Lecture Notes in Mathematics 2086

T0181885

For further volumes:
http://www.springer.com/series/304

Sébastien Boucksom • Philippe Eyssidieux
Vincent Guedj

Editors

An Introduction to the Kähler–Ricci Flow

 Springer

Editors
Sébastien Boucksom
Institut de Mathématiques de Jussieu
CNRS-Université Pierre et Marie Curie
Paris, France

Philippe Eyssidieux
Institut Fourier and Institut Universitaire
  de France
Université Joseph Fourier
Saint-Martin d'Hères, France

Vincent Guedj
Institut de Mathématiques de Toulouse
  and Institut Universitaire de France
Université Paul Sabatier
Toulouse, France

ISBN 978-3-319-00818-9   ISBN 978-3-319-00819-6 (eBook)
DOI 10.1007/978-3-319-00819-6
Springer Cham Heidelberg New York Dordrecht London

Lecture Notes in Mathematics ISSN print edition: 0075-8434
                ISSN electronic edition: 1617-9692

Library of Congress Control Number: 2013946026

Mathematics Subject Classification (2010): 35K96, 32Q15, 32Q20, 32Q25, 14E30

Printed on acid-free paper

Springer is part of Springer Science+Business Media (www.springer.com)

Preface

In 2010–2011 we organized several meetings of our ANR project MACK. These consisted of series of lectures centered around the Kähler–Ricci flow, which took place, respectively, in

- IMT (Toulouse, France), February 2010: mini course by H.-D. Cao:
  An introduction to the Kähler–Ricci flow on Fano manifolds;
- LATP (Marseille, France), March 2010: mini course by J. Song:
  The Kähler–Ricci flow on complex surfaces;
- CIRM (Luminy, France), February 2011: mini course B. Weinkove:
  An introduction to the Kähler–Ricci flow;
- CIRM (Luminy, France), February 2011: mini course J. Song:
  Kähler–Ricci flow and the Minimal Model Program;
- IMT (Toulouse, France), June 2011: mini course by D.-H. Phong:
  The normalized Kähler–Ricci flow on Fano manifolds;
- IMT (Toulouse, France), June 2011: mini course by S. Boucksom and V. Guedj:
  Regularizing properties of the Kähler–Ricci flow;
- FSSM (Marrakech, Morocco), October 2011: mini course by C. Imbert:
  Introduction to fully nonlinear parabolic equations;
- FSSM (Marrakech, Morocco), October 2011: mini course by V. Guedj:
  Convergence of the Kähler–Ricci flow on Kähler–Einstein Fano manifolds.

There were other lectures on more algebraic aspects (e.g., an introduction to the Minimal Model Program and finite generation of the canonical ring by S. Druel at the CIRM), or on elliptic problems (e.g., Moser–Trudinger inequalities by B. Berndtsson in Marrakech). Some of the speakers have produced a set of lecture notes, working hard to make them accessible to non-experts. This volume presents them in a unified way.

It is a pleasure to thank all the participants of these meetings for their enthusiasm and for creating a very pleasant atmosphere of work. Special thanks of course to the following lecturers:

- Sébastien Boucksom (CNRS and IMJ, Paris, France);
- Huai Dong Cao (Lehigh University, Bethlehem, USA);
- Vincent Guedj (IUF and IMT, Toulouse, France)
- Cyril Imbert (CNRS and Université Paris-Est Créteil, France);
- Duong Hong Phong (Columbia University, USA);
- Jian Song (Rutgers University, Piscataway, USA);
- Ben Weinkove (University of California, San Diego, USA).

We acknowledge financial support from the French ANR project MACK. We also would like to thank the CIRM and their staff for providing wonderful conditions of work during the thematic month "Complex and Riemannian geometry," as well as the LATP and the IMT for providing "professeur invité" positions (resp. for H.-D. Cao, J. Song, and D.-H. Phong).

Our last meeting in Marrakech was also extremely useful, we thank the organizers (Said Asserda and Ahmed Zeriahi), the other speakers (B. Berndtsson, S. Boucksom, A. Broustet, J.-P. Demailly, S. Diverio, N.C. Nguyen, S. Lamy) as well as all the participants for their help in creating a successful event.

Paris, France   S. Boucksom
Grenoble, France   P. Eyssidieux
Toulouse, France   V. Guedj
1 July 2012

Contents

Chapter 1
Introduction

Sébastien Boucksom, Philippe Eyssidieux, and Vincent Guedj

Abstract This book is the first comprehensive reference on the Kähler–Ricci flow. It provides an introduction to fully non-linear parabolic equations, to the Kähler–Ricci flow in general and to Perelman's estimates in the Fano case, and also presents the connections with the Minimal Model program.

1.1 Motivation

1.1.1 Some Historical Remarks

According to Weil [Weil57], the notion of a Kähler manifold, introduced by Kähler in 1933 [Käh33], became important through the work of Hodge [Hod41] that put on a firm footing the theorems of Lefschetz on the topology of complex projective manifolds [Lef24]. As Hodge remarks, the introduction of a metric on a complex projective manifold is a somewhat artificial operation. However artificial, this operation turned out to be very fruitful. Kodaira developed Hodge's ideas into his famous theorem giving a differential geometric characterization of

S. Boucksom (✉)
Institut de Mathématiques de Jussieu, CNRS-Université Pierre et Marie Curie, 4 place Jussieu, 75251, Paris, France
e-mail: boucksom@math.jussieu.fr

P. Eyssidieux
Institut Fourier and Institut Universitaire de France, Université Joseph Fourier, 100 rue des Maths, 38402, Saint-Martin d'Hères, France
e-mail: Philippe.Eyssidieux@ujf-grenoble.fr

V. Guedj
Institut de Mathématiques de Toulouse and Institut Universitaire de France, Université Paul Sabatier, 118 route de Narbonne, 31062, Toulouse Cedex 9, France
e-mail: vincent.guedj@math.univ-toulouse.fr

S. Boucksom et al. (eds.), *An Introduction to the Kähler–Ricci Flow*,
Lecture Notes in Mathematics 2086, DOI 10.1007/978-3-319-00819-6_1,
© Springer International Publishing Switzerland 2013

complex projective manifolds; this, together with Hörmander's L^2-estimates for the $\bar{\partial}$-equation, now form the basis of a unified approach to complex algebraic geometry and complex analysis (see for instance the book [Dem09]).

Hodge's remark raised the question of constructing canonical metrics on complex projective manifolds (or more generally on compact manifolds of Kähler type). Obvious candidates for such canonical metrics are Kähler–Einstein metrics that were actually introduced in [Käh33]. A more sophisticated guess for canonical Kähler metrics (in a given Kähler class) is Calabi's theory of extremal metrics. Their investigation is a very active field nowadays, which lies outside the scope of these lecture notes except for the Kähler–Einstein case.

1.1.2 Kähler–Einstein Metrics

A Kähler–Einstein metric on a complex manifold X is a Kähler metric g whose Ricci tensor is proportional to the metric tensor. The Kähler assumption is unnecessary for this definition to make sense, and Einstein metrics are indeed classical objects in Riemannian geometry (see [Bes87]). They were introduced in Lorentzian geometry by Einstein, the proportionality constant being known as the cosmological constant in that context.

If (X, ω)[1] is a compact Kähler–Einstein manifold of complex dimension n, the cosmological constant is essentially

$$\bar{s} = \int_X c_1(X)\{\omega\}^{n-1} / \int_X \{\omega\}^n,$$

a topological invariant of the pair $(X, \{\omega\})$ where $\{\omega\} \in H^{1,1}(X, \mathbb{R})$ is the cohomology class of ω.

If $\bar{s}$ is negative, the manifold is canonically polarized (i.e. its canonical bundle K_X is ample), and $\{\omega\} \in \mathbb{R}_{<0}c_1(X)$. Canonically polarized manifolds form a rather special class of varieties of general type: in dimension one, a compact Riemann surface is canonically polarized if and only if its genus g satisfies $g \geq 2$ and carries a unique Kähler–Einstein metric of curvature -1, its hyperbolic metric. By a theorem of Aubin and Yau, every canonically polarized complex projective manifold carries a unique Kähler–Einstein metric (up to scaling).

This celebrated work uses the reduction of the Kähler–Einstein equation to a complex Monge–Ampère equation (a scalar fully nonlinear elliptic equation). In fact, Kähler's original article already pointed out that solutions to a complex Monge–Ampère equation could furnish solutions to Einstein equations (which are not scalar) and ends with the formulation of the general problem of studying this equation. He also introduced the locally symmetric metric of the complex ball as a Kähler–Einstein metric.

---

[1]We follow the convention to specify a Kähler metric g on a complex manifold by the associated closed (1,1)-form ω.

A Kähler–Einstein metric satisfies $\bar{s} = 0$ if and only if $c_1(X) = 0$. One then has to specify the Kähler class of the Kähler–Einstein metric and it is a celebrated theorem of Yau that, given X a compact Kähler manifold such that $c_1(X) = 0$, every Kähler class contains a unique Ricci flat metric [Yau78].

If $\bar{s}$ is positive, the underlying manifold X is Fano and $\{\omega\} \in \mathbb{R}_{>0}c_1(X)$. In dimension one, a Fano manifold is a projective line and its Kähler–Einstein metrics are the Fubini–Study metrics. The Fano case is well known to be harder, and an algebro-geometric characterization of Kähler–Einstein Fano manifolds is still unknown and an intense subject of study. The case of surfaces was settled by Tian [Tian90] but the three-dimensional case is still open at the time of this writing.[2]

1.1.3 The Ricci Flow Approach

In Riemannian geometry, Hamilton [Ham82] introduced the Ricci flow

$$\frac{\partial g}{\partial t} = -2\text{Ric}(g),$$

and the development of his ideas gave rise to Perelman's proof of the Poincaré conjecture in three-dimensional topology. Bando observed that the Kähler condition is preserved under Hamilton's Ricci flow, hereby defining the main topic of these lecture notes: the Kähler–Ricci flow. To achieve this, Bando wrote out a scalar parabolic equation satisfied by the Kähler potential of the Kähler–Ricci flow. This is a parabolic version of the complex Monge–Ampère equation solved in [Yau78].

The convergence of the Kähler–Ricci flow to the canonical Kähler–Einstein metric on a compact Kähler manifold X with $c_1(X) < 0$ or $c_1(X) = 0$ was established by Cao [Cao85], and through the work of many authors the Kähler–Ricci flow became a major tool in Kähler geometry.

1.1.4 A New Hope?

It turns out that most results on the Kähler–Ricci flow on general type manifolds have been proved assuming that the Minimal Model Program (MMP) works. Although the canonical singular Kähler–Einstein metric [EGZ09] can be constructed unconditionnally [BEGZ10], its regularity properties can only be established using the existence of a minimal model proved in [BCHM10].

---

[2]The equivalence between K-polystability and the existence of a Kähler–Einstein metric has recently been announced by Chen–Donaldson–Sun and Tian, independently.

In a similar fashion, Song and Tian [ST12] rely on the existence of flips to establish long time existence of the Kähler–Ricci flow with surgeries on projective varieties with non-negative Kodaira dimension, and the behaviour of the Iitaka fibration as predicted by the abundance conjecture to establish convergence to a canonical current.

An approach to the MMP via Kähler–Einstein geometry had been advocated by Tsuji in the pre-BCHM era, and Song–Tian's convergence theorem suggests a way to construct the Iitaka fibration as the kernel foliation of the limit of the Kähler–Ricci flow. As far as the MMP for Kähler non-algebraic manifolds is concerned, the Kähler–Ricci flow is actually one of the few tools that could be used. There is obviously a long way before these dreams come true.

1.2 Contents

The ambition of these notes is to produce a reference for the foundations of the Kähler–Ricci flow and a guide to some recent developments. The lack of such a reference appeared clearly during the workshops that were organized by the editors of the present volume.

1.2.1 Organisation

The volume is divided into five chapters, as follows.

1.2.2 Chapter 2

The Kähler–Ricci flow being equivalent to a scalar fully nonlinear parabolic partial differential equation, it is important to have an overview of the general theory of such equations.

Chapter 2 contains a contribution of C. Imbert and L. Silvestre fulfilling this need, concentrating on three fundamental problems: Schauder estimates (regularity theory in Hölder classes), viscosity solutions (existence and uniqueness of weak solutions), and Harnack inequalities.

1.2.3 Chapter 3

This chapter contains a contribution of Song and Weinkove that surveys the fundamental estimates in the Kähler–Ricci flow and its long time existence theory. This lays the basis of the fascinating analytification of the Minimal Model Program that was conjectured by Tian as a Kähler analogue of Perelman's approach to Thurston's Geometrization Conjecture and established in part in [ST09]. Song and Weinkove go on discussing their recent contributions in this direction [SW10].

1.2.4   Chapter 4

In order to connect the Kähler–Ricci flow to the Minimal Model program, it is necessary to be able to work on varieties with mild singularities (terminal and $\mathbb{Q}$-factorial, at least), since minimal models of algebraic varieties with non-negative Kodaira dimension have these kind of singularities in dimension greater than 2.

Chapter 4 contains a contribution of Boucksom and Guedj presenting in detail the construction of the Kähler–Ricci flow on such mildly singular varieties, following Song and Tian's work [ST09]. After passing to a resolution of singularities, the result boils down to an existence and uniqueness result for certain degenerate parabolic complex Monge–Ampère equations. An illustration of the regularizing properties of such equations is proposed along the way, following [SzTo11].

1.2.5   Chapter 5

This chapter contains a contribution of Cao that surveys the Kähler–Ricci flow on Fano manifolds (long time existence, Li–Yau–Hamilton inequalities, etc.), culminating with an exposition of Perelman's estimates.

The latter are of central use in the study of the long time behavior of the (correctly normalized) Kähler–Ricci flow of Fano manifolds.

1.2.6   Chapter 6

This final chapter, written by Guedj, explains the proof of Perelman's convergence theorem: on a Kähler–Einstein Fano manifold without non trivial holomorphic vector fields, the normalized Kähler–Ricci flow converges in C^∞ topology to the unique Kähler–Einstein metric.

As far as weak convergence is concerned, an alternative approach due to [BBEGZ11] is presented, which can also be used on singular Fano varieties.

References

[BBEGZ11] R. Berman, S. Boucksom, P. Eyssidieux, V. Guedj, A. Zeriahi, Kähler-Einstein metrics and the Kähler–Ricci flow on log Fano varieties (2011). Preprint [arXiv.1111.7158v2]

[Bes87] A. Besse, in *Einstein Manifolds*. Erg. der Math. 3. Folge, vol. 10 (Springer, Berlin, 1987)

[BCHM10] C. Birkar, P. Cascini, C. Hacon, J. McKernan, Existence of minimal models for varieties of log general type. J. Am. Math. Soc. **23**(2), 405–468 (2010)

[BEGZ10] S. Boucksom, P. Eyssidieux, V. Guedj, A. Zeriahi, Monge-Ampère equations in big cohomology classes. Acta Math. **205**, 199–262 (2010)

[Cao85] H.D. Cao, Deformation of Kähler metrics to Kähler-Einstein metrics on compact Kähler manifolds. Invent. Math. **81**(2), 359–372 (1985)

[Dem09] J.P. Demailly, Complex analytic and differential geometry (2009), OpenContent book available at http://www-fourier.ujf-grenoble.fr/~demailly/manuscripts/agbook.pdf

[EGZ09] P. Eyssidieux, V. Guedj, A. Zeriahi, Singular Kähler-Einstein metrics. J. Am. Math. Soc. **22**, 607–639 (2009)

[Ham82] R.S. Hamilton, Three-manifolds with positive Ricci curvature. J. Differ. Geom. **17**(2), 255–306 (1982)

[Hod41] W. Hodge, *The Theory and Applications of Harmonic Integrals* (Cambridge University Press, Cambridge, 1941)

[Käh33] E. Kähler, Über eine bemerkenswerte Hermitesche metrik. Abh. Math. Semin. Univ. Hambg. **9**, 173–186 (1933)

[Lef24] S. Lefschetz, *L'Analysis Situs et la Géométrie Algébrique* (Gauthier-Villars, Paris, 1924)

[ST09] J. Song, G. Tian, The Kähler–Ricci flow through singularities (2009). Preprint [arXiv:0909.4898]

[ST12] J. Song, G. Tian, Canonical measures and Kähler–Ricci flow. J. Am. Math. Soc. **25**, 303–353 (2012)

[SW10] J. Song, B. Weinkove, Contracting exceptional divisors by the Kähler–Ricci flow (2010). Preprint (arXiv:1003.0718 [math.DG])

[SzTo11] G. Székelyhidi, V. Tosatti, Regularity of weak solutions of a complex Monge-Ampère equation. Anal. PDE **4**, 369–378 (2011)

[Tian90] G. Tian, On Calabi's conjecture for complex surfaces with positive first Chern class. Invent. Math. **101**(1), 101–172 (1990)

[Weil57] A. Weil, *Introduction à l'étude des variétés kählériennes* (Hermann, Paris, 1957)

[Yau78] S.T. Yau, On the Ricci curvature of a compact Kähler manifold and the complex Monge-Ampère equation. I. Comm. Pure Appl. Math. **31**(3), 339–411 (1978)

Chapter 2
An Introduction to Fully Nonlinear Parabolic Equations

Cyril Imbert and Luis Silvestre

Abstract These notes contain a short exposition of selected results about parabolic equations: Schauder estimates for linear parabolic equations with Hölder coefficients, some existence, uniqueness and regularity results for viscosity solutions of fully nonlinear parabolic equations (including degenerate ones), the Harnack inequality for fully nonlinear uniformly parabolic equations.

2.1 Introduction

The literature about parabolic equations is immense and it is very difficult to have a complete picture of available results. Very nice books such as [LSU67, Kryl87, Dong91,Lieb96] are attempt to gather and order the most significant advances in this wide field. If now one restricts himself to fully nonlinear parabolic equations, the task is still almost impossible. Indeed, many results proved for parabolic equations were first proved for elliptic equations and these results are numerous. We recall that many problems come from geometry; the reader is referred to the survey paper [Kryl97] where Krylov gives historical and bibliographical landmarks.

In these notes, we will focus on three specific topics concerning parabolic equations: Schauder estimates for linear parabolic equations (following Safonov [Saf84] and the textbook by Krylov [Kryl96]), viscosity solutions for fully nonlinear parabolic equations (see e.g. [CIL92]) and the Harnack inequality for fully nonlinear uniformly parabolic equations.

C. Imbert (✉)
CNRS, UMR8050, Université Paris-Est Créteil Val-de-Marne, Centre de mathématiques,
UFR sciences et technologies, 61 avenue du Général de Gaulle, 94010 Créteil cedex, France
e-mail: cyril.imbert@u-pec.fr

L. Silvestre
Department of Mathematics, University of Chicago, Chicago, IL 60637, USA

S. Boucksom et al. (eds.), *An Introduction to the Kähler–Ricci Flow*,
Lecture Notes in Mathematics 2086, DOI 10.1007/978-3-319-00819-6_2,
© Springer International Publishing Switzerland 2013

2.1.1   Main Objects and Notation

Geometric Objects

We first consider a connected open bounded set $\Omega \subset \mathbb{R}^d$. We refer to such a set as a *domain*. A domain is $C^{2,\alpha}$ if, locally, the boundary of the domain can be represented as the graph of a function with two derivatives that are α-Hölder continuous.

Parabolic equations are considered in cylindrical domain of the form $(0, T) \times \Omega$. The *parabolic boundary* of $(0, T) \times \Omega$ is denoted by $\partial_p (0, T) \times \Omega$; we recall that it is defined as follows

$$\partial_p (0, T) \times \Omega = \{0\} \times \Omega \cup (0, T) \times \partial\Omega.$$

The open ball of $\mathbb{R}^d$ centered at x of radius ρ is denoted by $B_\rho(x)$. If $x = 0$, we simply write B_ρ. The following elementary cylindrical domains play a central role in the theory: for all $\rho > 0$ and $x \in \mathbb{R}^d$, we define

$$Q_\rho(t, x) = (t - \rho^2, t) \times B_\rho(x).$$

When we write Q_ρ, we mean $Q_\rho(0, 0)$. It is also convenient to write

$$Q_\rho(t, x) = (t, x) + Q_\rho$$

and

$$Q_\rho = \rho Q_1.$$

A Linear Operator

The general parabolic equation considered in Sect. 2.2 involves the following linear operator

$$Lu = \sum_{i,j} a_{ij}(t, x) \frac{\partial^2 u}{\partial x_i \partial x_j} + \sum_i b_i(t, x) \frac{\partial u}{\partial x_i} + c(t, x)u.$$

The set of $d \times d$ real symmetric matrices is denoted by $\mathbb{S}_d$. The identity matrix is denoted by I. For $A, B \in \mathbb{S}_d$, $A \geq B$ means that all the eigenvalues of $A - B$ are non-negative.

Unknown functions $u : (0, T) \times \Omega \to \mathbb{R}$ depend on two (set of) variables: $t \in \mathbb{R}$ and $x \in \mathbb{R}^d$. It is convenient to use a capital letter X to refer to $(t, x) \in \mathbb{R}^{d+1}$.

The time derivative of u is either denoted by $\frac{\partial u}{\partial t}$ or $\partial_t u$ or u_t. Du denotes the gradient of the function u with respect to the space variable x. $D^2 u$ denotes the Hessian matrix of the function u with respect to x.

The linear operator introduced above can be written as follows

$$Lu = \text{trace}(AD^2 u) + b \cdot Du + cu$$

where $A = (a_{ij})_{ij}$.

Hölder Spaces and Semi-norms

We say that $u \in C^{0,\alpha}(Q)$ for $Q \subset (0, T) \times \Omega$ if u is $\frac{\alpha}{2}$-Hölder continuous with respect to time t and α-Hölder continuous with respect to space x. The corresponding semi-norm is denoted by $[u]_{\alpha,Q}$. See Sect. 2.1.4 for details.

2.1.2 Fully Nonlinear Parabolic Equations

We first emphasize the fact that we will not consider systems of parabolic equations; in other words, we will focus on scalar parabolic equations. This means that the unknown function u will always be real valued. We also restrict ourselves to second order parabolic equations.

We consider parabolic equations posed in a domain $\Omega \subset \mathbb{R}^d$; hence, unknown functions u are defined in $(0, T) \times \Omega$ with $T \in [0, \infty]$. In order to construct solutions and prove uniqueness for instance, initial and boundary conditions should be imposed. However, we will very often not specify them.

Fully nonlinear parabolic equations appear in optimal control theory and geometry. Here are several significant examples.

- The Bellman equation

$$\partial_t u + \sup_{\alpha \in A} \left\{ -\sum_{i,j} a_{ij}^\alpha(x) \frac{\partial^2 u}{\partial x_i \partial x_j} + \sum_i b_i^\alpha(x) \frac{\partial u}{\partial x_i} \right\} + \lambda u = 0.$$

- The mean curvature equation

$$\partial_t u = \Delta u = \frac{D^2 u Du \cdot Du}{|Du|^2}.$$

- The parabolic Monge–Ampère equations proposed by Krylov in [Kryl76]

$$-\frac{\partial u}{\partial t}\det(D^2 u) = H^{d+1}$$

$$-\det(D^2 u) + \left[\frac{\partial u}{\partial t} + H\right]^{d+1} = 0 \qquad (2.1)$$

$$-\det\left(D^2 u - \frac{\partial u}{\partial t} I\right) = H^d$$

where $H = H(t, x, Du)$ is a nonlinear first order term.
- For the study of the Kähler–Ricci flow, one would like to study:

$$\frac{\partial u}{\partial t} = \ln(\det(D^2 u)). \qquad (2.2)$$

2.1.3   Aim of These Notes

Our goal is to construct solutions and study their regularity. One would like to construct classical solutions, that is to say solutions such that the derivatives appearing in the equation exist in the classical sense and satisfy the equation. But this is not always possible and it is sometimes (very often?) necessary to construct weak solutions. They are different notions of weak solutions; we will focus in these notes on so-called viscosity solutions. The advantage is that it is easy to construct such solutions. One can next try to prove that these solutions are regular.

Before 1988 (date of publication of [Jens88]), it was popular (necessary) to construct solutions of fully nonlinear elliptic (or parabolic) equations by using the *continuity method*. To apply it, it is necessary to get appropriate a priori estimates (on third derivatives for instance, or on the modulus of continuity of the second ones).

The situation changed dramatically when Jensen [Jens88] managed to apply the viscosity solution techniques of Crandall–Lions [CL81] to second order elliptic and parabolic equations. In particular, he understood how to adapt the so-called doubling variable techniques to prove uniqueness. Ishii also contributed to this major breakthrough. The reader is referred to the survey paper [CIL92] for further details.

Before presenting the viscosity solution techniques and some selected regularity results for these weak solutions, we will present shortly the classical Schauder approach to linear parabolic equations.

2.1.4   Spaces of Hölder Functions

Because we study parabolic equations, Hölder continuity of solutions refers to uniform continuity with respect to

$$\rho(X, Y) = \sqrt{|t - s|} + |x - y|$$

where $X = (t, x)$ and $Y = (s, y)$. In other words, solutions are always twice more regular with respect to the space variable than with respect to the time variable.

Remark 2.1.1 (Important). The reader should keep in mind that, following Krylov [Kryl96], we choose to write $u \in C^{0,\alpha}$ for functions that are α-Hölder continuous in x and $\frac{\alpha}{2}$-Hölder continuous in t. This choice is made first to emphasize the link between regularities with respect to time and space variables, second to simplify notation.

Let $Q \subset (0, T) \times \Omega$ and $\alpha \in (0, 1]$.

- $u \in C^{0,\alpha}(Q)$ means that there exists $C > 0$ s.t. for all $(t, x), (s, y) \in Q$, we have

$$|u(t, x) - u(s, y)| \leq C(|t - s|^{\frac{\alpha}{2}} + |x - y|^{\alpha}).$$

In other words, u is $\frac{\alpha}{2}$-Hölder continuous in t and α-Hölder continuous in x.

- $u \in C^{1,\alpha}(Q)$ means that u is $\frac{\alpha+1}{2}$-Hölder continuous in t and Du is α-Hölder continuous in x.
- $u \in C^{2,\alpha}(Q)$ means that $\frac{\partial u}{\partial t}$ is $\frac{\alpha}{2}$-Hölder continuous in t and $D^2 u$ is α-Hölder continuous in x.

We also consider the following norms and semi-norms.

$$[u]_{\alpha,Q} = \sup_{X,Y \in Q, X \neq Y} \frac{|u(X) - u(Y)|}{\rho(X, Y)}$$

$$|u|_{0,Q} = \sup_{X \in Q} |u(X)|$$

$$[u]_{2+\alpha,Q} = \left[\frac{\partial u}{\partial t}\right]_{\alpha,Q} + [D^2 u]_{\alpha,Q}$$

$$|u|_{2+\alpha,Q} = |u|_{0,Q} + \left|\frac{\partial u}{\partial t}\right|_{0,Q} + |Du|_{0,Q} + |D^2 u|_{0,Q} + [u]_{2+\alpha,Q}.$$

We will use repeatedly the following elementary proposition.

Proposition 2.1.2.

$$[uv]_{\alpha,Q} \leq |u|_{0,Q}[v]_{\alpha,Q} + |v|_{0,Q}[u]_{\alpha,Q}$$

and for $k = 0, 2,$

$$[u + v]_{k+\alpha,Q} \leq [u]_{k+\alpha,Q} + [v]_{k+\alpha,Q}.$$

The following proposition implies in particular that in order to control the norm $|u|_{2+\alpha,Q}$, it is enough to control $|u|_{0,Q}$ and $[u]_{2+\alpha,Q}$.

Proposition 2.1.3 (Interpolation inequalities). *For all* $\varepsilon > 0$, *there exists* $C(\varepsilon) > 0$ *s.t. for all* $u \in C^{2,\alpha}$,

$$\begin{cases} |\frac{\partial u}{\partial t}|_{0,Q} \leq \varepsilon[u]_{2+\alpha,Q} + C(\varepsilon)|u|_{0,Q}, \\ [Du]_{\alpha,Q} \leq \varepsilon[u]_{2+\alpha,Q} + C(\varepsilon)|u|_{0,Q}, \\ [u]_{\alpha,Q} \leq \varepsilon[u]_{2+\alpha,Q} + C(\varepsilon)|u|_{0,Q}. \end{cases} \tag{2.3}$$

The following proposition is a precise parabolic statement of the following elliptic fact: in order to control the Hölder modulus of continuity of the gradient of u, it is enough to make sure that, around each point, the function u can be perturbed linearly so that the oscillation of u in a ball of radius $r > 0$ is of order $r^{1+\alpha}$.

Proposition 2.1.4 (An equivalent semi-norm). *There exists* $C \geq 1$ *such that for all* $u \in C^{2,\alpha}(Q)$,

$$C^{-1}[u]'_{2+\alpha,Q} \leq [u]_{2+\alpha,Q} \leq C[u]'_{2+\alpha,Q}$$

where

$$[u]'_{2+\alpha,Q} = \sup_{X \in Q} \sup_{\rho > 0} \rho^{-2-\alpha} \inf_{P \in \mathcal{P}_2} |u - P|_{0,Q_\rho(X) \cap Q}$$

where

$$\mathcal{P}_2 = \{\alpha t + p \cdot x + \frac{1}{2}Xx \cdot x + c : \alpha, c \in \mathbb{R}, p \in \mathbb{R}^d, X \in \mathbb{S}_d\}.$$

The reader is referred to [Kryl96] for proofs of the two previous propositions.

2.2   Schauder Estimates for Linear Parabolic Equations

In this first section, we state a fundamental existence and uniqueness result for linear parabolic equations with Hölder continuous coefficients.

The proof of this theorem is rather long and presenting it completely is out of the scope of the present lectures notes. Instead, we would like to focus on two particular aspects: uniqueness and interior estimates.

The uniqueness of the solution is proved by using a maximum principle (Sect. 2.2.3), the existence can be obtained through the *continuity method*. This method relies on the proof of the "good" a priori estimate (2.4) on any $C^{2,\alpha}$ solution. This estimate is global in the sense that it deals with what happens at the interior of $(0, T) \times \Omega$ and at its boundary. In Sect. 2.2.5, we focus on what happens in the interior of the domain. Precisely, we present a complete proof of the interior Schauder estimate in the general case. It relies on Schauder estimates for parabolic equations with constant coefficients. The derivation of these estimates are presented in Sect. 2.2.4 by studying first the heat equation. We present here an argument due to Safonov circa 1984.

2.2.1   Linear Parabolic Equations

The standing example of linear parabolic equations with constant coefficients is the *heat equation*

$$\frac{\partial u}{\partial t} - \Delta u = f$$

where f is a *source term*. The general form of a linear parabolic equation with variable coefficients is the following

$$\frac{\partial u}{\partial t} - \sum_{i,j} a_{ij}(X) \frac{\partial^2 u}{\partial x_i \partial x_j} - \sum_i b_i(X) \frac{\partial u}{\partial x_i} - c(X) u = 0$$

where

$$c \le 0$$

and $A(X) = (a_{ij}(X))_{i,j}$ is a symmetric matrix satisfying one of the following assumptions

- (Degenerate ellipticity) for all X, $A(X) \ge 0$;
- (Strict ellipticity) there exists $\lambda > 0$ s.t. for all X, $1 A(X) \ge \lambda I$;
- (Uniform ellipticity) there exists $\Lambda \ge \lambda > 0$ s.t. for all X, $\lambda I A(X) \le \Lambda I$.

We recall that I denotes the identity matrix and if $A, B \in \mathbb{S}_d$, $A \ge B$ means that all the eigenvalues of $A - B$ are non-negative.

It is convenient to consider the linear differential operator L defined as follows

$$Lu = \sum_{i,j} a_{ij}(X) \frac{\partial^2 u}{\partial x_i \partial x_j} + \sum_i b_i(X) \frac{\partial u}{\partial x_i} + c(X) u.$$

2.2.2 A Fundamental Existence and Uniqueness Result

In this subsection, we state a fundamental existence and uniqueness result for linear parabolic equation with Hölder continuous coefficients. Such a result together with its proof can be found in various forms in several classical monographs such as [LSU67, Kryl96]. We choose here to present the version given in [Kryl96].

In the following statement, $\mathbb{R}_+^{d+1}$ denotes $[0, +\infty) \times \mathbb{R}^d$.

Theorem 2.2.1. *If Ω is a $C^{2,\alpha}$ domain and the coefficients $A, b, c \in C^\alpha$ $((0, T) \times \Omega)$ and $f \in C^\alpha(\mathbb{R}_+^{d+1})$, $g \in C^{2+\alpha}((0, T) \times \Omega)$, $h \in C^{2,\alpha}(\mathbb{R}^d)$, and g and h are compatible (see Remark 2.2.3 below), then there exists a unique solution $u \in C^{2,\alpha}(Q)$ of*

$$\begin{cases} \frac{\partial u}{\partial t} - \Delta u = f & \text{in } (0, T) \times \Omega \\ u = g & \text{on } (0, +\infty) \times \partial\Omega \\ u = h & \text{on } \{0\} \times \bar{\Omega}. \end{cases}$$

In addition,

$$|u|_{2+\alpha,(0,T)\times\Omega} \leq C(|f|_{\alpha,\mathbb{R}_+^{d+1}} + |g|_{2+\alpha,(0,T)\times\Omega} + |h|_{2+\alpha,\mathbb{R}^d}) \qquad (2.4)$$

where $C = C(d, \lambda, K, \alpha, \rho_0, \operatorname{diam}(\Omega))$ and $K = |A|_{\delta,(0,T)\times\Omega} + |b|_{\delta,(0,T)\times\Omega} + |c|_{\delta,(0,T)\times\Omega}$ and ρ_0 is related to the $C^{2,\alpha}$ regularity of the boundary of Ω.

Remark 2.2.2. The inequality (2.4) is called the (global) Schauder a priori estimate.

Remark 2.2.3. The fact that data g and h are *compatible* has to do with conditions ensuring that a solution which is regular up to the boundary can be constructed. Since we will not address these problems, we refer the interested reader to [LSU67, Kryl96] for a precise definition.

2.2.3 Maximum and Comparison Principles

Maximum principles are powerful tools to study elliptic and parabolic equations. There are numerous statements which are not equivalent. We choose the following one.

Theorem 2.2.4 (Maximum principle). *Consider a bounded continuous function $u : (0, T) \times \Omega \to \mathbb{R}$ such that $\frac{\partial u}{\partial t}$ exists at each point of $(0, T) \times \Omega$ and Du, D^2u exist and are continuous in $(0, T) \times \Omega$.*

If

$$\frac{\partial u}{\partial t} - Lu \leq 0 \text{ in } (0, T) \times \Omega$$

$$u \leq 0 \text{ on } \partial_p (0, T) \times \Omega$$

then $u \leq 0$ in $(0, T) \times \Omega$.

Remark 2.2.5. The set $\partial_p (0, T) \times \Omega$ is the parabolic boundary of the cylindrical domain $(0, T) \times \Omega$. Its definition is recalled in the section devoted to notation.

Proof. Fix $\gamma > 0$ and consider the function $v(t, x) = u(t, x) - \frac{\gamma}{T-t}$. Assume that v is not non-positive. Then its maximum M on $(0, T) \times \Omega$ is positive. It is reached, and it cannot be attained for $t = 0$ or $x \in \partial\Omega$ since $v \leq u \leq 0$ on $\partial_p(0, T) \times \Omega$. It can neither be attained for $t = T$ since $v \to -\infty$ as $t \to T-$. We conclude that the maximum is attained for some $t \in (0, T)$ and $x \in \Omega$. In particular,

$$0 = \frac{\partial v}{\partial t}(t, x) = \frac{\partial u}{\partial t}(t, x) - \frac{\gamma}{(T-t)^2}$$

$$0 = Dv(t, x) = Du(t, x)$$

$$0 \geq D^2 v(t, x) = D^2 u(t, x).$$

Remark that since A is (uniformly) elliptic, the linear operator satisfies

$$Lu(t, x) = \text{trace}(AD^2 u) + b \cdot Du + cu = \text{trace}(AD^2 u) + cu \leq \text{trace}(AD^2 u) \leq 0$$

since $u(t, x) \geq v(t, x) > 0$, $c \leq 0$, $A \geq 0$ and $D^2 u(t, x) \leq 0$. We now use the fact that u satisfies $\frac{\partial u}{\partial t} - Lu \leq 0$ in $(0, T) \times \Omega$ to get the desired contradiction:

$$\frac{\gamma}{(T-t)^2} = \frac{\partial u}{\partial t}(t, x) \leq Lu(t, x) \leq 0.$$

Since γ is arbitrary, the proof is complete.   □

We now state two corollaries. The first one will be the starting point of the second section (Sect. 2.3). In the framework of linear equation, it is a direct consequence of the previous result.

Corollary 2.2.6 (Comparison principle I). *Consider two bounded continuous functions u and v which are differentiable with respect to time and such that first and second derivatives with respect to space are continuous. If*

$$\frac{\partial u}{\partial t} - Lu \leq f \ in \ (0, T) \times \Omega \tag{2.5}$$

$$\frac{\partial v}{\partial t} - Lv \geq f \ in \ (0, T) \times \Omega$$

and $u \leq v$ in $\partial_p Q$, then $u \leq v$ in $(0, T) \times \Omega$.

Remark 2.2.7. Remark that this corollary implies that as soon as u satisfies (2.5), it lies below any solution of $\frac{\partial u}{\partial t} - Lu = f$. This is the reason why it is referred to as a subsolution of the equation $\frac{\partial u}{\partial t} - Lu = f$. In the same way, v lies above any solution and is referred to as a supersolution.

Remark 2.2.8. In view of the previous remark, we can reformulate the result of the previous corollary as follows: if a subsolution lies below a supersolution at the parabolic boundary then it lies below in the whole cylindrical domain.

The next result contains a first estimate for solutions of linear parabolic equations.

Corollary 2.2.9 (A first estimate). *Consider a bounded continuous solution u of $\frac{\partial u}{\partial t} - Lu = f$ in $(0, T) \times \Omega$. Assume moreover that it is differentiable with respect to time and continuously twice differentiable with respect to space. Then*

$$|u|_{0,(0,T)\times\Omega} \leq T|f|_{0,(0,T)\times\Omega} + |g|_{0,\partial_p(0,T)\times\Omega}.$$

Sketch of proof. Consider $v^{\pm} = u \pm (|g|_{0,\partial_p(0,T)\times\Omega} + t|f|_{0,(0,T)\times\Omega})$ and check that v^+ is a supersolution and v^- is a subsolution. Then the previous corollary yields the desired result. □

2.2.4 Schauder Estimate for the Heat Equation

2.2.4.1 Statement and Corollary

The "interior" Schauder estimate for the heat equation takes the following form.

Theorem 2.2.10. *Let $\alpha \in (0, 1)$ and consider a C^{∞} function $u : \mathbb{R}^{d+1} \to \mathbb{R}$ with compact support and define $f = \frac{\partial u}{\partial t} - \Delta u$. Then there exists a constant $C > 0$ only depending on dimension and α such that*

$$[u]_{2+\alpha,\mathbb{R}^{d+1}} \leq C [f]_{\alpha,\mathbb{R}^{d+1}}.$$

It is then easy to derive a similar "interior" Schauder estimate for linear uniformly parabolic equation with constant coefficients and no lower order term.

Corollary 2.2.11. *Let $\alpha \in (0, 1)$ and assume that $A \equiv A_0$ in $\mathbb{R}^{d+1}$ and $b \equiv 0$, $c \equiv 0$. Then there exists a constant $C > 0$ only depending on dimension and α such that for any C^∞ function u with compact support*

$$[u]_{2+\alpha,\mathbb{R}^{d+1}} \leq C[f]_{\alpha,\mathbb{R}^{d+1}}$$

where $f = \frac{\partial u}{\partial t} - Lu$.

Sketch of proof. The proof consists in performing an appropriate change of coordinates. Precisely, we choose $P \in \mathbb{S}_d$ such that $A_0 = P^2$ and consider $v(t, x) = u(t, Px)$. Then check that $\Delta v = \text{trace}(A_0 D^2 u) = Lu$ and use Theorem 2.2.10. □

2.2.4.2 Two Useful Facts

Before proving Theorem 2.2.10, we recall two facts about the heat equation. We recall first that a solution $u \in C^\infty$ of

$$\frac{\partial u}{\partial t} - \Delta u = f,$$

with compact support included in $(0, +\infty) \times \mathbb{R}^d$, can be represented as

$$u(t, x) = \int_0^t \int_{\mathbb{R}^d} G(s, y) f(t - s, x - y) ds \, dy$$

where

$$G(t, x) = \frac{1}{(4\pi t)^{d/2}} e^{-\frac{|x|^2}{4t}}.$$

We write in short hand

$$u = G \star f,$$

keeping in mind that G should be extended by 0 for $t < 0$ in order to make this rigorous. This formula can be justified using Fourier analysis for instance.

Fact 1. *For any $0 \leq \rho \leq R$,*

$$|G \star \mathbf{1}_{Q_R(Z_0)}|_{0, Q_\rho(Z_0)} \leq CR^2$$

where $\mathbf{1}_{Q_R(Z_0)}(Z) = 1$ if $Z \in Q_R(Z_0)$ and 0 if not.

Fact 2. *There exists a constant $C > 0$ such that any solution of $\frac{\partial h}{\partial t} = \Delta h$ in $Q_R(0)$ satisfies*

$$\left| \frac{\partial^n}{\partial t^n} D^\alpha h(0) \right| \leq C \frac{|h|_{0,Q_R(0)}}{R^{2n+|\alpha|}}$$

where $\alpha = (\alpha_1, \ldots, \alpha_n)$, $|\alpha| = \sum_i \alpha_i$ and $D^\alpha h = \frac{\partial^{\alpha_1}}{\partial x_1^{\alpha_1}} \cdots \frac{\partial^{\alpha_d}}{\partial x_d^{\alpha_d}} h$.

This second fact can be proved by using Bernstein's techniques. See [Kryl96, Chap. 8, p. 116].

2.2.4.3   Proof of the Schauder Estimate

The following proof is due to Safonov circa 1984. It is presented in [Kryl96]. Krylov says in [Kryl97] that "[he] believes this proof should be part of a general knowledge for mathematicians even remotely concerned with the theory of PDEs".

Recall that the $C^{2,\alpha}$ regularity can be established "pointwise". Indeed, in view of Proposition 2.1.4, it is enough to be able to find a polynomial P which is linear in time and quadratic in space such that the oscillation of the difference between u and P decreases as $\rho^{2+\alpha}$ in a box of size ρ. The natural candidate for P is the "second order" Taylor polynomial of the function itself. The idea of Safonov is to perturb this natural candidate in order to reduce to the case where $f \equiv 0$.

Proof of Theorem 2.2.10. Without loss of generality, we can assume that the compact support of u is included in $(0, +\infty) \times \mathbb{R}^d$.

Take $X_0 \in \mathbb{R}^{d+1}$, $\rho > 0$ and $K \geq 1$ to be specified later. Let Q denote $Q_{(K+1)\rho}(X_0)$ and take $\zeta \in C^\infty(\mathbb{R}^{d+1})$ with compact support and such that $\zeta \equiv 1$ in Q.

We consider the "second order" Taylor polynomial associated with a function w at a point $X = (t, x)$

$$T_X w(s, y) = w(X) + w_t(X)(s-t) + Dw(X) \cdot (y-s) + \frac{1}{2} D^2 w(X)(y-x) \cdot (y-x).$$

We now consider

$$g = (\zeta T_{X_0} u)_t - \Delta(\zeta T_{X_0} u).$$

In view of properties of ζ,

$$g \equiv f(X_0) \text{ in } Q.$$

Keeping this piece of information in mind, we can write for $X \in Q$,

$$u - T_{X_0} u = u - \zeta T_{X_0} u = G \star (f - g)$$
$$= h + r$$

with

$$h = G \star ((f - g)\mathbf{1}_{Q^c}) \quad \text{and} \quad r = G \star ((f - f(X_0))\mathbf{1}_Q)$$

where $Q^c = \mathbb{R}^{d+1} \setminus Q$. Remark in particular that

$$h_t - \Delta h = 0 \text{ in } Q.$$

Now we estimate

$$|u - T_{X_0} u - T_{X_0} h|_{0, Q_\rho(X_0)} \leq |h - T_{X_0} h|_{0, Q_\rho(X_0)} + |r|_{0, Q_\rho(X_0)} \tag{2.6}$$

and we study the two terms of the right hand side.
   We use Fact 1 to get first

$$|r|_{0, Q_\rho(X_0)} \leq [f]_{\alpha, Q}(K + 1)^\alpha \rho^\alpha |G \star \mathbf{1}_Q|_{0, Q_\rho(X_0)}$$
$$\leq C(K + 1)^{2+\alpha} \rho^{2+\alpha} [f]_{\alpha, Q}. \tag{2.7}$$

We now write for $X \in Q_\rho(X_0)$,

$$h(X) = h(X_0) + h_t(\theta, x)(t - t_0) + Dh(X_0) \cdot (x - x_0) + \frac{1}{2} D^2 h(\Theta)(x - x_0) \cdot (x - x_0)$$

for some $\theta \in (t_0, t)$ and $\Theta = (t_0, y_0) \in Q_\rho(X_0)$. Hence, we have

$$h(X) - T_{X_0} h(X) = (h_t(\theta, x) - h_t(X_0))(t - t_0)$$
$$+ \frac{1}{2}(D^2 h(\Theta) - D^2 h(X_0))(x - x_0) \cdot (x - x_0)$$

from which we deduce

$$|h(X) - T_{X_0} h(X)| \leq \rho^2 |h_t(\theta, x) - h_t(X_0)| + \rho^2 |D^2 h(\Theta) - D^2 h(X_0)|. \tag{2.8}$$

We now use Fact 2 in order to get

$$
\begin{aligned}
|h - T_{X_0}h|_{0,Q_\rho(X_0)} &\leq \rho^2 \left(\rho^2 \left| \frac{\partial^2}{\partial t^2} h \right|_{0,Q_\rho(X_0)} + \rho \left| \frac{\partial}{\partial t} Dh \right|_{0,Q_\rho(X_0)} \right) \\
&\quad + C\rho^3 |D^3 h|_{0,Q_\rho(X_0)} \\
&\leq C(\rho^4 (K\rho)^{-4} + \rho^3 (K\rho)^{-3} + \rho^3 (K\rho)^{-3}) |h|_{0,Q} \\
&\leq C(K^{-4} + 2K^{-3}) |h|_{0,Q} \\
&\leq CK^{-3} |h|_{0,Q}
\end{aligned}
$$

by choosing $K \geq 1$. We next estimate $|h|_{0,Q}$ as follows

$$
\begin{aligned}
|h|_{0,Q} &\leq |u - T_{X_0}u - r|_{0,Q} \leq |u - T_{X_0}u|_{0,Q} + |r|_{0,Q} \\
&\leq C(K+1)^{2+\alpha} \rho^{2+\alpha} ([u]_{2+\alpha,Q} + |[f]_{\alpha,Q})
\end{aligned}
$$

where we used (2.8) for u instead of h and we used (2.7). Then, we have

$$
|h - T_{X_0}h|_{0,Q_\rho(X_0)} \leq C \frac{(K+1)^{2+\alpha}}{K^3} \rho^{2+\alpha} ([u]_{2+\alpha,Q} + [f]_{\alpha,Q}). \tag{2.9}
$$

Combining (2.6), (2.7) and (2.9), we finally get

$$
\begin{aligned}
\rho^{-(2+\alpha)} |u - T_{X_0}u - T_{X_0}h|_{0,Q_\rho(X_0)} &\leq C(K+1)^{2+\alpha} [f]_{\alpha,Q} \\
&\quad + C \frac{(K+1)^{2+\alpha}}{K^3} ([u]_{2+\alpha,Q} + [f]_{\alpha,Q}).
\end{aligned}
$$

In view of Proposition 2.1.4, it is enough to choose $K \geq 1$ large enough so that

$$
C \frac{(K+1)^{2+\alpha}}{K^3} \leq \frac{1}{2}
$$

to conclude the proof of the theorem.   □

2.2.5 Schauder Estimate in the Case of Variable Coefficients

Theorem 2.2.12. *Consider a function $u \in C^{2,\alpha}((0,T) \times \mathbb{R}^d)$ for some $\alpha \in (0,1)$. Then there exists $C = C(d,\alpha)$ such that*

$$
[u]_{2+\alpha,(0,T)\times\mathbb{R}^d} \leq C \left([f]_{\alpha,(0,T)\times\mathbb{R}^d} + |u|_{0,(0,T)\times\mathbb{R}^d} \right)
$$

where $f = \frac{\partial u}{\partial t} - Lu$.

Remark 2.2.13 (Notation). In the remaining of this subsection, it is convenient to write semi-norms as $[\cdot]_{k+\alpha}$ instead of $[\cdot]_{k+\alpha,(0,T)\times\mathbb{R}^d}$, $k = 0, 2$. In the same way, $|\cdot|_0$ stands for $|\cdot|_{0,(0,T)\times\mathbb{R}^d}$.

Remark 2.2.14. Recall that by Corollary 2.2.9, one has

$$|u|_0 \leq T|u_t - Lu|_0 + |u(0,\cdot)|_{0,\mathbb{R}^d}.$$

Before giving a rigorous proof, we would like first to explain the main idea.

Main idea of the proof of Theorem 2.2.12. Assume first that there are no lower order terms ($c \equiv 0$ and $b \equiv 0$).

In a neighbourhood of $X_0 \in \mathbb{R}^{d+1}$, the coefficients of the linear operator L are frozen: the linear operator with constant coefficients is denoted by L_0. If X is close to X_0, then L is not very far from L_0 and this can be measured precisely thanks to the Hölder continuity of coefficients.

Use first Corollary 2.2.11:

$$[u]_{2+\alpha} \leq C[u_t - L_0u]_\alpha \leq C[u_t - Lu]_\alpha + C[Lu - L_0u]_\alpha.$$

Now control $[Lu - L_0u]_\alpha$ thanks to $[u]_{2+\alpha}$ and conclude.

Next, lower order terms are treated by using interpolation inequalities.   □

Let us now make this precise and rigorous.

Proof of Theorem 2.2.12. We first assume that $b \equiv 0$ and $c \equiv 0$. Let f denote $\frac{\partial u}{\partial t} - Lu$.

Let $\varepsilon \in (0, T/2)$ and $\gamma \leq \varepsilon/2$ be a positive real number to be fixed later and consider X_1 and X_2 such that

$$[u_t]_{\alpha,(\varepsilon,T-\varepsilon)\times\mathbb{R}^d} \leq 2\rho(X_1, X_2)^{-\alpha}|u_t(X_1) - u_t(X_2)|$$

where we recall that $\rho(X_1, X_2) = \sqrt{|t_1 - t_2|} + |x_1 - x_2|$ if $X_i = (t_i, x_i)$, $i = 1, 2$.
If $\rho(X_1, X_2) \geq \gamma$, then we use interpolation inequalities (2.3) in order to get

$$[u_t]_{\alpha,(\varepsilon,T-\varepsilon)\times\mathbb{R}^d} \leq 2\gamma^{-\alpha}|u_t|_0$$

$$\leq \frac{1}{4}[u]_{2+\alpha} + C(\gamma)|u|_0.$$

If $\rho(X_1, X_2) < \gamma$, we consider $\zeta \in C^\infty(\mathbb{R}^{d+1})$ with compact support such that $\zeta(X) = 1$ if $\rho(X, 0) \leq 1$ and $\zeta(X) = 0$ if $\rho(X, 0) \geq 2$. We next define $\xi(t, x) = \zeta(\gamma^{-2}(t - t_1), \gamma^{-1}(x - x_1))$. In particular, $\xi(X) = 1$ if $\rho(X, X_1) \leq \gamma$ and $\xi(X) = 0$ if $\rho(X, X_1) \geq 2\gamma$.

Now we use Corollary 2.2.11 in order to get

$$[u_t]_{\alpha,(\varepsilon,T-\varepsilon)\times\mathbb{R}^d} \le 2\rho(X_1, X_2)^{-\alpha}|u_t(X_1) - u_t(X_2)|$$
$$\le 2[(u\xi)]_{2+\alpha}$$
$$\le 2C[(u\xi)_t - L(X_1)(u\xi)]_\alpha$$
$$\le 2C[(u\xi)_t - L(u\xi)]_\alpha + 2C[(L(X_1) - L)(u\xi)]_\alpha. \qquad (2.10)$$

We estimate successively the two terms of the right hand side of the last line. First, we write

$$(u\xi)_t - L(u\xi) = \xi f + u(\xi_t - L\xi) - 2ADu \cdot D\xi$$

since $L(u\xi) = uL\xi + \xi Lu + 2ADu \cdot D\xi$. Using interpolation inequalities (2.3), this implies

$$[(u\xi)_t - L(u\xi)]_\alpha \le C(\gamma)([f]_\alpha + [u]_\alpha + [Du]_\alpha)$$
$$\le \gamma^\alpha[u]_{2+\alpha} + C(\gamma)([f]_\alpha + |u|_0). \qquad (2.11)$$

We next write

$$(L(X_1) - L)(u\xi) = \text{trace}[(A(X_1) - A(X))D^2(u\xi)]$$

and for X such that $\rho(X_1, X) \le 2\gamma$, we thus get thanks to interpolation inequalities (2.3)

$$[(L(X_1) - L)(u\xi)]_\alpha \le C\gamma^\alpha[D^2(u\xi)]_\alpha + C|D^2(u\xi)|_0$$
$$\le C\gamma^\alpha[u]_{2+\alpha} + C(\gamma)|u|_0. \qquad (2.12)$$

Combining (2.10)–(2.12), we finally get in the case where $\rho(X_1, X_2) \le \gamma$,

$$[u_t]_{\alpha,(\varepsilon,T-\varepsilon)\times\mathbb{R}^d} \le C\gamma^\alpha[u]_{2+\alpha} + C(\gamma)([f]_\alpha + |u|_0).$$

We conclude that we have in both cases

$$[u_t]_{\alpha,(\varepsilon,T-\varepsilon)\times\mathbb{R}^d} \le (C\gamma^\alpha + 1/4)[u]_{2+\alpha} + C(\gamma)([f]_\alpha + |u|_0).$$

We can argue in a similar way to get

$$[D^2u]_{\alpha,(\varepsilon,T-\varepsilon)\times\mathbb{R}^d} \le (C\gamma^\alpha + 1/4)[u]_{2+\alpha} + C(\gamma)([f]_\alpha + |u|_0).$$

Adding these two inequalities yield

$$[u]_{2+\alpha,(\varepsilon,T-\varepsilon)\times\mathbb{R}^d} \leq (C\gamma^\alpha + 1/2)[u]_{2+\alpha} + C(\gamma)([f]_\alpha + |u|_0).$$

Now choose γ such that $C\gamma^\alpha \leq 1/4$ and get

$$[u]_{2+\alpha,(\varepsilon,T-\varepsilon)\times\mathbb{R}^d} \leq \frac{3}{4}[u]_{2+\alpha} + C([f]_\alpha + |u|_0).$$

Taking the supremum over $\varepsilon \in (0, T/2)$ allows us to conclude in the case where $b \equiv 0$ and $c \equiv 0$.

If now $b \neq 0$ and $c \neq 0$, we apply the previous result and get

$$[u]_{2+\alpha} \leq C([f + b \cdot Du + cu]_\alpha + |u|_0).$$

Use now interpolation inequalities once again to conclude.   □

2.3 Viscosity Solutions: A Short Overview

Viscosity solutions were first introduced by Crandall and Lions [CL81]. This notion of weak solution enabled to characterize the value function of an optimal control problem as the unique solution of the corresponding first order Hamilton–Jacobi equation. An example of such an equation is the following one

$$\frac{\partial u}{\partial t} + \frac{1}{2}|Du|^2 + V(x) = 0 \tag{2.13}$$

for some continuous function V. The viscosity solution theory is also by now a fundamental tool for the study of nonlinear elliptic and parabolic equations.

2.3.1 Definition and Stability of Viscosity Solutions

2.3.1.1 Degenerate Ellipticity

We recall that linear parabolic equations in non-divergence form have the following general form

$$\frac{\partial u}{\partial t} - Lu = f$$

with

$$Lu = \text{trace}(AD^2u) + b \cdot Du + cu$$

with $A \geq 0$ (in the sense of symmetric matrices).

We now consider very general *nonlinear* parabolic equation of the form

$$\frac{\partial u}{\partial t} + F(t, x, Du, D^2u) = 0 \qquad (2.14)$$

where we assume that the *nonlinearity* $F : (0, T) \times \Omega \times \mathbb{R}^d \times \mathbb{S}_d \to \mathbb{R}$ is continuous and satisfies the following condition

$$A \leq B \Rightarrow F(t, x, p, A) \geq F(t, x, p, B). \qquad (2.15)$$

In other words, the nonlinearity F is non-increasing with respect to the matrix variable. We say that F is *degenerate elliptic*.

Remark 2.3.1. In the case of parabolic Monge–Ampère equations such as (2.1) or (2.2), the nonlinearity is well-defined and degenerate elliptic only on a subset of $\mathbb{S}_d$; precisely, it is only defined either on the subset $\mathbb{S}_d^+$ of semi-definite symmetric matrices or on the subset $\mathbb{S}_d^{++}$ of definite symmetric matrices. Hence, solutions should be convex or strictly convex.

2.3.1.2 Semi-continuity

Consider an open set $Q \subset \mathbb{R}^{d+1}$. We recall that u is lower semi-continuous at (t, x) if, for all sequences $(s_n, y_n) \to (t, x)$,

$$u(t, x) \leq \liminf_{n \to \infty} u(s_n, y_n).$$

In the same way, one can define upper semi-continuous functions. Very often, the previous inequality is written

$$u(t, x) \leq \liminf_{(s,y) \to (t,x)} u(s, y).$$

If u is bounded from below in a neighbourhood of Q, one can define the *lower semi-continuous envelope* of u in Q as the largest lower semi-continuous function lying below u. It is denoted by u_*. Similarly, the upper semi-continuous envelope u^* of a locally bounded from above function u can be defined.

2.3.1.3   Definition(s)

In this paragraph, we give the definition of a viscosity solution of the fully nonlinear parabolic equation (2.14). We give a first definition in terms of test functions. We then introduce the notion of subdifferentials and superdifferentials with which an equivalent definition can be given (see Remark 2.3.8 below).

In order to motivate the definition of a viscosity solution, we first derive necessary conditions for smooth solutions of (2.14).

Consider an open set $Q \subset \mathbb{R}^{d+1}$ and a function $u : Q \to \mathbb{R}$ which is C^1 with respect to t and C^2 with respect to x. Consider also a function ϕ with the same regularity and assume that $u \leq \phi$ in a neighbourhood of $(t, x) \in Q$ and $u = \phi$ at (t, x). Then

$$\frac{\partial \phi}{\partial t}(t, x) = \frac{\partial u}{\partial t}(t, x)$$

$$D\phi(t, x) = Du(t, x)$$

$$D^2\phi(t, x) \geq D^2u(t, x).$$

Using the degenerate ellipticity of the nonlinearity F, we conclude that

$$\frac{\partial \phi}{\partial t}(t, x) + F(t, x, D\phi(t, x), D^2\phi(t, x))$$

$$\leq \frac{\partial u}{\partial t}(t, x) + F(t, x, Du(t, x), D^2u(t, x)) = 0.$$

A similar argument can be used to prove that if $u \geq \phi$ in a neighbourhood of (t, x) with $u(t, x) = \phi(t, x)$ then the reserve inequality holds true. These facts motivate the following definitions.

Definition 2.3.2 (Test functions). A *test function* on the set Q is a function $\phi : Q \to \mathbb{R}$ which is C^1 with respect to t and C^2 with respect to x.

Given a function $u : Q \to \mathbb{R}$, we say that the test function ϕ *touches u from above (resp. below) at* (t, x) if $u \leq \phi$ (resp. $u \geq \phi$) in a neighbourhood of (t, x) and $u(t, x) = \phi(t, x)$.

Remark 2.3.3. If $u - \phi$ reaches a local maximum (resp. minimum) at (t_0, x_0), then $\phi + [u(t_0, x_0) - \phi(t_0, x_0)]$ touches u from above (resp. below).

Definition 2.3.4 (Viscosity solutions). Consider a function $u : Q \to \mathbb{R}$ for some open set Q.

- u is a *subsolution* of (2.14) if u is upper semi-continuous and if, for all $(t, x) \in Q$ and all test functions ϕ touching u from above at (t, x),

$$\frac{\partial \phi}{\partial t}(t, x) + F(t, x, D\phi(t, x), D^2\phi(t, x)) \leq 0.$$

- u is a *supersolution* of (2.14) if u is lower semi-continuous and if, for all $(t, x) \in Q$ and all test functions ϕ touching u from below at (t, x),

$$\frac{\partial \phi}{\partial t}(t, x) + F(t, x, D\phi(t, x), D^2\phi(t, x)) \geq 0.$$

- u is a *solution* of (2.14) if it is both a sub- and a supersolution.

Remark 2.3.5. Remark that a viscosity solution of (2.14) is a continuous function.

When proving uniqueness of viscosity solutions, it is convenient to work with the following objects.

Definition 2.3.6 (Second order sub-/super-differentials). The following set

$$\mathcal{P}^{\pm}(u)(t, x) = \{(\alpha, p, X) \in \mathbb{R} \times \mathbb{R}^d \times \mathbb{S}_d :$$

$$(\alpha, p, X) = (\partial_t \phi(t, x), D\phi(t, x), D^2\phi(t, x))$$

$$\text{s.t. } \phi \text{ touches } u \text{ from above (resp. below) at } (t, x)\}$$

is *the super-(resp. sub-)differential* of the function u at the point (t, x).

Remark 2.3.7. Here is an equivalent definition: $(\alpha, p, X) \in \mathcal{P}^+ u(t, x)$ if and only if

$$u(s, y) \geq u(t, x) + \alpha(s-t) + p \cdot (y-x) + \frac{1}{2}X(x-y) \cdot (x-y) + o\left(|s-t| + |y-x|^2\right)$$

for (s, y) in a neighbourhood of (t, x). A similar characterization holds for $\mathcal{P}^-$.

Remark 2.3.8. The definition of a viscosity solution can be given using sub- and super-differentials of u. Indeed, as far as subsolutions are concerned, in view of Definitions 2.3.4 and 2.3.6, u is a viscosity subsolution of (2.14) in the open set Q if and only if for all $(t, x) \in Q$ and all $(\alpha, p, X) \in \mathcal{P}^+ u(t, x)$,

$$\alpha + F(t, x, p, X) \leq 0.$$

When proving uniqueness, the following limiting versions of the previous objects are used.

Definition 2.3.9 (Limiting super-/sub-differentials).

$$\overline{\mathcal{P}}^{\pm}(u)(t, x) = \{(\alpha, p, X) \in \mathbb{R} \times \mathbb{R}^d \times \mathbb{S}_d : \exists (t_n, x_n) \to (t, x) \text{ s.t.}$$

$$(\alpha_n, p_n, X_n) \to (\alpha, p, X), u(t_n, x_n) \to u(t, x),$$

$$(\alpha_n, p_n, X_n) \in \mathcal{P}^{\pm} u(t_n, x_n)\}$$

Remark 2.3.10. Since F is assumed to be continuous, the reader can remark that u is a viscosity subsolution of (2.14) in Q if and only if for all $(t, x) \in Q$ and all $(\alpha, p, X) \in \overline{\mathcal{P}}^+ u(t, x)$,

$$\alpha + F(t, x, p, X) \leq 0.$$

An analogous remark can be made for supersolutions.

2.3.1.4   First Properties

In this section, we state without proofs some important properties of sub- and supersolutions. Proofs in the elliptic case can be found in [CIL92] for instance. These proofs can be readily adapted to the parabolic framework.

Proposition 2.3.11 (Stability properties).

- *Let $(u_\alpha)_\alpha$ be a family of subsolutions of (2.14) in Q such that the upper semi-continuous envelope u of $\sup_\alpha u_\alpha$ is finite in Q. Then u is also a subsolution of (2.14) in Q.*
- *If $(u_n)_n$ is a sequence of subsolutions of (2.14), then the upper relaxed-limit u of the sequence defined as follows*

$$\bar{u}(t, x) = \limsup_{(s,y) \to (t,x), n \to \infty} u_n(s, y) \tag{2.16}$$

is everywhere finite in Q, then it is a subsolution of (2.14) in Q.

Remark 2.3.12. An analogous proposition can be stated for supersolutions.

2.3.2   The Perron Process

In this subsection, we would like to give an idea of the general process that allows one to construct solutions for fully nonlinear parabolic equations.

2.3.2.1   General Idea

The Perron process is well known in harmonic analysis and potential analysis. It has been adapted to the case of fully nonlinear elliptic equations in non-divergence form by Ishii [Ish87].

The general idea is the following one: assume that one can construct a subsolution u^- and a supersolution u^+ to a nonlinear parabolic equation of the form (2.14) such that $u^- \leq u^+$. Using Proposition 2.3.11, we can construct a maximal subsolution u lying between u^- and u^+. Then a general argument allows one to prove that the lower semi-continuous envelope of the maximal subsolution u is in fact a supersolution.

Remark 2.3.13. Before making the previous argument a little bit more precise, we would like to point out that the function u constructed by this general method is not a solution in the sense of Definition 2.3.4. It is a so-called *discontinuous (viscosity) solution* of (2.14). We decided to stick to continuous viscosity solution in these lecture notes and to state the result of the Perron process as in Lemma 2.3.15 below. See also Sect. 2.3.2.3.

Example 2.3.14. In many important cases, $u^\pm$ are chosen in the following form: $u_0(x) \pm Ct$ where u_0 is the smooth initial datum and C is a large constant, precisely:

$$C \geq \sup_{x \in \mathbb{R}^d} |F(0, x, Du_0(x), D^2 u_0(x))|.$$

If non-smooth/unbounded initial data are to be considered, discontinuous stability arguments can be used next.

2.3.2.2 Maximal Subsolution and Bump Construction

We now give more details about the general process to construct a "solution". We consider a cylindrical domain $Q = (0, T) \times \Omega$ for some domain $\Omega \subset \mathbb{R}^d$.

Lemma 2.3.15. *Assume that $u^\pm$ is a super-(resp. sub-) solution of (2.14) in Q. Then there exists a function $u : Q \to \mathbb{R}$ such that $u^- \leq u \leq u^+$ and u^* is a subsolution of (2.14) and u_* is a supersolution of (2.14).*

Proof. Consider

$$S = \{v : Q \to \mathbb{R} \text{ s.t. } u^- \leq v \leq u^+ \text{ and } v^* \text{ subsolution of (2.14)}\}.$$

By Proposition 2.3.11, we know that the upper semi-continuous envelope u^* of the function

$$u = \sup_{v \in S} v$$

is a subsolution of (2.14).

We next prove that the lower semi-continuous envelope u_* of u is a supersolution of (2.14) in Q. Arguing by contradiction, one can assume that there exists $(\alpha, p, X) \in \mathcal{P}^- u_*(t, x)$ such that

$$\alpha + F(t, x, p, X) =: -\theta < 0. \tag{2.17}$$

Remark that at (t, x), we have necessarily

$$u_*(t, x) < u^+(t, x).$$

Indeed, if this is not the case, then $(\alpha, p, X) \in \mathcal{P}^- u^+(t, x)$ and (2.17) cannot be true since u^+ is a supersolution of (2.14). Up to modifying the constant θ, we can also assume that

$$u_*(t, x) - u^+(t, x) \le -\theta < 0. \tag{2.18}$$

Without loss of generality, we can also assume that $(t, x) = (0, 0)$ and $u_*(t, x) = 0$. Let us consider the following "paraboloid"

$$P(s, y) = \tau s + p \cdot y + \frac{1}{2} X y \cdot y + \delta - \gamma \left(\frac{1}{2} |y|^2 + |s| \right)$$

with δ and γ to be chosen later. Compute next

$$\frac{\partial P}{\partial s}(s, y) + F(s, y, DP(s, y), D^2 P(s, y))$$

$$= \tau - \gamma \frac{s}{|s|} + F(s, y, p + X y - \gamma y, X - \gamma I)$$

(if $s = 0$, $\frac{s}{|s|}$ should be replaced with any real number $\sigma \in [-1, 1]$). Hence, for r and γ small enough, we have

$$\frac{\partial P}{\partial s} + F(s, y, DQ, D^2 Q) \le -\frac{\theta}{2} < 0$$

for all $(s, y) \in V_r$. Moreover, since $(\tau, p, X) \in \mathcal{P}^- u_*(t, x)$, we have

$$u_*(s, y) \ge \tau s + p \cdot y + \frac{1}{2} X y \cdot y + o(|y|^2 + |s|)$$

$$\ge P(s, y) - \delta + \gamma \left(\frac{1}{2} |y|^2 + |s| \right) + o(|y|^2 + |s|).$$

Choose now $\delta = \frac{\gamma r}{4}$ and consider $(s, y) \in V_r \setminus V_{r/2}$:

$$u_*(s, y) \ge P(s, y) - \frac{\gamma r}{4} + \frac{\gamma r}{2} + o(r) = P(s, y) + \frac{\gamma r}{4} + o(r).$$

Consequently, for r small enough,

$$u(s, y) - P(s, y) \ge \frac{\gamma r}{8} > 0 \text{ in } V_r \setminus V_{r/2},$$

$$P(s, y) < u^+(s, y) \text{ in } V_r$$

where we used (2.18) to get the second inequality.

We next consider

$$U(s, y) = \begin{cases} \max\{u(s, y), P(s, y)\} & \text{if } (s, y) \in V_r, \\ u(s, y) & \text{if not.} \end{cases}$$

On one hand, we remark that the function U^* is still a subsolution of (2.14) and $U \geq u \geq u_-$ and $U \leq u^+$. Consequently, $U \in S$ and in particular, $U \leq u$. On the other hand, $\sup_{\mathbb{R}^+ \times \mathbb{R}^d}\{U - u\} \geq \delta$; indeed, consider $(t_n, x_n) \to (0, 0)$ such that $u(t_n, x_n) \to u_*(0, 0) = 0$ and write

$$\lim_{n \to \infty} U(t_n, x_n) - u(t_n, x_n) \geq \lim_{n \to \infty} P(t_n, x_n) - u(t_n, x_n) = \delta > 0.$$

This contradicts the fact that $U \leq u$. The proof of the lemma is now complete. □

2.3.2.3   Continuous Solutions from Comparison Principle

As mentioned above, the maximal subsolution u^* is not necessarily continuous; hence, its lower semi-continuous envelope u_* does not coincide necessarily with it. In particular, we cannot say that u is a solution in the sense of Definition 2.3.4 (cf. Remark 2.3.13 above).

We would get a (continuous viscosity) solution if $u^* = u_*$. On one hand, u^* is upper semi-continuous by construction and on the other hand $u_* \leq u^*$ by definition of the semi-continuous envelopes. Hence, u is a solution of (2.14) if and only if $u^* \leq u_*$ in Q. Since u^* is a subsolution of (2.14) in Q and u_* is a supersolution of (2.14) in Q, it is thus enough that (2.14) satisfies a *comparison principle* and that the barriers $u^{\pm}$ satisfy some appropriate inequality on the parabolic boundary. More precisely, we would like on one hand that

Comparison principle. *If u is a subsolution of (2.14) in Q and v is a supersolution of (2.14) in Q and $u \leq v$ on the parabolic boundary $\partial_p Q$, then $u \leq v$ in Q.*

and on the other hand, we would like that $u^* \leq u_*$ on $\partial_p Q$. This boundary condition would be true if

$$(u^+)^* \leq (u^-)_* \text{ on } \partial_p Q.$$

We emphasize that the lower and upper semi-continuous envelopes appearing in the previous inequality are performed with respect to time and space.

Example 2.3.16. If for instance $Q = (0, T) \times \mathbb{R}^d$, then barriers should satisfy

$$(u^+)^*(0, x) \leq (u^-)_*(0, x) \text{ for } x \in \mathbb{R}^d.$$

This condition is fullfilled for such a Q if $u^{\pm} = u_0 \pm Ct$ (see Example 2.3.14).

In the next subsection, we will present general techniques for proving comparison principles. The reader should be aware of the fact that, in many practical cases, general theorems from the viscosity solution theory do not apply to the equation under study. In those cases, one has to adapt the arguments presented below in order to take into account the specific difficulties implied by the specific equation. The reader is referred to [CIL92] for a large review of available tools.

2.3.3   Introduction to Comparison Principles

In this subsection, we present classical techniques to prove comparison principles in some typical cases.

2.3.3.1   First Order Equations

In this paragraph, we first study first order Hamilton–Jacobi equations of the following form

$$\frac{\partial u}{\partial t} + H(x, Du) = 0. \tag{2.19}$$

As we will see, a comparison principle holds true if H satisfies the following structure condition: for all $x, y, p \in \mathbb{R}^d$,

$$|H(x, p) - H(y, p)| \le C|x - y|. \tag{2.20}$$

In order to avoid technicalities and illustrate main difficulties, we assume that $x \mapsto H(x, p)$ is $\mathbb{Z}^d$-periodic; hence, solutions should also be $\mathbb{Z}^d$-periodic for $\mathbb{Z}^d$-periodic initial data.

Theorem 2.3.17 (Comparison principle II). *Consider a continuous $\mathbb{Z}^d$-periodic function u_0. If u is a $\mathbb{Z}^d$-periodic subsolution of (2.19) in $(0, T) \times \mathbb{R}^d$ and v is a $\mathbb{Z}^d$-periodic supersolution of (2.19) in $(0, T) \times \mathbb{R}^d$ such that $u(0, x) \le u_0(x) \le v(0, x)$ for all $x \in \mathbb{R}^d$, then $u \le v$ in $(0, T) \times \mathbb{R}^d$.*

Proof. The beginning of the proof is the same as in the proof of Theorem 2.2.4: we assume that

$$M = \sup_{t \in (0,T), x \in \mathbb{R}^d} \left\{ u(t, x) - v(t, x) - \frac{\gamma}{T - t} \right\} > 0.$$

Here, we cannot use the equation directly, since it is not clear wether $u - v$ satisfies a nonlinear parabolic equation or not (recall that the equation is nonlinear). Hence, we should try to duplicate the (time and space) variables.

Doubling Variable Technique

Consider

$$
M_\varepsilon = \sup_{t,s\in(0,T),x,y\in\mathbb{R}^d} \left\{ u(t,x) - v(s,y) - \frac{(t-s)^2}{2\varepsilon} - \frac{|x-y|^2}{2\varepsilon} - \frac{\eta}{T-t} \right\}.
$$

Remark that $M_\varepsilon \geq M > 0$. This supremum is reached since u is upper semi-continuous and v is lower semi-continuous and both functions are $\mathbb{Z}^d$-periodic. Let $(t_\varepsilon, s_\varepsilon, x_\varepsilon, y_\varepsilon)$ denote a maximizer. Then we have

$$
\frac{(t_\varepsilon - s_\varepsilon)^2}{2\varepsilon} + \frac{|x_\varepsilon - y_\varepsilon|^2}{2\varepsilon} \leq u(t_\varepsilon, x_\varepsilon) - v(s_\varepsilon, y_\varepsilon) \leq |u^+|_0 + |v_-|_0
$$

where we recall that $|w|_0 = \sup_{(t,x)\in(0,T)\times\mathbb{R}^d} |w(t,x)|$. In particular, up to extracting subsequences, $t_\varepsilon \to t$, $s_\varepsilon \to t$ and $x_\varepsilon \to x$, $y_\varepsilon \to y$ and $t_\varepsilon - s_\varepsilon = O(\sqrt{\varepsilon})$ and $x_\varepsilon - y_\varepsilon - O(\sqrt{\varepsilon})$.

Assume first that $t = 0$. Then

$$
0 < M \leq \limsup_{\varepsilon\to0} M_\varepsilon \leq \limsup_\varepsilon u(t_\varepsilon, x_\varepsilon) - \liminf_\varepsilon v(s_\varepsilon, y_\varepsilon)
$$

$$
\leq u(0,x) - v(0,x) \leq 0.
$$

This is not possible. Hence $t > 0$.

Since $t > 0$, for ε small enough, $t_\varepsilon > 0$ and $s_\varepsilon > 0$. Now remark that the function ϕ_u

$$
(t,x) \mapsto v(s_\varepsilon, y_\varepsilon) + \frac{(t-s_\varepsilon)^2}{2\varepsilon} + \frac{|x-y_\varepsilon|^2}{2\varepsilon} + \frac{\eta}{T-t}
$$

is a test function such that $u - \phi_u$ reaches a maximum at $(t_\varepsilon, x_\varepsilon)$. Hence (recall Remark 2.3.3),

$$
\frac{\eta}{(T-t_\varepsilon)^2} + \frac{t_\varepsilon - s_\varepsilon}{\varepsilon} + H(x_\varepsilon, p_\varepsilon) \leq 0
$$

with $p_\varepsilon = \frac{x_\varepsilon - y_\varepsilon}{\varepsilon}$. Similarly, the function ϕ_v

$$
(s,y) \mapsto u(t_\varepsilon, x_\varepsilon) - \frac{(s-t_\varepsilon)^2}{2\varepsilon} - \frac{|y-x_\varepsilon|^2}{2\varepsilon} - \frac{\eta}{T-t_\varepsilon}
$$

is a test function such that $v - \phi_v$ reaches a minimum at $(s_\varepsilon, y_\varepsilon)$; hence

$$
\frac{t_\varepsilon - s_\varepsilon}{\varepsilon} + H(y_\varepsilon, p_\varepsilon) \leq 0
$$

with the same p_ε! Substracting the two viscosity inequalities yields

$$\frac{\eta}{(T - t_\varepsilon)^2} \leq H(y_\varepsilon, p_\varepsilon) - H(x_\varepsilon, p_\varepsilon).$$

In view of (2.20), we conclude that

$$\frac{\eta}{T^2} \leq C|x_\varepsilon - y_\varepsilon| = O(\sqrt{\varepsilon}).$$

Letting $\varepsilon \to 0$ yields the desired contradiction.   □

Remark 2.3.18. Condition (2.20) is satisfied by (2.13) if the potential V is Lipschitz continuous. On the contrary, if $H(x, p) = c(x)|p|$, then the Hamilton–Jacobi equation is the so-called eikonal equation and it does not satisfy (2.20) even if c is globally Lipschitz. Such an Hamiltonian satisfies

$$|H(x, p) - H(y,)| \leq C(1 + |p|)|x - y|. \tag{2.21}$$

For such equations, the penalization should be studied in greater details in order to prove that

$$\frac{|x_\varepsilon - y_\varepsilon|^2}{2\varepsilon} \to 0 \text{ as } \varepsilon \to 0.$$

With this piece of information in hand, the reader can check that the same contradiction can be obtained for Lipschitz c's. See for instance [Bar194] for details.

Since we will use once again this additional fact about penalization, we state it now in a lemma.

Lemma 2.3.19. *Consider* $\tilde{u}(t, x) = u(t, x) - \eta(T - t)^{-1}$. *Assume that*

$$M_\varepsilon = \sup_{\substack{x,y \in \mathbb{R}^d \\ t,s \in (0,T)}} \tilde{u}(t, x) - v(s, y) - \frac{|x - y|^2}{2\varepsilon} - \frac{|t - s|^2}{2\varepsilon}$$

is reached at $(x_\varepsilon, y_\varepsilon, t_\varepsilon, s_\varepsilon)$. *Assume moreover that* $(x_\varepsilon, y_\varepsilon, t_\varepsilon, s_\varepsilon) \to (x, y, t, s)$ *as* $\varepsilon \to 0$. *Then*

$$\frac{|x_\varepsilon - y_\varepsilon|^2}{\varepsilon} \to 0 \text{ as } \varepsilon \to 0.$$

Remark 2.3.20. The reader can check that the previous lemma still holds true if $v(s, y)$ is replaced with $v(t, y)$ and if the term $\varepsilon^{-1}|t - s|^2$ is removed.

Proof. Remark first that $\varepsilon \mapsto M_\varepsilon$ is non-decreasing and $M_\varepsilon \geq M := \sup_{\mathbb{R}^d} (\tilde{u} - v)$. Hence, as $\varepsilon \to 0$, M_ε converges to some limit $l \geq M$. Moreover,

$$M_{2\varepsilon} \geq \tilde{u}(t_\varepsilon, x_\varepsilon) - v(s_\varepsilon, y_\varepsilon) - \frac{|x_\varepsilon - y_\varepsilon|^2}{4\varepsilon} - \frac{|t_\varepsilon - s_\varepsilon|^2}{4\varepsilon}$$

$$\geq M_\varepsilon + \frac{|x_\varepsilon - y_\varepsilon|^2}{4\varepsilon} + \frac{|t_\varepsilon - s_\varepsilon|^2}{4\varepsilon}.$$

Hence,

$$\frac{|x_\varepsilon - y_\varepsilon|^2}{4\varepsilon} + \frac{|t_\varepsilon - s_\varepsilon|^2}{4\varepsilon} \leq M_{2\varepsilon} - M_\varepsilon \to l - l = 0. \qquad \square$$

2.3.3.2 Second Order Equations with No x Dependance

In this subsection we consider the following equation

$$\frac{\partial u}{\partial t} + H(x, Du) - \Delta u = 0 \tag{2.22}$$

still assuming that $x \mapsto H(x, p)$ is $\mathbb{Z}^d$-periodic and satisfies (2.20). The classical parabolic theory implies that there exists smooth solutions for such an equation. However, we illustrate viscosity solution techniques on this (too) simple example.

Theorem 2.3.21 (Comparison principle III). *Consider a continuous $\mathbb{Z}^d$-periodic function u_0. If u is a $\mathbb{Z}^d$-periodic subsolution of (2.22) in $(0, T) \times \mathbb{R}^d$ and v is a $\mathbb{Z}^d$-periodic supersolution of (2.19) in $(0, T) \times \mathbb{R}^d$ such that $u(0, x) \leq u_0(x) \leq v(0, x)$ for all $x \in \mathbb{R}^d$, then $u \leq v$ in $(0, T) \times \mathbb{R}^d$.*

Remark 2.3.22. A less trivial example would be

$$\frac{\partial u}{\partial t} + H(x, Du) - \text{trace}(A_0 D^2 u) = 0$$

for some degenerate matrix $A_0 \in \mathbb{S}_d$, $A_0 \geq 0$. We prefer to keep it simple and study (2.22).

First attempt of proof. We follow the proof of Theorem 2.3.17. If one uses the two test functions ϕ_u and ϕ_v to get viscosity inequalities, this yields

$$\frac{1}{(T - t_\varepsilon)^2} + \frac{t_\varepsilon - s_\varepsilon}{\varepsilon} + H(x_\varepsilon, p_\varepsilon) \leq \text{trace}(\varepsilon^{-1} I),$$

$$\frac{t_\varepsilon - s_\varepsilon}{\varepsilon} + H(y_\varepsilon, p_\varepsilon) \geq - \text{trace}(\varepsilon^{-1} I).$$

Substracting these two inequalities, we get

$$\frac{1}{T^2} \leq O(\sqrt{\varepsilon}) + \frac{2d}{\varepsilon}$$

and it is not possible to get a contradiction by letting $\varepsilon \to 0$.   □

In the previous proof, we lost a very important piece of information about second order derivatives; indeed, assume that u and v are smooth. As far as first order equations are concerned, using the first order optimality condition

$$Du(t_\varepsilon, x_\varepsilon) - p_\varepsilon = 0 \quad \text{and} \quad -Dv(s_\varepsilon, y_\varepsilon) + p_\varepsilon = 0$$

is enough. But for second order equations, one has to use second order optimality condition

$$\begin{pmatrix} Du(t_\varepsilon, x_\varepsilon) & 0 \\ 0 & -Dv(s_\varepsilon, y_\varepsilon) \end{pmatrix} \leq \begin{pmatrix} \varepsilon^{-1}I & -\varepsilon^{-1}I \\ -\varepsilon^{-1}I & \varepsilon^{-1}I \end{pmatrix}.$$

It turns out that for *semi-continuous functions*, the previous inequality still holds true up to an arbitrarily small error in the right hand side.

Uniqueness of viscosity solutions for second order equations where first obtained by Lions [Lions83] by using probabilistic methods. The analytical breakthrough was achieved by Jensen [Jens88]. Ishii's contribution was also essential [Ish89]. In particular, he introduced the matrix inequalities contained in the following lemma. See [CIL92] for a detailed historical survey.

We give a first version of Jensen–Ishii's lemma for the specific test function $(2\varepsilon)^{-1}|x - y|^2$.

Lemma 2.3.23 (Jensen–Ishii's lemma I). *Let U and V be two open sets of $\mathbb{R}^d$ and I an open interval of $\mathbb{R}$. Consider also a bounded subsolution u of (2.14) in $I \times U$ and a bounded supersolution v of (2.14) in $I \times V$. Assume that $u(t, x) - v(t, y) - \frac{|x-y|^2}{2\varepsilon}$ reaches a local maximum at $(t_0, x_0, y_0) \in I \times U \times V$. Letting p denote $\varepsilon^{-1}(x_0 - y_0)$, there exists $\tau \in \mathbb{R}$ and $X, Y \in \mathbb{S}_d$ such that*

$$(\tau, p, X) \in \overline{\mathcal{P}}^+ u(t_0, x_0), (\tau, p, Y) \in \overline{\mathcal{P}}^- v(t_0, y_0)$$

$$-\frac{2}{\varepsilon} \begin{pmatrix} I & 0 \\ 0 & I \end{pmatrix} \leq \begin{pmatrix} X & 0 \\ 0 & -Y \end{pmatrix} \leq \frac{3}{\varepsilon} \begin{pmatrix} I & -I \\ -I & I \end{pmatrix}. \tag{2.23}$$

Remark 2.3.24. As a matter of fact, it is not necessary to assume that u and v are sub- and supersolution of an equation of the form (2.14). We chose to present first the result in this way to avoid technicalities. Later on, we will need the standard version of this lemma, so we will state it. See Lemma 2.3.30 below.

Remark 2.3.25. Such a result holds true for more general test functions $\phi(t, x, y)$ than $(2\varepsilon)^{-1}|x - y|^2$. However, this special test function is a very important one and many interesting results can be proven with it. We will give a more general version of this important result, see Lemma 2.3.30.

Remark 2.3.26. The attentive reader can check that the matrix inequality (2.23) implies in particular $X \leq Y$.

Remark 2.3.27. This lemma can be used as a black box and one does so very often. But we mentioned above that some times, one has to work more to get a uniqueness result for some specific equation. In this case, it could be necessary to consider more general test functions, or even to open the black box and go through the proof to adapt it in a proper way.

With such a lemma in hand, we can now prove Theorem 2.3.21.

Proof of Theorem 2.3.21. We argue as in the proof of Theorem 2.3.17 but we do not duplicate the time variable since it is embedded in Lemma 2.3.23. Instead, we consider

$$M_\varepsilon = \sup_{\substack{x,y \in \mathbb{R}^d \\ t \in (0,T)}} \left\{ u(t, x) - v(t, y) - \frac{|x - y|^2}{2\varepsilon} - \frac{\eta}{T - t} \right\},$$

let $(t_\varepsilon, x_\varepsilon, y_\varepsilon)$ denote a maximiser and apply Lemma 2.3.23 with $\tilde{u}(t, x) = u(t, x) - \frac{\eta}{T-t}$ and v and we get τ, X, Y such that

$$\left(\tau + \frac{\eta}{(T - t)^2}, p_\varepsilon, X \right) \in \overline{\mathcal{P}}^+ u(t_\varepsilon, x_\varepsilon), (\tau, p_\varepsilon, Y) \in \overline{\mathcal{P}}^- v(t_\varepsilon, y_\varepsilon), \quad X \leq Y$$

(see Remark 2.3.26 above). Hence, we write the two viscosity inequalities

$$\frac{\gamma}{(T - t)^2} + \tau + H(x_\varepsilon, p_\varepsilon) \leq \text{trace } X$$

$$\tau + H(y_\varepsilon, p_\varepsilon) \geq \text{trace } Y \geq \text{trace } X$$

and we substract them in order to get the desired contradiction

$$\frac{\gamma}{T^2} \leq O(\sqrt{\varepsilon}).$$

The proof is now complete.   □

2.3.3.3 Second Order Equations with x Dependance

In this paragraph, we prove a comparison principle for the following *degenerate elliptic* equation

$$\frac{\partial u}{\partial t} + H(x, Du) - \text{trace}(\sigma(x)\sigma^T(x)D^2u) = 0 \qquad (2.24)$$

under the following assumptions

- $x \mapsto H(x, p)$ is $\mathbb{Z}^d$-periodic and satisfies (2.21);
- $\sigma : \mathbb{R}^d \to \mathbb{M}_{d,m}(\mathbb{R})$ is Lipschitz continuous and $\mathbb{Z}^d$-periodic, $m \leq d$.

Here, $\mathbb{M}_{d,m}(\mathbb{R})$ denotes the set of real $d \times m$-matrices. We make precise that σ^T denotes the transpose matrix of the $d \times m$-matrix σ.

The following theorem is, to some respects, the nonlinear counterpart of the first comparison principle we proved in Sect. 2.2 (see Corollary 2.2.6). Apart from the nonlinearity of the equation, another significant difference with Corollary 2.2.6 is that (2.24) is degenerate elliptic and not uniformly elliptic.

Theorem 2.3.28 (Comparison principle IV). *Consider a continuous $\mathbb{Z}^d$-periodic function u_0. If u is a $\mathbb{Z}^d$-periodic subsolution of (2.22) in $(0, T) \times \mathbb{R}^d$ and v is a $\mathbb{Z}^d$-periodic supersolution of (2.19) in $(0, T) \times \mathbb{R}^d$ such that $u(0, x) \leq u_0(x) \leq v(0, x)$ for all $x \in \mathbb{R}^d$, then $u \leq v$ in $(0, T) \times \mathbb{R}^d$.*

Proof. We argue as in the proof of Theorem 2.3.21. The main difference lies after writing viscosity inequalities thanks to Jensen–Ishii's lemma. Indeed, one gets

$$\frac{\eta}{T^2} \leq -H(x_\varepsilon, p_\varepsilon) + H(y_\varepsilon, p_\varepsilon) + \text{trace}(\sigma(x_\varepsilon)\sigma^T(x_\varepsilon)X) - \text{trace}(\sigma(y_\varepsilon)\sigma^T(y_\varepsilon)Y)$$

$$\leq C\left(1 + \frac{|x_\varepsilon - y_\varepsilon|}{\varepsilon}\right)|x_\varepsilon - y_\varepsilon|$$

$$+ \text{trace}(\sigma(x_\varepsilon)\sigma^T(x_\varepsilon)X) - \text{trace}(\sigma(y_\varepsilon)\sigma^T(y_\varepsilon)Y).$$

The first term can be handled thanks to Lemma 2.3.19. But one cannot just use $X \leq Y$ obtained from the matrix inequality (2.23) to handle the second one. Instead, consider an orthonormal basis $(e_i)_i$ of $\mathbb{R}^m$ and write

$$\text{trace}(\sigma(x_\varepsilon)\sigma^T(x_\varepsilon)X) - \text{trace}(\sigma(y_\varepsilon)\sigma^T(y_\varepsilon)Y)$$

$$= \text{trace}(\sigma^T(x_\varepsilon)X\sigma(x_\varepsilon)) - \text{trace}(\sigma^T(y_\varepsilon)Y\sigma(y_\varepsilon))$$

$$= \sum_{i=1}^{m}(X\sigma(x_\varepsilon)e_i \cdot \sigma(x_\varepsilon)e_i - Y\sigma(y_\varepsilon)e_i \cdot \sigma(y_\varepsilon)e_i)$$

$$\leq \frac{3}{\varepsilon}\sum_{i=1}^{m}|\sigma(x_\varepsilon)e_i - \sigma(y_\varepsilon)e_i|^2;$$

we applied (2.23) to vectors of the form $(\sigma(x_\varepsilon)e_i, \sigma(y_\varepsilon)e_i) \in \mathbb{R}^d \times \mathbb{R}^d$ to get the last line. We can now use the fact that σ is Lipschitz continuous and get

$$\operatorname{trace}(\sigma(x_\varepsilon)\sigma^T(x_\varepsilon)X) - (\sigma(y_\varepsilon)\sigma^T(y_\varepsilon)Y) \le C\frac{|x_\varepsilon - y_\varepsilon|^2}{\varepsilon}.$$

We thus finally get

$$\frac{\eta}{T^2} \le C|x_\varepsilon - y_\varepsilon| + C\frac{|x_\varepsilon - y_\varepsilon|^2}{\varepsilon}.$$

We can now get the contradiction $\eta < 0$ by using Lemma 2.3.19 and letting $\varepsilon \to 0$. The proof is now complete.   □

2.3.4 Hölder Continuity Through the Ishii–Lions Method

In this subsection, we want to present a technique introduced by Ishii and Lions in [IL90] in order to prove Hölder continuity of solutions of very general fully nonlinear elliptic and parabolic equations. On one hand, it is much simpler than the proof we will present in the next section; on the other hand, it cannot be used to prove further regularity such as Hölder continuity of the gradient.

The fundamental assumptions is that the equation is uniformly elliptic (see below for a definition). For pedagogical purposes, we do not want to prove a theorem for the most general case. Instead, we will look at (2.24) for $\mathbb{S}_d$-valued σ's and special H's

$$\frac{\partial u}{\partial t} + c(x)|Du| - \operatorname{trace}(\sigma(x)\sigma(x)D^2u) = 0 \qquad (2.25)$$

Assumptions (A)

- c is bounded and Lipschitz continuous in Q;
- $\sigma : Q \to \mathbb{S}_d$ is bounded and Lipschitz continuous in x and constant in t;
- There exists $\lambda > 0$ such that for all $X = (t, x) \in Q$,

$$A(x) := \sigma(x)\sigma(x) \ge \lambda I.$$

Under these assumptions, the equation is *uniformly elliptic*, i.e. there exist two positive numbers $0 < \lambda \le \Lambda$, called *ellipticity constants*, such that

$$\forall X = (t, x) \in Q, \quad \lambda I \le A(x) \le \Lambda I. \qquad (2.26)$$

Theorem 2.3.29. *Under Assumptions (A) on H and σ, any viscosity solution u of (2.25) in an open set $Q \subset \mathbb{R}^{d+1}$ is Hölder continuous in time and space.*

When proving Theorem 2.3.29, we will need to use Jensen–Ishii's lemma for a test function which is more general than $(2\varepsilon)^{-1}|x - y|^2$. Such a result can be found in [CIL92].

Lemma 2.3.30 (Jensen–Ishii's Lemma II). *Let U and V be two open sets of $\mathbb{R}^d$ and I an open interval of $\mathbb{R}$. Consider also a bounded subsolution u of (2.14) in $I \times U$ and a bounded supersolution v of (2.14) in $I \times V$. Assume that $u(t, x) - v(t, y) - \phi(x - y)$ reaches a local maximum at $(t_0, x_0, y_0) \in I \times U \times V$. Letting p denote $D\phi(x_0 - y_0)$, for all $\beta > 0$ such that $\beta Z < I$, there exists $\tau \in \mathbb{R}$ and $X, Y \in \mathbb{S}_d$ such that*

$$(\tau, p, X) \in \overline{\mathcal{P}}^+ u(t_0, x_0), (\tau, p, Y) \in \overline{\mathcal{P}}^- v(t_0, y_0)$$

$$-\frac{2}{\beta} \begin{pmatrix} I & 0 \\ 0 & I \end{pmatrix} \leq \begin{pmatrix} X & 0 \\ 0 & -Y \end{pmatrix} \leq \begin{pmatrix} Z^\beta & -Z^\beta \\ -Z^\beta & Z^\beta \end{pmatrix} \tag{2.27}$$

where $Z = D^2\phi(x_0 - y_0)$ and $Z^\beta = (I - \beta Z)^{-1} Z$.

We can now turn to the proof of Theorem 2.3.29.

Proof of Theorem 2.3.29. We first prove that u is Hölder continuous with respect to x. Without loss of generality, we can assume that Q is bounded. We would like to prove that for all $X_0 = (t_0, x_0) \in Q$ and $(t, x), (t, y) \in Q$,

$$u(t, x) - u(t, y) \leq L_1 |x - y|^\alpha + L_2 |x - x_0|^2 + L_2 (t - t_0)^2$$

for $L_1 = L_1(X_0)$ and $L_2 = L_2(X_0)$ large enough. We thus consider

$$M = \sup_{(t,x),(t,y)\in Q} \{u(t, x) - u(t, y) - \phi(x - y) - \Gamma(t, x)\}$$

with $\phi(z) = L_1 |z|^\alpha$ and $\Gamma(t, x) = L_2 |x - x_0|^2 + L_2 (t - t_0)^2$ and we argue by contradiction: we assume that for all $\alpha \in (0, 1)$, $L_1 > 0$, $L_2 > 0$, we have $M > 0$.

Since Q is bounded, M is reached at a point denoted by $(\bar{t}, \bar{x}, \bar{y})$. The fact that $M > 0$ implies first that $\bar{x} \neq \bar{y}$. It also implies

$$\begin{cases} |\bar{x} - \bar{y}| \leq \left(\frac{2|u|_{0,Q}}{L_1}\right)^{\frac{1}{\alpha}} =: A < d(X_0, \partial Q), \\ |\bar{X} - X_0| < \sqrt{\frac{2|u|_{0,Q}}{L_2}} =: R_2 \leq \frac{d(X_0, \partial Q)}{2} \end{cases} \tag{2.28}$$

if L_1 and L_2 are chosen so that

$$L_1 > \frac{2|u|_{0,Q}}{(d(X_0, \partial Q))^\alpha}, \quad L_2 \geq \frac{8|u|_{0,Q}}{(d(X_0, \partial Q))^2}.$$

In particular we have $\bar{x}, \bar{y} \in \Omega$. We next apply Jensen–Ishii's Lemma 2.3.30 to $\tilde{u}(t, x) = u(t, x) - \Gamma(t, x)$ and $v(s, y)$. Then there exists $\tau \in \mathbb{R}$ and $X, Y \in \mathbb{S}_d$ such that

$$(\tau + 2L_2(\bar{t} - t_0), \bar{p} + 2L_2(\bar{x} - x_0), X + 2L_2 I) \in \overline{\mathcal{P}}^+ u(\bar{t}, \bar{x}), \quad (\tau, \bar{p}, Y) \in \overline{\mathcal{P}}^- u(\bar{t}, \bar{y})$$

where $\bar{p} = D\phi(\bar{x} - \bar{y})$ and $Z = D^2\phi(\bar{x} - \bar{y})$ and (2.27) holds true. In particular, $X \leq Y$. We can now write the two viscosity inequalities

$$2L_2(\bar{t} - t_0) + \tau + H(\bar{x}, \bar{p} + 2L_2(\bar{x} - x_0)) \leq \text{trace}(A(\bar{x})(X + 2L_2 I))$$

$$\tau + H(\bar{y}, \bar{p}) \geq \text{trace}(A(\bar{y})Y)$$

and combine them with (2.28) and (2.26) to get

$$-CL_2 \leq 2L_2(\bar{t} - t_0) \leq c(\bar{y})|\bar{p}| - c(\bar{x})|\bar{p} + 2L_2(\bar{x} - x_0)|$$
$$+ CL_2 + \text{trace}(A(\bar{x})X) - \text{trace}(A(\bar{y})Y). \tag{2.29}$$

We next estimate successively the difference of first order terms and the difference of second order terms. As far as first order terms are concerned, we use that c is bounded and Lipschitz continuous and (2.28) to get

$$c(\bar{y})|\bar{p}| - c(\bar{x})|\bar{p} + 2L_2(\bar{x} - x_0)| \leq C|\bar{x} - \bar{y}||\bar{p}| + CL_2|\bar{x} - x_0|$$
$$\leq C|\bar{x} - \bar{y}||\bar{p}| + CL_2. \tag{2.30}$$

As far as second order terms are concerned, we use (2.26) to get

$$\text{trace}(A(\bar{x})X) - \text{trace}(A(\bar{y})Y) \leq \text{trace}(A(\bar{x})(X - Y)) + \text{trace}((A(\bar{x}) - A(\bar{y}))Y)$$

$$\leq \lambda \, \text{trace}(X - Y)$$

$$+ \sum_i (\sigma(\bar{x})Y\sigma(\bar{x})e_i \cdot e_i - \sigma(\bar{y})Y\sigma(\bar{y})e_i \cdot e_i)$$

$$\leq \lambda \, \text{trace}(X - Y) + C\|Y\||\bar{x} - \bar{y}|.$$

We should next estimate $|\bar{p}|$, $\text{trace}(X - Y)$ and $\|Y\|$. In order to do so, we compute $D\phi$ and $D^2\phi$. It is convenient to introduce the following notation

$$a = \bar{x} - \bar{y}, \qquad \hat{a} = \frac{a}{|a|}, \qquad \varepsilon = |a|.$$

$$\bar{p} = D\phi(a) = L_1 \alpha |a|^{\alpha-2} a \tag{2.31}$$

$$Z = D^2\phi(a) = L_1 \alpha (|a|^{\alpha-2} I + (\alpha - 2)|a|^{\alpha-4} a \otimes a)$$

$$= \gamma^{-1}(I - (2 - \alpha)\hat{a} \otimes \hat{a}). \tag{2.32}$$

with $\gamma = (L_1\alpha)^{-1}\varepsilon^{2-\alpha}$. The reader can remark that if one chooses $\beta = \gamma/2$, then

$$Z^\beta = (I - \beta Z)^{-1} Z = \frac{2}{\gamma}\left(I - 2\frac{2-\alpha}{3-\alpha}\hat{a} \otimes \hat{a}\right). \tag{2.33}$$

Since Y is such that $-\frac{1}{\beta}I \leq -Y \leq Z^\beta$, we conclude that

$$\|Y\| \leq \frac{2}{\gamma}.$$

We next remark that (2.27) and (2.33) imply that all the eigenvalues of $X - Y$ are non-positive and that one of them is less than

$$4Z^\beta \hat{a} \cdot \hat{a} = -\frac{8}{\gamma}\frac{1-\alpha}{3-\alpha}.$$

Hence

$$\text{trace}(X - Y) \leq -\frac{8}{\gamma}\frac{1-\alpha}{3-\alpha}.$$

Finally, second order terms are estimated as follows

$$\text{trace}(A(\bar{x})X) - \text{trace}(A(\bar{y})Y) \leq -\frac{C}{\gamma} + C\frac{\varepsilon}{\gamma} \leq -\frac{C}{2\gamma} \tag{2.34}$$

(choosing L_1 large enough so that $\varepsilon \leq 1/2$). Combining now (2.29), (2.30) and (2.34) and recalling the definition of γ and ε, we finally get

$$-CL_2 \leq C\varepsilon^\alpha - \frac{CL_1}{\varepsilon^{2-\alpha}} \leq \frac{C}{L_1} - CL_1^{\frac{2}{\alpha}}.$$

Since L_2 is fixed, it is now enough to choose L_1 large enough to get the desired contradiction. The proof is now complete. □

2.4 Harnack Inequality

In this section, we consider the following special case of (2.14)

$$\frac{\partial u}{\partial t} + F(x, D^2 u) = f \tag{2.35}$$

for some uniformly elliptic nonlinearity F (see below for a definition) and some continuous function f. The goal of this section is to present and prove the Harnack inequality (Theorem 2.4.35). This result states that the supremum of a non-negative solution of (2.35) can be controlled from above by its infimum times a universal constant plus the L^{d+1}-norm of the right hand side f. The estimates that will be obtained do not depend on the regularity of F with respect to x.

We will see that it is easy to derive the Hölder continuity of solutions from the Harnack inequality, together with an estimate of the Hölder semi-norm.

The Harnack inequality is a consequence of both the L^ε-estimate (Theorem 2.4.15) and of the local maximum principle (Proposition 2.4.34). Since this local maximum principle is a consequence of the L^ε-estimate, the heart of the proof of the Harnack inequality thus lies in proving that a (small power of) non-negative supersolution is integrable, see Theorem 2.4.15 below.

The proof of the L^ε estimate relies on various measure estimates of the solution. These estimates are obtained through the use of a *maximum principle* due to Krylov in the parabolic case.

The proof of the L^ε estimate also involves many different sets, cylinders and cubes. The authors are aware of the fact that it is difficult to follow the corresponding notation. Some pictures are provided and the authors hope they are helpful with this respect.

Pucci's Operators

Given ellipticity constants $0 < \lambda \leq \Lambda$, we consider

$$P^+(M) = \sup_{\lambda I \leq A \leq \Lambda I} \{-\text{trace}(AM)\},$$

$$P^-(M) = \inf_{\lambda I \leq A \leq \Lambda I} \{-\text{trace}(AM)\}.$$

Some model fully nonlinear parabolic equations are

$$\frac{\partial u}{\partial t} + P^+(D^2 u) = f, \tag{2.36}$$

$$\frac{\partial u}{\partial t} + P^-(D^2 u) = f. \tag{2.37}$$

Remark that those nonlinear operators only depend on ellipticity constants λ, Λ and dimension d. They are said *universal*. Similarly, constants are said *universal* if they only depend on λ, Λ and d.

Uniform Ellipticity

Throughout the remaining of this section, we make the following assumptions on F: for all $X, Y \in \mathbb{S}_d$ and $x \in \Omega$,

$$P^-(X - Y) \leq F(x, X) - F(x, Y) \leq P^+(X - Y).$$

This condition is known as the *uniform ellipticity* of F. Remark that this condition implies in particular that F is degenerate elliptic in the sense of Sect. 2.3.1.1 (see Condition 2.15).

2.4.1 A Maximum Principle

In order to state and prove the maximum principle, it is necessary to define first the parabolic equivalent of the convex envelope of a function, which we will refer to as the monotone envelope.

2.4.1.1 Monotone Envelope of a Function

Definition 2.4.1 (Monotone envelope). If Ω is a convex set of $\mathbb{R}^d$ and (a, b) is an open interval, then the *monotone envelope* of a lower semi-continuous function $u : (a, b) \times \Omega \to \mathbb{R}$ is the largest function $v : (a, b) \times \Omega \to \mathbb{R}$ lying below u which is non-increasing with respect to t and convex with respect to x. It is denoted by $\Gamma(u)$.

Combining the usual definition of the convex envelope of a function with the non-increasing envelope of a function of one real variable, we obtain a first representation formula for $\Gamma(u)$.

Lemma 2.4.2 (Representation formula I).

$$\Gamma(u)(t, x) = \sup\{\xi \cdot x + h : \xi \cdot x + h \leq u(s, x) \text{ for all } s \in (a, t], x \in \Omega\}.$$

The set where $\Gamma(u)$ coincides with u is called the *contact set*; it is denoted by C_u. The following lemma comes from convex analysis, see e.g. [HUL].

Lemma 2.4.3. *Consider a point* (t, x) *in the contact set* C_u *of* u. *Then* $\xi \cdot x + h = \Gamma(u)(t, x)$ *if and only if* ξ *lies in the convex subdifferential* $\partial u(t, x)$ *of* $u(t, \cdot)$ *at* x *and* $-h$ *equals the convex conjugate* $u^*(t, x)$ *of* $u(t, \cdot)$ *at* x.

Recall that a convex function is locally Lipschitz continuous and in particular a.e. differentiable, for a.e. contact points, $(\xi, h) = (Du(t, x), u(t, x) - x \cdot Du(t, x))$. This is the reason why we next consider for $(t, x) \in (a, b) \times \Omega$ the following function

$$G(u)(t, x) = (Du(t, x), u(t, x) - x \cdot Du(t, x)).$$

The proof of the following elementary lemma is left to the reader.

Lemma 2.4.4. *If* u *is* $C^{1,1}$ *with respect to* x *and Lipschitz continuous with respect to* t, *then the function* $G : (a, b) \times \Omega \to \mathbb{R}^{d+1}$ *is Lipschitz continuous in* (t, x) *and for a.e.* $(t, x) \in (a, b) \times \Omega$,

$$\det D_{t,x} G(u) = u_t \det D^2 u.$$

We now give a second representation formula for $\Gamma(u)$ which will help us next to describe viscosity subdifferentials of the monotone envelope (see Lemma 2.4.6 below).

Lemma 2.4.5 (Representation formula II).

$$\Gamma(u)(t, x) = \inf \left\{ \sum_{i=1}^{d+1} \lambda_i u(s_i, x_i) : \sum_{i=1}^{d+1} \lambda_i x_i = x, s_i \in [a, t], \right.$$

$$\left. \sum_{i=1}^{d+1} \lambda_i = 1, \lambda_i \in [0, 1] \right\}. \tag{2.38}$$

In particular, if

$$\Gamma(u)(t_0, x_0) = \sum_{i=1}^{d+1} \lambda_i u(t_i^0, x_i^0),$$

then

- *for all* $i = 1, \ldots, d + 1$, $\Gamma(u)(t_i, x_i) = u(t_i, x_i)$;
- $\Gamma(u)$ *is constant with respect to* t *and linear with respect to* x *in the convex set* $\operatorname{co}\{(t, x_i^0), (t_i^0, x_i^0), i = 1, \ldots d + 1\}$.

Proof. Let $\tilde{\Gamma}(u)$ denote the function defined by the right hand side of (2.38). First, we observe that $\tilde{\Gamma}(u)$ lies below u and is non-increasing with respect to t and convex with respect to x. Consider now another function v lying below u which is non-increasing with respect to t and convex with respect to x. We then have

$$u(t, x) \geq \tilde{\Gamma}(u)(t, x) \geq \tilde{\Gamma}(v)(t, x) \geq v(t, x).$$

The proof is now complete.   □

We next introduce the notion of *harmonic sum of matrices*. For $A_1, A_2 \in \mathbb{S}_d$ such that $A_1 + A_2 \geq 0$, we consider

$$(A_1 \square A_2)\zeta \cdot \zeta = \inf_{\zeta_1 + \zeta_2 = \zeta} \{A_1 \zeta_1 \cdot \zeta_1 + A_2 \zeta_2 \cdot \zeta_2\}.$$

The reader can check that if A_1 and A_2 are not singular, $A_1 \square A_2 = (A_1^{-1} + A_2^{-1})^{-1}$. We can now state and prove

Lemma 2.4.6. *Let* $(\alpha, p, X) \in \mathcal{P}^- \Gamma(u)(t_0, x_0)$ *and*

$$\Gamma(u)(t_0, x_0) = \sum_{i=1}^{d+1} \lambda_i u(t_i^0, x_i^0). \tag{2.39}$$

Then for all $\varepsilon > 0$ *such that* $I + \varepsilon X > 0$, *there exist* $(\alpha_i, X_i) \in (-\infty, 0] \times \mathbb{S}_d$, $i = 1, \ldots, d + 1$, *such that*

$$\begin{cases} (\alpha_i, p, X_i) \in \overline{\mathcal{P}}^- u(t_i^0, x_i^0) \\ \sum_{i=1}^{d+1} \lambda_i \alpha_i = \alpha \\ X_\varepsilon \leq \lambda_1^{-1} X_1 \square \cdots \square \lambda_{d+1}^{-1} X_{d+1} \end{cases} \tag{2.40}$$

where $X_\varepsilon = X \square \varepsilon^{-1} I = (I + \varepsilon X)^{-1} X$.

Proof. We first define for two arbitrary functions $v, w : \mathbb{R}^d \to \mathbb{R}$,

$$v \overset{x}{\square} w(x) = \inf_{y \in \mathbb{R}^d} v(x - y) + w(y).$$

For a given function $v : [0, +\infty) \times \mathbb{R}^d \to \mathbb{R}$, we also consider the non-increasing envelope $M[v]$ of v:

$$M[v](t, x) = \inf_{s \in [0,t]} v(s, x).$$

We now can write

$$\Gamma(u)(t, x) = \overset{x}{\underset{1 \leq i \leq d+1}{\square}} M[u_i](t, x)$$

where

$$u_i(t, x) = \lambda_i u\left(t, \frac{x}{\lambda_i}\right).$$

Consider also $t_i^0 \in [0, t_0]$ such that

$$M[u_i](t_0, x_i^0) = u_i(t_i^0, x_i^0) = \lambda_i u\left(t_i^0, \frac{x_i^0}{\lambda_i}\right).$$

Lemma 2.4.6 is a consequence of the two following ones.

Lemma 2.4.7. *Consider* $(\alpha, p, X) \in \mathcal{P}^- V(t_0, x_0)$ *where*

$$V(t, x) = \underset{1 \le i \le d+1}{\overset{x}{\square}} v_i(t, x)$$

$$V(t_0, x_0) = \sum_{i=1}^{d+1} v_i(t_0, x_i^0).$$

Then for all $\varepsilon > 0$ *such that* $I + \varepsilon X > 0$, *there exist* $(\beta_i, Y_i) \in \mathbb{R} \times \mathbb{S}_d$ *such that we have*

$$(\beta_i, p, Y_i) \in \overline{\mathcal{P}^-} v_i(t_0, x_i^0)$$

$$\sum_{i=1}^{d+1} \beta_i = \alpha$$

$$X_\varepsilon \le \square_{i=1}^{d+1} Y_i.$$

Proof. We consider a test function ϕ touching V from below at (t_0, x_0) such that

$$(\alpha, p, X) = (\partial_t \phi, D\phi, D^2\phi)(t_0, x_0).$$

We write for (t, x_i) in a neighborhood of (t_0, x_i^0),

$$\phi\left(t, \sum_{i=1}^{d+1} x_i\right) - \phi\left(t_0, \sum_{i=1}^{d+1} x_i^0\right) \le \sum_{i=1}^{d+1} v_i(t, x_i) - \sum_{i=1}^{d+1} v_i(t_0, x_i^0).$$

Following [ALL97, Imb06], we conclude through Jensen–Ishii's lemma for $d + 1$ functions and general test functions (see Lemma 2.5.6 in appendix) that for all $\varepsilon > 0$ such that $I + d\varepsilon X > 0$, there exist $(\beta_i, Y_i) \in \mathbb{R} \times \mathbb{S}_d, i = 1, \ldots, d + 1$ such that

$$(\beta_i, p, Y_i) \in \overline{\mathcal{P}^-} v_i(t_0, x_i^0)$$

$$\sum_{i=1}^{d+1} \beta_i = \alpha$$

and

$$\begin{pmatrix} X & \dots & X \\ \vdots & \ddots & \vdots \\ X & \dots & X \end{pmatrix}_\varepsilon \leq \begin{pmatrix} Y_1 & 0 & \dots & 0 \\ 0 & \ddots & \ddots & \vdots \\ \vdots & \ddots & \ddots & 0 \\ 0 & \dots & 0 & Y_{d+1} \end{pmatrix}$$

where, for any matrix A, $A_\varepsilon = (I + \varepsilon A)^{-1} A$. A small computation (presented e.g. in [Imb06, p. 796]) yields that the previous matrix inequality is equivalent to the following one

$$X_{d\varepsilon} \zeta \cdot \zeta \leq \sum_{i=1}^{d+1} Y_i \zeta_i \cdot \zeta_i$$

where $\zeta = \sum_{i=1}^{d+1} \zeta_i$. Taking the infimum over decompositions of ζ, we get the desired matrix inequality. $\qquad\square$

Lemma 2.4.8. *Consider $s_1 \in [0, s_0]$ such that*

$$M[v](s_0, y_0) = v(s_1, y_0).$$

Then for all $(\beta, q, Y) \in \mathcal{P}^- M[v](s_0, y_0)$,

$$(\beta, q, Y) \in \mathcal{P}^- v(s_1, y_0) \quad and \quad \beta \leq 0.$$

Proof. We consider the test function ϕ associated with (β, q, Y) and we write for h and δ small enough

$$\phi(s_0 + h, y_0 + \delta) - \phi(s_0, y_0) \leq M[v](s_0 + h, y_0 + \delta) - M[v](s_0, y_0)$$
$$\leq v(s_1 + h, y_0 + \delta) - v(s_1, y_0).$$

This implies $(\beta, q, Y) \in \mathcal{P}^- v(s_1, y_0)$. Moreover, choosing $\delta = 0$, we get

$$\phi(s_0 + h, y_0) \leq \phi(s_0, y_0)$$

and $\beta \leq 0$ follows. $\qquad\square$

The proof is now complete. $\qquad\square$

2.4.1.2 Statement

The following result is the first key result in the theory of regularity of fully non-linear parabolic equations. It is the parabolic counterpart of the famous Alexandroff estimate, also called Alexandroff–Bakelman–Pucci (ABP) estimate, see [CafCab] for more details about this elliptic estimate. The following one was first proved for linear equations by Krylov [Kryl76] and then extended by Tso [Tso85]. The following result appears in [Wang92a].

Theorem 2.4.9 (Maximum principle). *Consider a supersolution of* (2.36) *in* $Q_\rho = Q_\rho(0,0)$ *such that* $u \geq 0$ *on* $\partial_p(Q_\rho)$. *Then*

$$\sup_{Q_\rho} u^- \leq C\rho^{\frac{d}{d+1}} \left(\int_{u=\Gamma(u)} (f^+)^{d+1} \right)^{\frac{1}{d+1}} \tag{2.41}$$

where C *is universal and* $\Gamma(u)$ *is the monotone envelope of* $\min(0, u)$ *extended by* 0 *to* $Q_{2\rho}$.

Remark 2.4.10. This is a maximum principle since, if $f \leq 0$, then u cannot take negative values.

Proof. We prove the result for $\rho = 1$ and the general one is obtained by considering $v(t, x) = u(\rho^2 t, \rho x)$. Moreover, replacing u with $\min(0, u)$ and extending it by 0 in $Q_2 \setminus Q_1$, we can assume that $u = 0$ on $\partial_p Q_1$ and $u \equiv 0$ in $Q_2 \setminus Q_1$.

We are going to prove the three following lemmas. Recall that $G(u)$ is defined page 44.

Lemma 2.4.11. *The function* $\Gamma(u)$ *is* $C^{1,1}$ *with respect to* x *and Lipschitz continuous with respect to* t *in* Q_1. *In particular,* $G\Gamma(u) := G(\Gamma(u))$ *is Lipschitz continuous with respect to* (t, x).

The second part of the statement of the previous lemma is a consequence of Lemma 2.4.4 above. We will prove the previous lemma together with the following one.

Lemma 2.4.12. *The partial derivatives* $(\partial_t \Gamma(u), D^2 \Gamma(u))$ *satisfy for a.e.* $(t, x) \in Q_1 \cap C_u$,

$$-\partial_t \Gamma(u) + \lambda \Delta(\Gamma(u)) \leq f^+(x)$$

where $C_u = \{u = \Gamma(u)\}$.

The key lemma is the following one.

Lemma 2.4.13. *If* M *denotes* $\sup_{Q_1} u^-$, *then*

$$\{(\xi, h) \in \mathbb{R}^{d+1} : |\xi| \leq M/2 \leq -h \leq M\} \subset G\Gamma(u)(Q_1 \cap C_u)$$

where $C_u = \{u = \Gamma(u)\}$.

Before proving these lemmas, let us derive the conclusion of the theorem. Using successively Lemma 2.4.13, the area formula for Lipschitz maps (thanks to Lemma 2.4.11) and Lemma 2.4.4, we get

$$
\begin{aligned}
CM^{d+1} &= |\{(\xi, h) \in \mathbb{R}^{d+1} : |\xi| \le M/2 \le -h \le M\}| \\
&\le |G\Gamma(u)(Q_1 \cap C_u)| \\
&\le \int_{Q_1 \cap C_u} |\det G\Gamma(u)| \\
&\le \int_{Q_1 \cap C_u} -\partial_t \Gamma(u) \det(D^2 \Gamma(u)).
\end{aligned}
$$

Now using the geometric–arithmetic mean inequality and Lemma 2.4.12, we get

$$
\begin{aligned}
CM^{d+1} &\le \lambda^{-d} \int_{Q_1 \cap C_u} -\partial_t \Gamma(u) \det(\lambda D^2 \Gamma(u)) \\
&\le \frac{1}{\lambda^d (d+1)^{d+1}} \int_{Q_1 \cap C_u} (-\partial_t \Gamma(u) + \lambda \Delta(\Gamma(u))^{d+1} \\
&\le C \int_{Q_1 \cap C_u} (f^+)^{d+1}
\end{aligned}
$$

where C's are universal. $\square$

We now turn to the proofs of Lemmas 2.4.11–2.4.13.

Proof of Lemmas 2.4.11 and 2.4.12. In order to prove that $\Gamma(u)$ is Lipschitz continuous with respect to t and $C^{1,1}$ with respect to x, it is enough to prove that there exists $C > 0$ such that

$$
\forall (t,x) \in Q_2, \ \forall (\alpha, p, X) \in \mathcal{P}^-\Gamma(u)(t,x), \qquad \begin{cases} -\alpha \le C \\ X \le CI. \end{cases} \tag{2.42}
$$

Indeed, since $\Gamma(u)$ is non-increasing with respect to t and convex with respect to x, (2.42) yields that $\Gamma(u)$ is Lipschitz continuous with respect to t and $C^{1,1}$ with respect to x. See Lemma 2.5.8 in appendix for more details.

In order to prove (2.42), we first consider $(\alpha, p, X) \in \mathcal{P}^-\Gamma(u)(t,x)$ such that $X \ge 0$. Recall (cf. Lemma 2.4.6 above) that $\alpha \le 0$. We then distinguish two cases.

Assume first that $\Gamma(u)(t,x) = u(t,x)$. In this case, $(\alpha, p, X) \in \mathcal{P}^- u(t,x)$ and since u is a supersolution of (2.36), we have

$$
\alpha - \lambda \operatorname{trace}(X) = \alpha + P^+(X) \ge f(x) \ge -C
$$

where $C = |f|_{0;Q_1}$. Hence, we get (2.42) since $X \geq 0$ implies that $X \leq \text{trace}(X)I$. We also remark that the same conclusion holds true if $(\alpha, p, X) \in \overline{\mathcal{P}}^-\Gamma(u)(t, x)$ such that $X \geq 0$.

Assume now that $\Gamma(u)(t, x) < u(t, x)$. In this case, there exist $\lambda_i \in [0, 1]$, $i = 1, \ldots, d + 1$, and $x_i \in Q_2$, $i = 1, \ldots, d + 1$, such that (2.39) holds true with (t_0, x_0) and (t_i^0, x_i^0) replaced with (t, x) and (t_i, x_i). If $(t_i, x_i) \in Q_2 \setminus Q_1$ for two different i's, then Lemma 2.4.5 implies that $M = 0$ which is false. Similarly, $t_i > -1$ for all i. Hence, there is at most one index i such that $(t_i, x_i) \in Q_2 \setminus Q_1$ and in this case $(t_i, x_i) \in \partial_p Q_2$ and $t_i > -1$. In particular, $|x_i| = 2$. We thus distinguish two subcases.

Assume first that $(t_{d+1}, x_{d+1}) \in \partial_p Q_2$ with $t_{d+1} > -1$ and $(t_i, x_i) \in Q_1$ for $i = 1, \ldots, d$. In particular $|x_{d+1}| = 2$ and since $x \in Q_1$, we have $\lambda_{d+1} \leq \frac{2}{3}$. This implies that there exists λ_i such that $\lambda_i \geq (3d)^{-1}$. We thus can assume without loss of generality that $\lambda_1 \geq (3d)^{-1}$. Then from Lemma 2.4.6, we know that for all $\varepsilon > 0$ such that $I + \varepsilon X > 0$, there exist $(\alpha_i, X_i) \in \mathbb{R} \times \mathbb{S}_d$, $i = 1, \ldots, d + 1$ such that (2.40) holds true. In particular,

$$X_\varepsilon \leq \frac{1}{\lambda_1} X_1 \leq 3d X_1.$$

Since $(\alpha_1, p, X_1) \in \overline{\mathcal{P}}^- u(t_1, x_1)$ and $\Gamma(u)(t_1, x_1) = u(t_1, x_1)$, we know from the discussion above that $X_1 \leq CI$. Hence for all ε small enough,

$$X_\varepsilon \leq 3d CI.$$

Letting $\varepsilon \to 0$ allows us to conclude that $X \leq 3d CI$ in the first subcase. As far as α is concerned, we remark that $\alpha_{d+1} = 0$ and $-\alpha_i \leq C$ for all $i = 1, \ldots, d + 1$ so that

$$-\alpha = \sum_{i=1}^{d+1} \lambda_i(-\alpha_i) \leq C.$$

Assume now that all the points (t_i, x_i), $i = 1, \ldots, d + 1$, are in Q_1. In this case, we have for all i that $-\alpha_i \leq C$ and $X_i \leq CI$ which implies

$$-\alpha = \sum_{i=1}^{d+1} \lambda_i(-\alpha_i) \leq C,$$

$$X_\varepsilon \leq \square_{i=1}^{d+1} \lambda_i^{-1} CI = CI.$$

We thus proved (2.42) in all cases where $X \geq 0$. Consider now a general subdifferential $(\alpha, p, X) \in \mathcal{P}^-\Gamma(u)(t, x)$. We know from Lemma 2.5.9 in appendix that there exists a sequence (α_n, p_n, X_n) such that

$$(\alpha_n, p_n, X_n) \in \mathcal{P}^- \Gamma(u)(t_n, x_n)$$

$$(t_n, x_n, \alpha_n, p_n) \to (t, x, \alpha, p)$$

$$X \le X_n + o_n(1), X_n \ge 0.$$

From the previous discussion, we know that

$$\alpha = \alpha_n + o_n(1) \le (C + 1)$$

$$X \le X_n + o_n(1) \le (C + 1)I$$

for all n. The proof is now complete. □

Proof of Lemma 2.4.13. The supersolution $u \le 0$ is lower semi-continuous and the minimum $-M < 0$ in Q_2 is thus reached at some $(t_0, x_0) \in Q_1$ (since $u \equiv 0$ outside Q_1). Now pick (ξ, h) such that

$$|\xi| \le M/2 \le -h \le M.$$

We consider $P(y) = \xi \cdot y + h$. We remark that $P(y) < 0$ for $y \in Q_1$, hence $P(y) < u(0, y)$ in Q_1. Moreover, since $|x_0| < 1$,

$$P(x_0) - u(t_0, x_0) = \xi \cdot x_0 + h + M > h - |\xi| + M \ge 0$$

hence $\sup_{y \in Q_2}(P(y) - u(t_0, y)) \ge 0$. We thus choose

$$t_1 = \sup\{t \ge 0 : \forall s \in [0, t], \sup_{Q_2}(P(y) - u(s, y)) < 0\}.$$

We have $0 \le t_1 \le t_0$ and

$$0 = \sup_{Q_2}(P(y) - u(t_1, y)) = P(y_1) - u(t_1, y_1).$$

In particular, $\xi = Du(t_1, y_1)$ and $h = u(t_1, x_1) - \xi \cdot x_1$, that is to say, $(\xi, h) = G(u)(t_1, y_1)$ with $(t_1, y_1) \in C_u$. □

2.4.2 The L^ε-Estimate

This subsection is devoted to the important "L^ε estimate" given in Theorem 2.4.15. This estimate is sometimes referred to as the weak Harnack inequality.

Theorem 2.4.15 claims that the L^ε-"norm" in a neighbourhood $\tilde{K}_1$ of $(0, 0)$ of a non-negative (super-)solution u of the model equation (2.36) can be controlled by its infimum over a neighbourhood $\tilde{K}_2$ of $(1, 0)$ plus the L^{d+1}-norm of f.

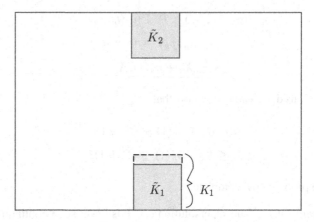

Fig. 2.1 The two neighbourhoods $\tilde{K}_1$ and $\tilde{K}_2$

Remark 2.4.14. Since ε can be smaller than 1, the integral of u^ε is in fact not the (ε-power of) a norm.

We introduce the two neighbourhoods mentioned above (see Fig. 2.1).

$$\tilde{K}_1 = (0, R^2/2) \times (-R, R)^d,$$
$$\tilde{K}_2 = (1 - R^2, 1) \times (-R, R)^d.$$

Theorem 2.4.15 (L^ε **estimate**). *There exist universal positive constants R, C and ε, such that for all non-negative supersolution u of*

$$\frac{\partial u}{\partial t} + P^+(D^2 u) \geq f \quad in\ (0, 1) \times B_{\frac{1}{R}}(0),$$

the following holds true

$$\left(\int_{\tilde{K}_1} u^\varepsilon \right)^{\frac{1}{\varepsilon}} \leq C \left(\inf_{\tilde{K}_2} u + \|f\|_{L^{d+1}((0,1) \times B_{\frac{1}{R}}(0))} \right). \tag{2.43}$$

The proof of this theorem is difficult and lengthy; this is the reason why we explain the main steps of the proof now.

First, one should observe that it is possible to assume without loss of generality that $\inf_{\tilde{K}_2} u \leq 1$ and $\|f\|_{L^{d+1}((0,1) \times B_{\frac{1}{R}}(0))} \leq \varepsilon_0$ (for some universal constant ε_0 to be determined) and to prove

$$\int_{\tilde{K}_1} u^\varepsilon(t, x)\,dx \leq C$$

where $\varepsilon > 0$ and $C > 0$ are universal. We recall that a constant is said to be universal if it only depends on ellipticity constants λ and Λ and dimension d. Getting such an estimate is equivalent to prove that

$$|\{u > t\} \cap \tilde{K}_1| \leq C t^{-\varepsilon}$$

(see page 69 for more details). To get such a decay estimate, it is enough to prove that

$$|\{u > N^k\} \cap \tilde{K}_1| \leq C N^{-k\varepsilon}$$

for some universal constant $N > 1$. This inequality is proved by induction thanks to a covering lemma (see Lemma 2.4.27 below). This amounts to cut the set $\{u > N^k\} \cap \tilde{K}_1$ in small pieces (the dyadic cubes) and make sure that the pieces where u is very large ($u \geq t, t \gg 1$) have small measures.

This will be a consequence of a series of measure estimates obtained from a basic one. The proof of the basic measure estimate is a consequence of the maximum principle proved above and the construction of an appropriate barrier we will present soon. But we should first introduce the parabolic cubes we will use in the decomposition. We also present the choice of parameters we will make.

2.4.2.1   Parabolic Cubes and Choice of Parameters

We consider the following subsets of $Q_1(1, 0)$.

$$K_1 = (0, R^2) \times (-R, R)^d,$$
$$K_2 = (R^2, 10R^2) \times (-3R, 3R)^d,$$
$$K_3 = (R^2, 1) \times (-3R, 3R)^d.$$

The constant R will be chosen as follows

$$R = \min\left(\frac{1}{3\sqrt{d}}, 3 - 2\sqrt{2}, \frac{1}{\sqrt{10(m+1)}}\right) \qquad (2.44)$$

where m will be chosen in a universal way in the proof of the L^ε estimate.

2.4.2.2   A Useful Barrier

The following lemma will be used to derive the basic measure estimate. This estimate is the cornerstone of the proof of the L^ε estimate.

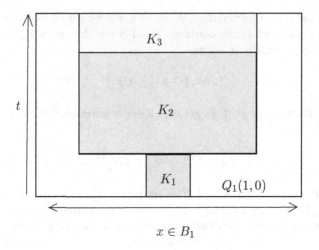

Fig. 2.2 The sets K_1, K_2 and K_3

Lemma 2.4.16. *For all $R \in (0, \min((3\sqrt{d})^{-1}, (10)^{-1/2}))$, there exists a nonnegative Lipschitz function $\phi : Q_1(1,0) \to \mathbb{R}$, C^2 with respect to x where it is positive, such that*

$$\frac{\partial \phi}{\partial t} + P^+(D^2\phi) \le g$$

for some continuous bounded function $g : Q_1(1,0) \to \mathbb{R}$ and such that

$$\text{supp } g \subset K_1$$
$$\phi \ge 2 \text{ in } K_3$$
$$\phi = 0 \text{ in } \partial_p Q_1(1,0).$$

Remark 2.4.17. Recall the definitions of K_1, K_2 and K_3 (see Fig. 2.2).

$$K_1 = (0, R^2) \times (-R, R)^d,$$
$$K_2 = (R^2, 10R^2) \times (-3R, 3R)^d,$$
$$K_3 = (R^2, 1) \times (-3R, 3R)^d.$$

The proof of the lemma above consists in constructing the function ϕ more or less explicitly. It is an elementary computation. However, it is an important feature of non divergence type equations that this type of computations can be made. Consider in contrast the situation of parabolic equations with measurable coefficients in divergence form. For that type of equations, a result like the one of Lemma 2.4.16 would be significantly harder to obtain.

Proof. We will construct a function φ which solves the equation

$$\varphi_t + P^+(D^2\varphi) \le 0 \tag{2.45}$$

in the whole cylinder $Q_1(1,0)$, such that φ is positive and unbounded near $(0,0)$ but $\varphi = 0$ in $\partial_p Q_1(1,0)$ away from $(0,0)$, and moreover $\varphi > 0$ in K_2. Note that if the equation were linear, φ could be its heat kernel in the cylinder. Once we have this function φ, we obtain ϕ simply by taking

$$\phi(t,x) = 2\frac{\varphi(t,x)}{\min_{K_2} \varphi} \text{ for } (t,x) \in \setminus K_1,$$

and making ϕ equal any other smooth function in K_1 which is zero on $\{t = 0\}$.

We now construct this function φ. We will provide two different formulas for $\varphi(t,x)$. The first one will hold for $t \in (0,T)$ for some $T \in (0,1)$. Then the second formula provides a continuation of the definition of φ on $[T,1]$.

For some constant $p > 0$ and a function $\Phi : \mathbb{R}^d \to \mathbb{R}$, we will construct the function φ in $(0,T)$ with the special form

$$\varphi(t,x) = t^{-p}\Phi\left(\frac{x}{\sqrt{t}}\right).$$

Let us start from understanding what conditions Φ must satisfy in order for φ to be a subsolution to (2.45).

$$0 \ge \varphi_t + P^+(D^2\varphi) = t^{-1-p}\left(-p\Phi\left(\frac{x}{\sqrt{t}}\right)\right.$$
$$\left.-\frac{1}{2}\frac{x}{\sqrt{t}}\cdot\nabla\Phi\left(\frac{x}{\sqrt{t}}\right) + P^+(D^2\Phi)\left(\frac{x}{\sqrt{t}}\right)\right).$$

Therefore, we need to find a function $\Phi : \mathbb{R}^d \to \mathbb{R}$ and some exponent p such that

$$-p\Phi(x) - \frac{1}{2}x\cdot\nabla\Phi(x) + P^+(D^2\Phi)(x) \le 0. \tag{2.46}$$

For some large exponent q, we choose Φ like this

$$\Phi(x) = \begin{cases} \text{something smooth and bounded between 1 and 2 if } |x| \le 3\sqrt{d}, \\ (6\sqrt{d})^q(2^q-1)^{-1}\left(|x|^{-q} - (6\sqrt{d})^{-q}\right) & \text{if } 3\sqrt{d} \le |x| \le 6\sqrt{d}, \\ 0 & \text{if } |x| \ge 6\sqrt{d}. \end{cases}$$

For $3\sqrt{d} < |x| < 6\sqrt{d}$, we compute explicitly the second and third terms in (2.46),

$$-\frac{1}{2}x \cdot \nabla \Phi(x) = (6\sqrt{d})^q (2^q - 1)^{-1}\frac{q}{2}|x|^{-q}$$

$$P^+(D^2\Phi)(x) = (6\sqrt{d})^q (2^q - 1)^{-1} q(\Lambda(d - 1) - \lambda(q + 1))|x|^{-q-2}.$$

By choosing q large enough so that $\lambda(q + 1) > \Lambda(d - 1) + 18d$, we get that

$$-\frac{1}{2}x \cdot \nabla \Phi(x) + P^+ \Phi(x) \le 0.$$

In order for (2.46) to hold in $B_{3\sqrt{d}}$, we just have to choose the exponent p large enough, since at those points $\Phi \ge 1$. Furthermore, since $\Phi \ge 0$ everywhere and $\Phi = 0$ outside $B_{6\sqrt{d}}$, then the inequality (2.46) holds in the full space $\mathbb{R}^d$ in the viscosity sense.

Since Φ is supported in $B_{6\sqrt{d}}$, then $\varphi = 0$ on $(0, T) \times \partial B_1$, for $T = (36d)^{-1}$. Thus, $\varphi = 0$ on the lateral boundary $(0, T) \times \partial B_1$. Moreover,

$$\lim_{t \to -1} \varphi(t, x) = 0,$$

uniformly in $B_1 \setminus B_\varepsilon$ for any $\varepsilon > 0$.

We have provided a value of φ up to time $T \in (0, 1)$. In order to continue φ in $[T, 1]$ we can do the following. Observe that by the construction of Φ, we have $P^+(D^2\varphi(T, x)) \le 0$ for $x \in B_1 \setminus B_{1/2}$ and $\varphi(x, T) \ge T^{-p}$ for $x \in B_{1/2}$. Therefore, let

$$C = \sup_{x \in B_1} \frac{P^+(D^2\varphi(T, x))}{\varphi(T, x)} < +\infty,$$

then we define $\varphi(t, x) = e^{-C(t-T)}\varphi(T, x)$ for all $t > T$, which is clearly a positive subsolution of (2.45) in $(T, 1] \times B_1$ with $\varphi = 0$ on $[T, 1] \times \partial B_1$.

The constructed function φ vanishes only on the set $\{(t, x) : t < T \text{ and } |x| \ge 6\sqrt{dt}\}$. Since the set $K_3 = (R^2, 1) \times (-3R, 3R)^d$ has no intersection with this set, then

$$\inf_{K_3} \varphi > 0.$$

This is all that was needed to conclude the proof.   □

2.4.2.3 The Basic Measure Estimate

As in the elliptic case, the basic measure estimate is obtained by combining the maximum principle of Theorem 2.4.9 and the barrier function constructed in Lemma 2.4.16. For the following proposition, we use the notation from Remark 2.4.17.

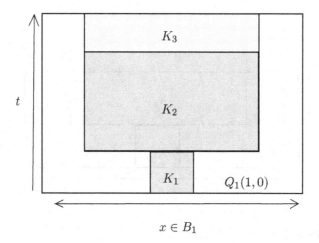

Fig. 2.3 Basic measure estimate in $Q_1(1,0)$

Proposition 2.4.18 (Basic measure estimate). *There exist universal constants $\varepsilon_0 \in (0,1)$, $M > 1$ and $\mu \in (0,1)$ such that for any non-negative supersolution of*

$$\frac{\partial u}{\partial t} + P^+(D^2 u) \geq f \text{ in } Q_1(1,0),$$

the following holds true: if

$$\begin{cases} \inf_{K_3} u \leq 1 \\ \|f\|_{L^{d+1}(Q_1(1,0))} \leq \varepsilon_0 \end{cases}$$

then

$$|\{u \leq M\} \cap K_1| \geq \mu |K_1|.$$

Remark 2.4.19. Since $K_2 \subset K_3$ (see Fig. 2.3), the result also holds true if $\inf_{K_3} u$ is replaced with $\inf_{K_2} u$. This will be used in order to state and prove the stacked measure estimate.

Remark 2.4.20. If u is a non-negative supersolution of

$$\frac{\partial u}{\partial t} + P^+(D^2 u) \geq f \text{ in } (0,T) \times B_1,$$

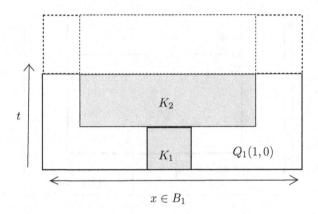

Fig. 2.4 A supersolution in a smaller cylinder

for some $T \in (R^2, 1)$ (see Fig. 2.4), we still get

$$\left.\begin{array}{r} \inf_{(R^2,T) \times (-3R,3R)^d} u \leq 1 \\ \|f\|_{L^{d+1}((0,T) \times B_1)} \leq \varepsilon_0 \end{array}\right\} \Rightarrow |\{u \leq M\} \cap K_1| \geq \mu |K_1|.$$

The reason is that such a solution could be extended to $Q_1(1, 0)$ (for example giving any boundary condition on $(T, 1) \times \partial B_1$ and making f quickly become zero for $t > T$), and then Proposition 2.4.18 can be applied to this extended function. This remark will be useful when getting the "stacked" measure estimate in the case where the stack of cubes reaches the final time.

Proof. Consider the function $w = u - \phi$ where ϕ is the barrier function from Lemma 2.4.16. Then w satisfies (in the viscosity sense)

$$\frac{\partial w}{\partial t} + P^+(D^2 w) \geq \frac{\partial u}{\partial t} + P^+(D^2 u) - \frac{\partial \phi}{\partial t} - P^+(D^2 \phi) \geq f - g.$$

Remark also that

- $w \geq u \geq 0$ on $\partial_p Q_1(1, 0)$;
- $\inf_{K_3} w \leq \inf_{K_3} u - 2 \leq -1$ so that $\sup_{K_3} w^- \geq 1$;
- $\{\Gamma(w) = w\} \subset \{w \leq 0\} \subset \{u \leq \phi\}$.

We recall that $\Gamma(w)$ denotes the monotone envelope of $\min(w, 0)$ extended by 0 to $Q_2(1, 0)$. We now apply the maximum principle (Theorem 2.4.9) and we get

$$1 \leq \sup_{K_3} w^- \leq \sup_{Q_1} w^- \leq C_{\max} \|f\|_{L^{d+1}(Q_1(1,0))} + C_{\max} \left(\int_{\{u \leq \phi\}} |g|^{d+1} \right)^{\frac{1}{d+1}}.$$

Remember now that supp $g \subset K_1$ and get

$$1 \leq C_{\max}\varepsilon_0 + C_{\max}|\{u \leq M\} \cap K_1|$$

with $M > \max(\sup_{K_1} \phi, 1)$. Choose now ε_0 so that $C_{\max}\varepsilon_0 \leq 1/2$ and get the result with $\mu = \frac{1}{1+2C_{\max}|K_1|}$. The proof is now complete. $\square$

Corollary 2.4.21 (Basic measure estimate scaled). *For the same constants ε_0, M and μ of Proposition 2.4.18 and any $x_0 \in \mathbb{R}^d$, $t_0 \in \mathbb{R}$ and $h > 0$, consider any nonnegative supersolution of*

$$\frac{\partial u}{\partial t} + P^+(D^2 u) \geq f \ in \ (t_0, x_0) + \rho Q_1(1, 0).$$

If

$$\|f\|_{L^{d+1}((t_0,x_0)+\rho Q_1(0,1))} \leq \varepsilon_0 \frac{h}{M\rho^{d/(d+1)}}$$

then

$$|\{u > h\} \cap \{(t_0, x_0) + \rho K_1\}| > (1-\mu)|(t_0, x_0) + \rho K_1| \Rightarrow u > \frac{h}{M} \ in \ (t_0, x_0) + \rho K_3.$$

Here, we recall that by ρK we mean $\{(\rho^2 t, \rho x) : (t, x) \in K\}$.

Remark 2.4.22. As in Remark 2.4.20, $(t_0, x_0) + \rho(0, 1) \times B_{\frac{1}{R}}(0)$ can be replaced with $(t_0, x_0) + \rho(0, T) \times B_{\frac{1}{R}}(0)$ for any $T \in (0, 1)$.

Proof. We consider the scaled function

$$v(t, x) = Mh^{-1}u(t_0 + \rho^2 t, x_0 + \rho x).$$

This function solves the equation

$$\frac{\partial v}{\partial t} + P^+(D^2 v) \geq \tilde{f} \ in \ Q_1(1, 0)$$

where $\tilde{f}(t, x) = Mh^{-1}\rho^2 f(t_0 + \rho^2 t, x_0 + \rho x)$. Note that

$$\|\tilde{f}\|_{L^{d+1}(Q_1(1,0))} = Mh^{-1}\rho^{d/(d+1)}\|f\|_{L^{d+1}((t_0,x_0)+\rho Q_1(1,0))} \leq \varepsilon_0.$$

We conclude the proof applying Proposition 2.4.18 to v. $\square$

2.4.2.4 Stacks of Cubes

Given $\rho \in (0, 1)$, we consider for all $k \in \mathbb{N}$, $k \geq 1$,

$$K_2^{(k)} = (\alpha_k R^2, \alpha_{k+1} R^2) \times (-3^k R, 3^k R)^d$$

where $\alpha_k = \sum_{i=0}^{k-1} 9^i = \frac{9^k - 1}{8}$.

The first stack of cubes that we can consider is the following one

$$\cup_{k \geq 1} K_2^{(k)}.$$

This stack is obviously not contained in $Q_1(1, 0)$ since time goes to infinity. It can spill out of $Q_1(1, 0)$ either on the lateral boundary or at the final time $t = 1$. We are going to see that at the final time, the "x-section" is contained in $(-3, 3)^d$.

We consider a scaled version of K_1 included in K_1 and we stack the corresponding $K_2^{(k)}$'s. The scaled versions of K_1, K_2 and $K_2^{(k)}$ are

$$\rho K_1 = (0, \rho^2 R^2) \times B_{\rho R}(0),$$

$$\rho K_2 = (\rho^2 R^2, 10\rho^2 R^2) \times B_{\rho R}(0),$$

$$\rho K_2^{(k)} = (\alpha_k \rho^2 R^2, \alpha_{k+1} \rho^2 R^2) \times (-3^k \rho R, 3^k \rho R)^d.$$

We now consider

$$L_1 = (t_0, x_0) + \rho K_1 \subset K_1$$

and

$$L_2^{(k)} = (t_0, x_0) + \rho K_2^{(k)}.$$

Lemma 2.4.23 (Stacks of cubes). *Choose* $R \leq \min(3 - 2\sqrt{2}, \sqrt{\frac{2}{5}}) = 3 - 2\sqrt{2}$. *For all* $L_1 = (t_0, x_0) + \rho K_1 \subset K_1$, *we have*

$$\tilde{K}_2 \subset \left(\cup_{k \geq 1} L_2^{(k)}\right) \cap (0, 1) \times (-3, 3)^d = \left(\cup_{k \geq 1} L_2^{(k)}\right) \cap \{0 < t < 1\}.$$

In particular, if moreover $R \leq (3\sqrt{d})^{-1}$,

$$\left(\cup_{k \geq 1} L_2^{(k)}\right) \subset (0, 1) \times B_{\frac{1}{R}}(0).$$

Moreover, the first $k^* = k$ *such that* $L_2^{(k+1)} \cap \{t = 1\} = \emptyset$ *satisfies*

$$\rho^2 R^2 \leq \frac{1}{\alpha_{k^*}}.$$

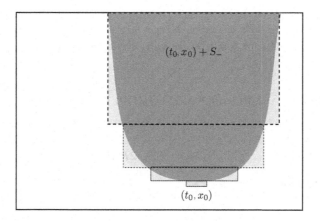

Fig. 2.5 Stacks of cubes

Proof. We first remark that the stack of cubes lies between two "square" paraboloids (see Fig. 2.5)

$$(t_0, x_0) + S_- \subset \cup_{k \geq 1} L_2^{(k)} \subset (t_0, x_0) + S_+$$

where

$$S_\pm = \cup_{s \geq s_\pm} \{p_\pm(s)\} \times (-s, s)^d$$

and $p_\pm(s_\pm) = \rho^2 R^2$ and $p_\pm(z) = a_\pm z^2 + b_\pm \rho^2 R^2$ are such that

$$p_+(3^k \rho R) = \alpha_k \rho^2 R^2$$
$$p_-(3^k \rho R) = \alpha_{k+1} \rho^2 R^2.$$

This is equivalent to

$$a_+ = \frac{1}{8} \quad \text{and} \quad a_- = \frac{9}{8} \quad \text{and} \quad b_+ = b_- = -\frac{1}{8} \quad \text{and} \quad s_\pm = \sqrt{\frac{9}{8}} \rho R.$$

Remark now that

$$[(t_0, x_0) + S_+] \cap Q_1(1, 0) \subset [0, 1] \times (-R - a_+^{-\frac{1}{2}}, R + a_+^{-\frac{1}{2}})^d.$$

We thus choose R such that $(R + a_+^{-\frac{1}{2}}) \leq 3$. This condition is satisfied if

$$R \leq 3 - 2\sqrt{2}.$$

Remark next that

$$(t_0, x_0) + S_- \supset \cap_{x \in (-R,R)^d} [(R^2, x) + S_-].$$

Hence

$$[(t_0, x_0) + S_-] \cap Q_1(1,0) \supset \tilde{K}_2$$

as soon as

$$a_+(2R)^2 \leq 1 - 2R^2.$$

It is enough to have

$$\frac{5}{2} R^2 = (4a_+ + 2) R^2 \leq 1.$$

Finally, the integer k^* satisfies

$$t_0 + \alpha_{k^*} R^2 \rho^2 \leq 1 < t_0 + \alpha_{k^*+1} R^2 \rho^2. \qquad \square$$

2.4.2.5 The Stacked Measure Estimate

In this paragraph, we apply repeatedly the basic measure estimate obtained above and get an estimate in the finite stacks of cubes we constructed in the previous paragraph.

Proposition 2.4.24 (Stacked measure estimate). *For the same universal constants $\varepsilon_0 \in (0,1)$, $M > 1$ and $\mu \in (0,1)$ from Proposition 2.4.18, the following holds true: consider a non-negative supersolution u of*

$$\frac{\partial u}{\partial t} + P^+(D^2 u) \geq f \text{ in } (0,1) \times B_{\frac{1}{R}}(0)$$

and a cube $L_1 = (t_0, x_0) + \rho K_1 \subset K_1$. Assume that for some $k \geq 1$ and $h > 0$

$$\|f\|_{L^{d+1}((0,1) \times B_{\frac{1}{R}}(0))} \leq \varepsilon_0 \frac{h}{M^k \rho^{d/(d+1)}}.$$

Then

$$|\{u > h\} \cap L_1| > (1 - \mu)|L_1| \Rightarrow \inf_{L_2^{(k)} \cap \{0 < t < 1\}} u > \frac{h}{M^k}.$$

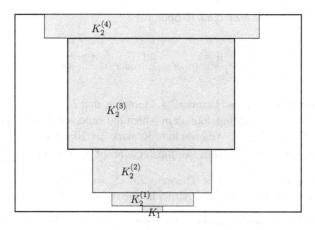

Fig. 2.6 Stacks of neighbourhoods $K_2^{(k)}$

Remark 2.4.25. Remember that $L_2^{(k)} = (t_0, x_0) + \rho K_2^{(k)}$ and see Fig. 2.6. Thanks to Lemma 2.4.23, we know that $L_2^{(k)} \cap \{0 < t < 1\} \subset (0, 1) \times B_{\frac{1}{R}}(0)$.

Proof. We prove the result by induction on k. Corollary 2.4.21 corresponds to the case $k = 1$ if we can verify that

$$\|f\|_{L^{d+1}((t_0,x_0)+\rho Q_1(1,0))} \leq \varepsilon_0 \frac{h}{M \rho^{d/(d+1)}}.$$

It is a consequence of the fact that $Q_1(1,0) \subset (0, 1) \times B_{\frac{1}{R}}(0)$.

For $k > 1$, the inductive hypothesis reads

$$\inf_{L_2^{(k-1)} \cap \{0 < t < 1\}} u > \frac{h}{M^{k-1}}.$$

If $L_2^{(k-1)}$ is not contained in $(0, 1) \times B_{\frac{1}{R}}(0)$, there is nothing to prove at rank k since $L_2^{(k)} \cap \{0 < t < 1\} = \emptyset$. We thus assume that $L_2^{(k-1)} \subset (0, 1) \times B_{\frac{1}{R}}(0)$.

In particular

$$\left|\{u > \frac{h}{M^{k-1}}\} \cap L_2^{(k-1)}\right| = |L_2^{(k-1)}|. \tag{2.47}$$

Note that $L_2^{(k-1)} = (t_1, 0) + \rho_1 K_1$ and $L_2^{(k)} = (t_1, 0) + \rho_1 K_2$ with $t_1 = t_0 + \alpha_{k-1} R^2 \rho^2$ and $\rho_1 = 3^{k-1}\rho$. In particular (2.47) implies

$$\left|\{u > \frac{h}{M^{k-1}}\} \cap \{(t_1, 0) + \rho_1 K_1\}\right| > (1 - \mu)|(t_1, 0) + \rho_1 K_1|.$$

So we apply Corollary 2.4.21 again to obtain

$$\inf_{L_2^{(k)} \cap \{0 < t < 1\}} u = \inf_{\{(t_1, 0) + \rho_1 K_2\} \cap \{0 < t < 1\}} u > \frac{h}{M^k}.$$

We can do so since $\rho_1 \geq \rho$ and Lemma 2.4.23 implies that $L_2^{(k)} \subset (0, 1) \times (-3, 3)^d$. In particular, the corresponding domain in which the supersolution is considered is contained in $(0, 1) \times B_{\frac{1}{R}}(0)$. We used here Remark 2.4.20 when $(t_1, 0) + \rho_1 K_2$ is not contained in $\{0 < t < 1\}$. Thus, we finish the proof by induction. □

Before turning to the proof of Theorem 2.4.15, we observe that the previous stacked measure estimate implies in particular the following result.

Corollary 2.4.26 (Straight stacked measure estimate). *Assume that $R \leq \frac{1}{\sqrt{10(m+1)}}$. Under the assumptions of Proposition 2.4.24 with $k = m$, for any cube $L_1 \subset K_1$*

$$|\{u > h\} \cap L_1| > (1 - \mu)|L_1| \Rightarrow u > \frac{h}{M^m} \text{ in } \overline{L_1}^{(m)} \subset Q_1(1, 0).$$

Proof. Apply Proposition 2.4.24 with $k = m$ and remark that $\overline{L_1}^{(m)} \subset L_2^{(m)}$. The fact that $\overline{L_1}^{(m)} \subset Q_1(1, 0)$ (see Fig. 2.7) comes from the fact that $10(m + 1) R^2 \leq 1$. □

2.4.2.6 A Stacked Covering Lemma

When proving the fundamental L^ε-estimate (sometimes called the weak Harnack inequality) for fully nonlinear elliptic equations, the Calderón–Zygmund decomposition lemma plays an important role (see [CafCab] for instance). It has to be adapted to the parabolic framework.

We need first some definitions. A *cube* Q is a set of the form $(t_0, x_0) + (0, s^2) \times (-s, s)^d$. A *dyadic cube* K of Q is obtained by repeating a finite number of times the following iterative process: Q is divided into 2^{d+2} cubes by considering all the translations of $(0, s^2/4) \times (0, s)^d$ by vectors of the form $(l(s^2/4), sk)$ with $k \in \mathbb{Z}^d$ and $l \in \mathbb{Z}$ included in Q. When a cube K_1 is split in different cubes including K_2, K_1 is called a *predecessor* of K_2.

Given $m \in \mathbb{N}$, and a dyadic cube K of Q, the set $\bar{K}^{(m)}$ is obtained by "stacking" m copies of its predecessor $\bar{K}$. More rigorously, if the predecessor $\bar{K}$ has the form $(a, b) \times L$, then we define $\bar{K}^{(m)} = (b, b + m(b - a)) \times L$. Figure 2.8 corresponds to the case $m = 3$.

Lemma 2.4.27 (Stacked covering lemma). *Let $m \in \mathbb{N}$. Consider two subsets A and B of a cube Q. Assume that $|A| \leq \delta|Q|$ for some $\delta \in (0, 1)$. Assume also the following: for any dyadic cube $K \subset Q$,*

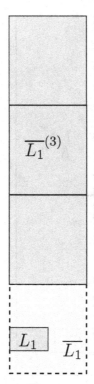

Fig. 2.7 L_1 and the predecessors $\overline{L_1}$ and $\overline{L_1}^{(3)}$

$$|K \cap A| > \delta|A| \Rightarrow \bar{K}^m \subset \mathcal{B}.$$

Then $|A| \leq \delta \frac{m+1}{m} |\mathcal{B}|$.

Remark 2.4.28. This lemma is implicitly used in [Wang92a] (see e.g. Lemma 3.23 of this paper) but details of the proof are not given.

The proof uses a Lebesgue's differentiation theorem with assumptions that are not completely classical, even if we believe that such a generalization is well-known. For the sake of completeness, we state and prove it in appendix (see Theorem 2.5.1 and Corollary 2.5.2).

Proof of Lemma 2.4.27. By iterating the process described to define dyadic cubes, we know that there exists a countable collection of dyadic cubes K_i such that

$$|K_i \cap A| \geq \delta|K_i| \quad \text{and} \quad |\bar{K}_i \cap A| \leq \delta|\bar{K}_i|$$

where $\bar{K}_i$ is a predecessor of K_i. We claim that thanks to Lebesgue's differentiation theorem (Corollary 2.5.2), there exists a set N of null measure such that

$$A \subset (\cup_{i=1}^{\infty} K_i) \cup N.$$

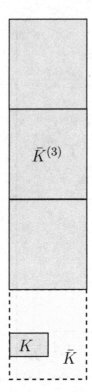

Fig. 2.8 A dyadic cube K and stacked predecessors $\bar{K}^{(m)}$

Indeed, consider $(t, x) \in A \setminus \cup_{i=1}^{\infty} K_i$. On one hand, since $(t, x) \in Q$, it belongs to a sequence of closed dyadic cubes of the form $L_j = (t_j, x_j) + [0, r_j^2] \times [-r_j, r_j]^d$ with $r_j \to 0$ as $j \to +\infty$ such that

$$|A \cap L_j| \leq \delta |L_j|$$

that is to say

$$\fint_{L_j} \mathbf{1}_A \leq \delta < 1.$$

On the other hand, for $(t, x) \in A \setminus \cup_{i=1}^{\infty} K_i$,

$$0 < 1 - \delta \leq 1 - \fint_{L_j} \mathbf{1}_A = \fint_{L_j} |\mathbf{1}_A - \mathbf{1}_A(t, x)|.$$

We claim that the right hand side of the previous equality goes to 0 as $j \to \infty$ as soon as $(t, x) \notin N$ where N is a set of null measure. Indeed, Corollary 2.5.2 implies that for (t, x) outside of such a set N,

$$\fint_{L_j} |\mathbf{1}_A - \mathbf{1}_A(t,x)| \le \fint_{\tilde{L}_j} |\mathbf{1}_A - \mathbf{1}_A(t,x)| \to 0$$

where $\tilde{L}_j = (t,x) + [0, 4r_j^2] \times [-2r_j, 2r_j]^d$. We conclude that $A \setminus \cup_i K_i \subset N$.

We can relabel predecessors $\bar{K}_i$ so that they are pairewise disjoint. We thus have $A \subset \cup_{i=1}^\infty K_i \cup N$ with $\bar{K}_i^m \subset B$ thanks to the assumption; in particular,

$$A \subset \cup_{i=1}^\infty K_i \cup N \subset \cup_{i=1}^\infty \bar{K}_i \cup \bar{K}_i^m \cup N$$

with $\cup_{i=1}^\infty \bar{K}_i^m \subset B$. Classically, we write

$$|A| \le \sum_{i \ge 1} |A \cap \bar{K}_i| \le \delta \sum_{i \ge 1} |\bar{K}_i| \le \delta |\cup_{i=1}^\infty \bar{K}_i|. \tag{2.48}$$

In order to conclude the proof of the lemma, it is thus enough to prove that for a countable collection $(\bar{K}_i)_i$ of disjoint cubes, we have

$$|\cup_{i=1}^\infty \bar{K}_i \cup \bar{K}_i^m| \le \frac{m}{m+1} |\cup_{i=1}^\infty \bar{K}_i^m|. \tag{2.49}$$

Indeed, combining (2.48) and (2.49) yields the desired estimate (keeping in mind that $\cup_i \bar{K}_i^m \subset B$).

Estimate (2.49) is not obvious since, even if the $\bar{K}_i$'s are pairwise disjoint, the stacked cubes $\bar{K}_i^m$ can overlap. In order to justify (2.49), we first write

$$\cup_{i=1}^\infty \bar{K}_i \cup \bar{K}_i^m = \cup_{j=1}^\infty J_j \times L_j$$

where L_j are disjoint cubes of $\mathbb{R}^d$ and J_j are open sets of $\mathbb{R}$ of the form

$$J = \cup_{k=1}^\infty (a_k, a_k + (m+1)h_k).$$

Remark that

$$\cup_{i=1}^\infty \bar{K}_i^m = \cup_{j=1}^\infty \tilde{J}_j \times L_j$$

where $\tilde{J}_j$ has the general form

$$\tilde{J} = \cup_{k=1}^\infty (a_k + h_k, a_k + (m+1)h_k).$$

Hence, the proof is complete once Lemma 2.4.29 below is proved. $\qquad\square$

Lemma 2.4.29. *Let $(a_k)_{k=1}^N$ and $(h_k)_{k=1}^N$ be two (possibly infinite) sequences of real numbers for $N \in \mathbb{N} \cup \{\infty\}$ with $h_k > 0$ for $k = 1, \dots, N$. Then*

$$\left| \cup_{k=1}^N (a_k, a_k + (m+1)h_k) \right| \le \frac{m}{m+1} \left| \cup_{k=1}^N (a_k + h_k, a_k + (m+1)h_k) \right|.$$

Proof. We first assume that N is finite. We write $\cup_{k=1}^N (a_k + h_k, a_k + (m+1)h_k)$ as $\cup_{l=1}^L I_l$ where I_l are disjoint open intervals. We can write them as

$$I_l = \cup_{k=1}^{N_l} (b_k + l_k, b_k + (m+1)l_k) = (\inf_{k=1,\dots,N_l} (b_k + l_k), \sup_{k=1,\dots,N_l} (b_k + (m+1)l_k)).$$

Pick k_l such that $\inf_{k=1,\dots,N_l} (b_k + l_k) = b_{k_l} + l_{k_l}$. In particular,

$$|I_l| = \sup_{k=1,\dots,N_l} (b_k + (m+1)l_k)) - \inf_{k=1,\dots,N_l} (b_k + l_k)$$

$$\ge m l_{k_l}.$$

Then

$$\left| \cup_{k=1}^N (a_k + h_k, a_k + (m+1)h_k) \right| \ge m \sum_l l_{k_l} = \frac{m}{m+1} \sum_l (m+1)l_{k_l}.$$

It is now enough to remark that $(m+1)l_{k_l}$ coincide with the length of one of the intervals $\{(a_k, a_k + (m+1)h_k)\}_k$ and they are distinct since so are the I_l's. The proof is now complete in the case where N is finite.

If now $N = \infty$, we get from the previous case that for any $N \in \mathbb{N}$,

$$\left| \cup_{k=1}^N (a_k, a_k + (m+1)h_k) \right| \le \frac{m}{m+1} \left| \cup_{k=1}^N (a_k + h_k, a_k + (m+1)h_k) \right|$$

$$\le \frac{m}{m+1} \left| \cup_{k=1}^\infty (a_k + h_k, a_k + (m+1)h_k) \right|.$$

It is now enough to let $N \to \infty$ to conclude. $\qquad\square$

2.4.2.7   Proof of the L^ε-Estimate

The proof of the L^ε estimate consists in obtaining a decay in the measure of the sets $\{u > M^k\} \cap \tilde{K}_1$ (see Fig. 2.9). As in the elliptic case, the strategy is to apply the covering Lemma 2.4.27 iteratively making use of Corollary 2.4.26. The main difficulty of the proof (which is not present in the elliptic case) comes from the fact that if K is a cube contained in $\tilde{K}_1$, then nothing prevents $\tilde{K}^{(m)}$ spilling out of K_1.

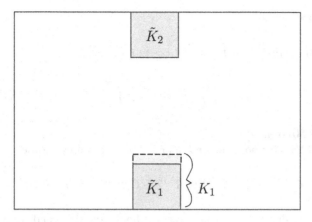

Fig. 2.9 The two neighbourhoods $\tilde{K}_1$ and $\tilde{K}_2$

Proof of Theorem 2.4.15. First, we can assume that

$$\inf_{\tilde{K}_2} u \le 1 \quad \text{and} \quad \|f\|_{L^{d+1}((0,1) \times B_{\frac{1}{R}}(0))} \le \varepsilon_0$$

(where ε_0 comes from Proposition 2.4.24) by considering

$$v^\delta(t,x) = \frac{u}{\inf_{\tilde{K}_2} u + \varepsilon_0^{-1} \|f\|_{L^{d+1}((0,1) \times B_{\frac{1}{R}}(0))} + \delta}.$$

We thus want to prove that there exits a universal constant $C > 0$ such that

$$\int_{\tilde{K}_1} u^\varepsilon(t,x)\, dt\, dx \le C. \tag{2.50}$$

In order to get (2.43), it is enough to find universal constants $m, k_0 \in \mathbb{N}$ and $B > 1$ such that for all $k \ge k_0$,

$$|\{u > M^{km}\} \cap (0, R^2/2 + C_1 B^{-k}) \times (-R, R)^d| \le C(1 - \mu/2)^k \tag{2.51}$$

where C is universal and M and μ comes from Proposition 2.4.24. Indeed, first for $t \in [M^{km}, M^{(k+1)m})$, we have

$$|\{u > t\} \cap (0, R^2/2 + C_1 B^{-k}) \times (-R, R)^d| \le C(1 - \mu/2)^k \le Ct^{-\varepsilon}$$

with $\varepsilon = -\frac{\ln(1-\mu/2)}{m \ln M} > 0$. We deduce that for all $t > 0$, we have

$$|\{u > t\} \cap \tilde{K}_1| \le Ct^{-\varepsilon}.$$

Now we use the formula

$$\int_{\tilde{K}_1} u^\varepsilon(t,x)\,dt\,dx = \varepsilon \int_0^\infty \tau^{\varepsilon-1} |\{u > \tau\} \cap \tilde{K}_1| d\tau$$

$$\leq \varepsilon |\tilde{K}_1| \int_0^1 \tau^{\varepsilon-1} d\tau + \varepsilon \int_1^\infty \tau^{\varepsilon-1} |\{u > \tau\} \cap \tilde{K}_1| d\tau$$

and we get (2.50) from (2.51).

We prove (2.51) by induction on k. For $k = k_0$, we simply choose

$$C \geq (1 - \mu/2)^{-k_0} |(0, R^2/2 + C_1 B^{-1}) \times (-R, R)^d|.$$

Now we assume that $k \geq k_0$, that the result holds true for k and we prove it for $k + 1$. In order to do so, we want to apply the covering Lemma 2.4.27 with

$$A = \{u > M^{(k+1)m}\} \cap (0, R^2/2 + C_1 B^{-k-1}) \times (-R, R)^d$$

$$\mathcal{B} = \{u > M^{km}\} \cap (0, R^2/2 + C_1 B^{-k}) \times (-R, R)^d$$

$$Q = K_1 = (0, R^2) \times (-R, R)^d$$

for some universal constants B and C_1 to be chosen later. We can choose k_0 (universal) so that $\mathcal{B} \subset K_1$. For instance

$$2C_1 B^{-k_0} \leq R^2.$$

The induction assumption reads

$$|\mathcal{B}| \leq C(1 - \mu/2)^k.$$

Lemma 2.4.30. We have $|A| \leq (1 - \mu)|Q|$.

Proof. Since, $\inf_{\tilde{K}_2} u \leq 1$, we have in particular $\inf_{K_3} u \leq 1$. The basic measure estimate (Proposition 2.4.18) then implies that

$$|A| \leq |\{u > M\} \cap K_1\}| \leq (1 - \mu)|K_1| = (1 - \mu)|Q|. \qquad \square$$

Lemma 2.4.31. *Consider any dyadic cube* $K = (t, x) + \rho K_1$ *of* Q. *If*

$$|K \cap \{u > M^{(k+1)m}\} \cap (0, R^2/2 + C_1 B^{-k-1}) \times (-R, R)^d\}| > (1-\mu)|K|, \quad (2.52)$$

then

$$\bar{K}^m \subset \{u > M^{km}\} \cap (0, R^2/2 + C_1 B^{-k}) \times (-R, R)^d$$

where $\bar{K}^m$ is defined at the beginning of Sect. 2.4.2.6.

Proof. We remark that the straight stacked measure estimate, Corollary 2.4.26, applied with $h = M^{(k+1)m} \geq M^m$, implies

$$\bar{K}^m \subset \{u > M^{km}\}.$$

We thus have to prove that

$$\bar{K}^m \subset [0, R^2/2 + C_1 B^{-k}] \times (-R, R)^d. \tag{2.53}$$

Because of (2.52), we have

$$K \cap (0, R^2/2 + C_1 B^{-k-1}) \times (-R, R)^d \neq \emptyset.$$

Hence

$$\bar{K}^m \subset [0, R^2/2 + C_1 B^{-k-1} + \text{height}(\bar{K}) + \text{height}(\bar{K}^m)] \times (-R, R)^d$$

where $\text{height}(L) = \sup\{t : \exists x, (t, x) \in L\} - \inf\{t : \exists x, (t, x) \in L\}$. Moreover,

$$\text{height}(K) = R^2 \rho^2$$
$$\text{height}(\bar{K}) = 4\,\text{height}(K)$$
$$\text{height}(\bar{K}^m) = m\,\text{height}(\bar{K}).$$

Hence, (2.53) holds true if

$$R^2/2 + C_1 B^{-k-1} + 4(m + 1)R^2\rho^2 \leq R^2/2 + C_1 B^{-k}$$

i.e.

$$R^2\rho^2 \leq \frac{C_1(B - 1)}{4(m + 1)} B^{-k-1}. \tag{2.54}$$

In order to estimate $R^2\rho^2$ we are going to use the stacked measure estimate given by Proposition 2.4.24 together with the fact that K is a cube for which (2.52) holds.

On one hand, Proposition 2.4.24 and (2.52) imply that as long as $l \leq (k + 1)m$, we have

$$u > M^{(k+1)m-l} \text{ in } L_2^{(l)} \cap \{0 < t < 1\};$$

in particular,

$$\inf_{\bigcup_{l=1}^{(k+1)m} L_2^{(l)} \cap \{0<t<1\}} u > 1.$$

On the other hand, using notation from Lemma 2.4.23,

$$\inf_{\bigcup_{l=1}^{k^*+1} L_2^{(l)} \cap \{0<t<1\}} u \le \inf_{\bar{K}_2} u \le 1$$

Hence $(k+1)m < k^* + 1$. Moreover, Lemma 2.4.23 implies

$$R^2\rho^2 \le (1-t_0)(\alpha_{k^*})^{-1} \le \frac{9}{9(k+1)m}.$$

Hence, we choose $B = 9^m$ and $C_1 = \frac{36(m+1)}{9^m-1}$.   □

We can now apply the covering lemma and conclude that

$$|A| \le \delta \frac{m+1}{m} |B|.$$

We choose m large enough (universal) such that

$$(1-\mu)\frac{m+1}{m} \le 1 - \mu/2.$$

Recalling that we chose μ such that $\frac{1}{\mu} = 1 + 2C_{\max} R^{d+2}$ (where $C_{\max}$ is the universal constant appearing in the maximum principle), the previous condition is equivalent to

$$m \ge 4C_{\max} R^{d+2}.$$

Since $R \le 1$, it is enough to choose $m \ge 4C_{\max}$.

Thanks to the induction assumption, we thus finally get

$$|\{u > M^{(k+1)m}\} \cap (0, R^2/2 + C_1 B^{-k-1}) \times (-R, R)^d| \le C(1-\mu/2)^{k+1}.$$

The proof is now complete.   □

2.4.3 Harnack Inequality

The main result of this subsection is the following theorem.

Theorem 2.4.32 (Harnack inequality). *For any non-negative function u such that*

$$\left.\begin{array}{l} \frac{\partial u}{\partial t} + P^+(D^2u) \ge -f \\ \frac{\partial u}{\partial t} + P^-(D^2u) \le f \end{array}\right\} \tag{2.55}$$

in Q_1, we have

$$\sup_{\tilde{K}_4} u \leq C(\inf_{Q_{R^2}} u + \|f\|_{L^{d+1}(Q_1)})$$

where $\tilde{K}_4 = (-R^2 + \frac{3}{8}R^4, -R^2 + \frac{1}{2}R^4) \times B_{\frac{R^2}{2\sqrt{2}}}(0)$.

Remark 2.4.33. The case where u solves (2.55) in Q_ρ instead of Q_1 follows by scaling. Indeed, consider $v(t, x) = u(\rho^2 t, \rho x)$ and change constants accordingly.

We will derive Theorem 2.4.32 combining Theorem 2.4.15 with the following proposition (which in turn also follows from Theorem 2.4.15).

Proposition 2.4.34 (Local maximum principle). *Consider a function u such that*

$$\frac{\partial u}{\partial t} + P^-(D^2 u) \leq f \text{ in } Q_1. \tag{2.56}$$

Then for all $p > 0$, we have

$$\sup_{Q_{1/2}} u \leq C\left(\left(\int_{Q_1}(u^+)^p\right)^{\frac{1}{p}} + \|f\|_{L^{d+1}(Q_1)}\right).$$

Proof. First we can assume that $u \geq 0$ by remarking that u^+ satisfies (2.56) with f replaced with $|f|$.

Let Ψ be defined by

$$\Psi(t, x) = h \max((1 - |x|)^{-2\gamma}, (1 + t)^{-\gamma})$$

where γ will be chosen later. We choose h minimal such that

$$\Psi \geq u \text{ in } Q_1.$$

In other words

$$h = \min_{(t,x) \in Q_1} \frac{u(t, x)}{\max((1 - |x|)^{-2\gamma}, (1 + t)^{-\gamma})}.$$

We want to estimate h from above. Indeed, we have

$$\sup_{Q_{\frac{1}{2}}} u \leq Ch$$

for some constant C depending on γ and $Q_{\frac{1}{2}}$.

In order to do estimate h, we consider a point (t_0, x_0) realizing the minimum in the definition of h. We consider

$$\delta^2 = \min((1 - |x_0|)^2, (1 + t_0)).$$

In particular

$$u(t_0, x_0) = h\delta^{-2\gamma}$$

and $Q_\delta(t_0, x_0) \subset Q_1$.

We consider next the function $v(t, x) = C - u(t, x)$ where

$$C = \sup_{Q_{\beta\delta}(t_0, x_0)} \Psi$$

for some parameter $\beta \in (0, 1)$ to be chosen later. Remark first that

$$h\delta^{-2\gamma} \le C \le h((1 - \beta)\delta)^{-2\gamma}.$$

Remark next that v is a supersolution of

$$\frac{\partial v}{\partial t} + P^+(D^2 v) + |f| \ge 0 \quad \text{in } Q_1$$

and $v \ge 0$ in $(t_0 - (R\beta\delta)^2, t_0) \times B_{\beta\delta}(x_0) \subset Q_{\beta\delta}(t_0, x_0)$. From the L^ε estimate (Theorem 2.4.15 properly scaled and translated), we conclude that

$$\int_L v^\varepsilon \le C(\beta\delta)^{d+2} \left(\inf_{(t_0 - \beta\delta, x_0) + \beta\delta\tilde{K}_2} v + (\beta\delta)^{\frac{d}{d+1}} \|f\|_{L^{d+1}(Q_1)} \right)^\varepsilon$$

where $L = (t_0 - \beta\delta, x_0) + \beta\delta\tilde{K}_1$. Moreover,

$$\inf_{(t_0 - \beta\delta, x_0) + \beta\delta\tilde{K}_2} v \le v(t_0, x_0)$$

$$= C - u(t_0, x_0)$$

$$\le h\left((1 - \beta)^{-2\gamma} - 1 \right)\delta^{-2\gamma}.$$

Hence, we have

$$\int_L v^\varepsilon \le C(\beta\delta)^{d+2} \left[h\left((1 - \beta)^{-2\gamma} - 1 \right)\delta^{-2\gamma} + (\beta\delta)^{\frac{d}{d+1}} \|f\|_{d+1} \right]^\varepsilon. \qquad (2.57)$$

We now consider the set

$$A = \left\{ (t, x) \in L : u(t, x) < \frac{1}{2} u(t_0, x_0) = \frac{1}{2} h \delta^{-2\gamma} \right\}.$$

We have

$$\int_A v^\varepsilon \geq |A| \left(h \delta^{-2\gamma} - \frac{1}{2} h \delta^{-2\gamma} \right)^\varepsilon = |A| \left(\frac{h \delta^{-2\gamma}}{2} \right)^\varepsilon.$$

We thus get from (2.57) the following estimate

$$|A| \leq C |L| \left[\left((1 - \beta)^{-2\gamma} - 1 \right)^\varepsilon + (\delta^{2\gamma} h^{-1})^\varepsilon (\beta \delta)^{\frac{d\varepsilon}{d+1}} \| f \|_{d+1}^\varepsilon \right].$$

Finally, we estimate $\int_{Q_1} u^\varepsilon$ from below as follows

$$\int_{Q_1} u^\varepsilon \geq \int_{L \backslash A} u^\varepsilon \geq (|L| - |A|) 2^{-\varepsilon} (h \delta^{-2\gamma})^\varepsilon.$$

Hence, choosing $\gamma = \frac{d+2}{2\varepsilon}$ and combining the two previous inequalities, we get

$$\beta^{2+d} C_1 h^\varepsilon = |L| 2^{-\varepsilon} (h \delta^{-2\gamma})^\varepsilon \leq \int_{Q_1} u^\varepsilon$$

$$+ \beta^{2+d} C_2 h^\varepsilon \left((1 - \beta)^{-2\gamma} - 1 \right)^\varepsilon$$

$$+ \beta^{2+d+\frac{d\varepsilon}{d+1}} C_2 \| f \|_{d+1}^\varepsilon.$$

We used $\delta \leq 1$. Choose now β small enough so that

$$C_2 \left((1 - \beta)^{-2\gamma} - 1 \right)^\varepsilon \leq C_1/2$$

and conclude in the case $p = \varepsilon$. The general case follows by interpolation. □

Theorem 2.4.32 is a direct consequence of the following one.

Theorem 2.4.35. *For any non-negative function u satisfying (2.55) in $(-1, 0) \times B_{\frac{1}{R}}(0)$, we have*

$$\sup_{\tilde{K}_3} u \leq C (\inf_{Q_R} u + \| f \|_{L^{d+1}((-1,0) \times B_{\frac{1}{R}}(0))})$$

where $\tilde{K}_3 = (-1 + \frac{3}{8} R^2, -1 + R^2/2) \times B_{\frac{R}{2\sqrt{2}}}(0)$ (see Fig. 2.10).

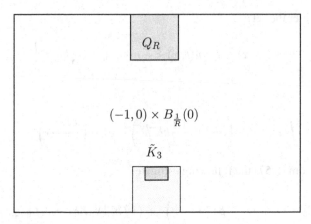

Fig. 2.10 The set $\widetilde{K}_3$

Proof of Theorem 2.4.35. On the one hand, from Theorem 2.4.15 (the L^ε estimate) applied to $u(t + 1, x)$ we know that

$$\left(\int_{(-1,-1+R^2/2)\times B_{R/\sqrt{2}}} u(x)^\varepsilon dx\right)^{1/\varepsilon} \leq C(\inf_{Q_R} u + \|f\|_{L^{d+1}(Q_1)}). \qquad (2.58)$$

On the other hand, we apply Proposition 2.4.34 to the scaled function $v(t,x) = u((t + 1 - R^2/2)/(R^2/2), \sqrt{2}x/R) \geq 0$ and $p = \varepsilon$ to obtain

$$\sup_{Q_{\frac{1}{2}}} v \leq C\left(\left(\int_{Q_1} v^\varepsilon\right)^{\frac{1}{\varepsilon}} + \|f\|_{L^{d+1}(Q_1)}\right).$$

Scaling back to the original variables, we get

$$\sup_{\widetilde{K}_3} u \leq C\left(\left(\int_{(-1,-1+R^2/2)\times B_{R/\sqrt{2}}} u^\varepsilon\right)^{\frac{1}{\varepsilon}} + \|f\|_{L^{d+1}(Q_1)}\right). \qquad (2.59)$$

Combining (2.58) with (2.59) we get

$$\sup_{\widetilde{K}_3} u \leq C\left(\inf_{Q_R} u + \|f\|_{L^{d+1}(Q_1)}\right),$$

which finishes the proof.   □

2.4.4 Hölder Continuity

An important consequence of Harnack inequality (Theorem 2.4.32) is the Hölder continuity of functions satisfying (2.55).

Theorem 2.4.36. *If u satisfies (2.55) in Q_ρ then u is α-Hölder continuous in Q_ρ and*

$$[u]_{\alpha, Q_{\rho/2}} \le C\rho^{-\alpha} \left(|u|_{0, Q_\rho} + \rho^{\frac{d}{d+1}} \|f\|_{L^{d+1}(Q_\rho)} \right).$$

Proof. We only deal with $\rho = 1$. We prove that if u satisfies (2.55) in Q_1 then u is α-Hölder continuous at the origin, i.e.

$$|u(t, x) - u(0, 0)| \le C \left(|u|_{0, Q_1} + \|f\|_{L^{d+1}(Q_1)} \right) (|x| + \sqrt{t})^\alpha. \tag{2.60}$$

To get such an estimate, it is enough to prove that the oscillation of the function u in Q_ρ decays as ρ^α; more precisely, we consider

$$M_\rho = \sup_{Q_\rho} u,$$

$$m_\rho = \inf_{Q_\rho} u,$$

$$\mathrm{osc}_{Q_\rho} u = M_\rho - m_\rho.$$

Then (2.60) holds true as soon as

$$\mathrm{osc}_{Q_\rho} u \le C \left(|u|_{0, Q_1} + \|f\|_{L^{d+1}(Q_1)} \right) \rho^\alpha. \tag{2.61}$$

Indeed, consider $(t, x) \in Q_\rho \setminus Q_{\rho/2}$ and estimate $|u(t, x) - u(0, 0)|$ from above by $\mathrm{osc}_{Q_\rho} u$ and $\rho/2$ from above by $|x|_\infty + \sqrt{t}$.

In order to prove (2.61), we consider the two functions $u - m_\rho \ge 0$ and $M_\rho - u \ge 0$ in Q_ρ. They both satisfy (2.55) in Q_ρ. From the Harnack inequality, we thus get

$$\sup_{\rho \tilde{K}_4} (u - m_\rho) \le C(\inf_{Q_{R^2\rho}} (u - m_\rho) + \rho^{\frac{d}{d+1}} \|f\|_{d+1})$$

$$\sup_{\rho \tilde{K}_4} (M_\rho - u) \le C(\inf_{Q_{R^2\rho}} (M_\rho - u) + \rho^{\frac{d}{d+1}} \|f\|_{d+1})$$

where $\rho \tilde{K}_4 \subset Q_\rho$ follows from $\tilde{K}_4 \subset (-1, 0) \times B_1$. We next add these two inequalities which yields

$$\mathrm{osc}_{Q_\rho} u \le C(\mathrm{osc}_{Q_\rho} u - \mathrm{osc}_{Q_{\gamma\rho}} u + \rho^{\frac{d}{d+1}} \|f\|_{d+1})$$

with $C > 1$ and where γ denotes R^2. Rearranging terms, we get

$$\mathrm{osc}_{Q_{\gamma\rho}} u \le \frac{C-1}{C} \mathrm{osc}_{Q_\rho} u + \rho^{\frac{d}{d+1}} \|f\|_{d+1}$$

where C is universal. Then an elementary iteration lemma allows us to achieve the proof of the theorem; see Lemma 2.5.13 in appendix with $h(\rho) = \mathrm{osc}_{Q_\rho} u$ and $\delta = (C-1)/C$ and $\beta = d/(d+1)$. □

Appendix: Technical Lemmas

A.1   Lebesgue's Differentiation Theorem

The purpose of this appendix is to prove a version of Lebesgue's differentiation theorem with parabolic cylinders. Recall that the usual version of the result says that if $f \in L^1(\Omega, dt \otimes dx)$ where Ω is a Borel set of $\mathbb{R}^{d+1}$, then for a.e. $(t, x) \in \Omega$,

$$\lim_{j\to\infty} \fint_{G_j} |f - f(t, x)| = 0$$

as long as the sequence of sets G_j satisfies the *regularity* condition:

$$G_j \subset B_j$$
$$|G_j| \ge c|B_j|$$

where B_j is a sequence of balls $B_{r_j}(t, x)$ with $r_j \to 0$.

A sequence of parabolic cylinders $Q_{r_j}(t, x)$ cannot satisfy the regularity condition because of the different scaling between space and time. Indeed $|Q_{r_j}(t, x)| = r_j^{d+2}$ which is an order of magnitude smaller than r_j^{d+1}.

Fortunately, the classical proof of Lebesgue's differentiation theorem can be repeated and works for parabolic cylinders as well, as it is shown below.

Theorem 2.5.1 (Lebesgue's differentiation theorem). *Consider an integrable function* $f \in L^1(\Omega, dt \otimes dx)$ *where* Ω *is an open set of* $\mathbb{R}^{d+1}$. *Then for a.e.* $(t, x) \in \Omega$,

$$\lim_{r\to 0+} \fint_{(t-r^2, t) \times B_r(x)} |f - f(t, x)| = 0$$

where $\fint_O g = \frac{1}{|O|} \int_O g$ *for any Borel set* $O \subset \mathbb{R}^{d+1}$ *and integrable function g.*

In the proof, we will in fact use the following corollary.

Corollary 2.5.2 (Generalized Lebesgue's differentiation theorem). *Let G_j be a family of sets which is regular in the following sense: there exists a constant $c > 0$ and $r_j \to 0$ such that*

$$G_j \subset (t - r_j^2, t) \times B_{r_j}(x),$$

$$|G_j| \geq c r_j^{d+2}.$$

Then, except for a set of measure zero which is independent of the choice of $\{G_j\}$, we have

$$\lim_{j \to +\infty} \fint_{G_j} |f - f(t, x)| = 0.$$

Remark 2.5.3. It is interesting to point out that if the parabolic cylinders were replaced by other families of sets not satisfying the *regularity* condition, the result of Lemma 2.5.5 may fail. For example if we take

$$\tilde{M} f(t, x) = \sup_{(a,b) \times B_r(y) \ni (t,x)} \fint_{(a,b) \times B_r(y) \cap \Omega} |f|$$

then Lemma 2.5.5 would fail for $\tilde{M} f$.

Proof of Corollary 2.5.2. We obtain Corollary 2.5.2 as an immediate consequence of Theorem 2.5.1 by noting that since $G_j \subset (t - r_j^2, t) \times B_{r_j}(x)$.

$$\fint_{G_j} |f - f(t, x)| \leq \frac{r^2 |B_r|}{|G_j|} \fint_{(t-r^2, t) \times B_r(x)} |f - f(t, x)|.$$

Thus, the result holds at all points where this right hand side goes to zero, which is a set of full measure by Theorem 2.5.1 and that $\frac{r^2 |B_r|}{|G_j|} \geq c > 0$.   □

In order to prove Theorem 2.5.1, we first need a version of Vitali's covering lemma.

Lemma 2.5.4 (Vitali's covering lemma). *Consider a bounded collection of cubes $(Q_\alpha)_\alpha$ of the form $Q_\alpha = (t_\alpha - r_\alpha^2, t_\alpha) \times B_{r_\alpha}(x_\alpha)$ and a set A such that $A \subset \cup_\alpha Q_\alpha$. Then there is a finite number of cubes $Q_1, \ldots, Q_N$ such that $A \subset \cup_{j=1}^{N} 5Q_j$ where $5Q_j = (t_\alpha - 25r_\alpha^2, t_\alpha) \times B_{5r_\alpha}(x_\alpha)$.*

Consider next the maximal function Mf associated with a function $f \in L^1(\Omega, dt \otimes dx)$

$$Mf(t, x) = \sup_{Q \ni (t,x)} \fint_{Q \cap \Omega} |f|$$

where the supremum is taken over cubes Q of the form $(s, y) + (-r^2, 0) \times B_r$.

Lemma 2.5.5 (The maximal inequality). *Consider* $f \in L^1(\Omega, dt \otimes dx)$, f
positive, and $\lambda > 0$, *we have*

$$|\{Mf > \lambda\}| \le \frac{C}{\lambda} \|f\|_{L^1}$$

for some constant C depending only on dimension d.

Proof. For all $x \in \{Mf > \lambda\}$, there exists $Q \ni x$ such that

$$\inf_Q f \ge \frac{\lambda}{2}|Q|.$$

Hence, the set $\{Mf > \lambda\}$ can be covered by cubes Q. From Vitali's covering lemma,
there exists a finite cover of $\{Mf > \lambda\}$ with some $5Q$'s:

$$\{Mf > \lambda\} \subset \cup_{j=1}^N 5Q_j$$

with Q_j that are disjoint and such that

$$\int_{Q_j \cap \Omega} f \ge \frac{\lambda}{2}|Q_j \cap \Omega|.$$

Hence

$$\int_\Omega f \ge \int_{\cup_j Q_j \cap \Omega} f = \sum_j \int_{Q_j \cap \Omega} f$$

$$\ge \frac{\lambda}{2}|\cup_j Q_j \cap \Omega| = \frac{\lambda}{2} \times \frac{1}{5^{d+2}}|\cup_j 5Q_j \cap \Omega| \ge \frac{\lambda}{C}|\{Mf > \lambda\}|$$

with $C = 2 \times 5^{d+2}$. $\qquad\qquad\qquad\qquad\qquad\qquad\qquad\qquad\qquad\qquad\qquad\qquad$ □

We can now prove Lebesgue's differentiation theorem (Theorem 2.5.1).

Proof of Theorem 2.5.1. We can assume without loss of generality that the set Ω
is bounded. We first remark that the result is true if f is continuous. If f is not
continuous, we consider a sequence $(f_n)_n$ of continuous functions such that

$$\|f - f_n\|_{L^1} \le \frac{C}{2^n}.$$

Moreover, up to a subsequence, we can also assume that for a.e. $(t, x) \in \Omega$,

$$f_n(t, x) \to f(t, x) \quad \text{as } n \to \infty.$$

Thanks to the maximal inequality (Lemma 2.5.5), we have in particular

$$|\{M(f - f_n) > \lambda\}| \le \frac{C}{\lambda 2^n}.$$

By Borel–Cantelli's Lemma, we conclude that for all $\lambda > 0$, there exists $n_\lambda \in \mathbb{N}$ such that for all $n \ge n_\lambda$,

$$M(f - f_n) \le \lambda \quad \text{a.e. in } \Omega.$$

We conclude that for a.e. $(t, x) \in \Omega$ and all $k \in \mathbb{N}$, there exists a strictly increasing sequence n_k such that for all $r > 0$ such that $Q_r(t, x) \subset \Omega$,

$$\fint_{Q_r(t,x)} |f - f_{n_k}| \le M(f - f_{n_k}) \le \frac{1}{k}.$$

Moreover, since f_n is continuous and Ω is bounded, there exists $r_k > 0$ such that for $r \in (0, r_k)$, we have

$$\fint_{Q_r(t,x)} |f_{n_k} - f_{n_k}(t, x)| \le \frac{1}{k}.$$

Moreover, for a.e. $(t, x) \in \Omega$,

$$|f_{n_k}(t, x) - f(t, x)| \to 0 \quad \text{as } k \to \infty.$$

These three facts imply that for a.e. $(t, x) \in \Omega$, for all $\varepsilon > 0$, there exists $r_\varepsilon > 0$ such that $r \in (0, r_\varepsilon)$,

$$\fint_{Q_r(t,x)} |f - f(t, x)| \le \varepsilon.$$

This achieves the proof of the lemma. $\square$

A.2 Jensen–Ishii's Lemma for N Functions

When proving Theorem 2.4.9 (more precisely, Lemma 2.4.6), we used the following generalization of Lemmas 2.3.23 and 2.3.30 whose proof can be found in [CIL92].

Lemma 2.5.6 (Jensen–Ishii's Lemma III). *Let U_i, $i = 1, \ldots, N$ be open sets of $\mathbb{R}^d$ and I an open interval of $\mathbb{R}$. Consider also lower semi-continuous functions $u_i : I \times U_i \to \mathbb{R}$ such that for all $v = u_i$, $i = 1, \ldots, N$, $(t, x) \in I \times U_i$, there exists $r > 0$ such that for all $M > 0$ there exists $C > 0$,*

$$\left.\begin{array}{r}(s, y) \in Q_r(t, x) \\ (\beta, q, Y) \in \mathcal{P}^- v(s, y) \\ |v(s, y)| + |q| + |Y| \le M\end{array}\right\} \Rightarrow -\beta \le C.$$

Let $x = (x_1, \dots, x_N)$ *and* $x_0 = (x_1^0, \dots, x_N^0)$. *Assume that* $\sum_{i=1}^N u_i(t, x_i) - \phi(t, x)$ *reaches a local minimum at* $(t_0, x_0) \in I \times \Pi_i U_i$. *If* α *denotes* $\partial_t \phi(t_0, x_0)$ *and* p_i *denotes* $D_{x_i} \phi(x_0)$ *and* A *denotes* $D^2 \phi(t_0, x_0)$, *then for any* $\beta > 0$ *such that* $I + \beta A > 0$, *there exist* $(\alpha_i, X_i) \in \mathbb{R} \times \mathbb{S}_d$, $i = 1, \dots, N$, *such that for all* $i = 1, \dots, N$,

$$(\alpha_i, p_i, X_i) \in \overline{\mathcal{P}}^- u(t_0, x_i^0)$$

$$\sum_{i=1}^N \alpha_i = \alpha$$

and

$$\frac{1}{\beta}\begin{pmatrix} I & 0 & \dots & 0 \\ 0 & \ddots & \ddots & \vdots \\ \vdots & \ddots & \ddots & 0 \\ 0 & \dots & 0 & I \end{pmatrix} \ge \begin{pmatrix} X_1 & 0 & \dots & 0 \\ 0 & \ddots & \ddots & \vdots \\ \vdots & \ddots & \ddots & 0 \\ 0 & \dots & 0 & X_N \end{pmatrix} \ge A_\beta$$

where $A_\beta = (I + \beta A)^{-1} A$.

Remark 2.5.7. The condition on the functions u_i is satisfied as soon as the u_i's are supersolutions of a parabolic equation. This condition ensures that some compactness holds true when using the doubling variable technique in the time variable. See [CIL92, Theorem 8.2, p. 50] for more details.

A.3   Technical Lemmas for Monotone Envelopes

When proving the maximum principle (Theorem 2.4.9), we used the two following technical lemmas.

Lemma 2.5.8. *Consider a convex set* Ω *of* $\mathbb{R}^d$ *and a lower semi-continuous function* $v : [a, b] \times \bar{\Omega} \to \mathbb{R}$ *which is non-increasing with respect to* $t \in (a, b)$ *and convex with respect to* $x \in \Omega$. *Assume that* v *is bounded from above and that for all* $(\alpha, p, X) \in \mathcal{P}^- v(t, x)$, *we have*

$$-\alpha \le C \quad and \quad X \le CI.$$

Then v is Lipschitz continuous with respect $t \in (a, b)$ and $C^{1,1}$ with respect to $x \in \Omega$.

Proof of Lemma 2.5.8. We assume without loss of generality that Ω is bounded. In this case, v is bounded from above and from below, hence is bounded. Next, we also get that v is Lipschitz continuous with respect to x in $[a, b] \times F$ for all closed convex set $F \subset \Omega$ such that $d(F, \partial\Omega) > 0$.

Step 1.

We first prove that v is Lipschitz continuous with respect to t: for all $(t_0, x_0) \in (a, b) \times \Omega$,

$$M = \sup_{s,t \in (a,b), x, y \in \Omega} \left\{ v(t, x) - v(s, y) - L|t - s| - \frac{L}{4\varepsilon}|x - y|^2 - L\varepsilon \right.$$
$$\left. - L_0|x - x_0|^2 - L_0(t - t_0)^2 \right\} \leq 0$$

for L large enough only depending on C and the Lipschitz constant of v with respect to x around (t_0, x_0) and for L_0 large enough. We argue by contradiction by assuming that $M > 0$. Consider $(\bar{s}, \bar{t}, \bar{x}, \bar{y})$ where the maximum M is reached. Remark first that

$$L_0|\bar{y} - x_0|^2 + L_0(\bar{s} - t_0)^2 + L|\bar{t} - \bar{s}| + \frac{L}{4\varepsilon}|\bar{x} - \bar{y}|^2 + L\varepsilon \leq v(\bar{t}, \bar{x}) - v(\bar{s}, \bar{y})$$
$$\leq 2|v|_{0,[a,b] \times \bar{\Omega}}.$$

In particular, we can choose L_0 and L large enough so that $(\bar{s}, \bar{y}), (\bar{t}, \bar{x}) \in (a, b) \times \Omega$. Remark next that $\bar{t} \neq \bar{s}$. Indeed, if $\bar{t} = \bar{s}$, then

$$0 < M \leq v(\bar{t}, \bar{x}) - v(\bar{t}, \bar{y}) - \frac{L}{4\varepsilon}|\bar{x} - \bar{y}|^2 - L\varepsilon$$

and choosing L larger than the Lipschitz constant of v with respect to x yields a contradiction. Hence the function v is touched from below at $(\bar{s}, \bar{y})$ by the test function

$$(s, y) \mapsto C_0 - \frac{L}{4\varepsilon}|\bar{x} - y|^2 - L|\bar{t} - s|$$

where C_0 is a constant depending on $(\bar{t}, \bar{x})$. In particular,

$$(L \operatorname{sign}(\bar{t} - \bar{s}), L(4\varepsilon)^{-1}(\bar{x} - \bar{y}), L(4\varepsilon)^{-1}I) \in \mathcal{P}^- v(\bar{s}, \bar{y}).$$

We thus should have $L \leq C$. Choosing $L > C$ yields also the desired contradiction.

Step 2.

In order to prove that for all $t \in (a, b)$, $u(t, \cdot)$ is $C^{1,1}$ with respect to x, it is enough to prove that for all $(p, X) \in D^{2,-} u(t, x)$ (see below), $X \leq CI$. Indeed, this implies that $u(t, \cdot) + \frac{C}{2} |\cdot|^2$ is concave [ALL97]. Since $u(t, \cdot)$ is convex, this implies that it is $C^{1,1}$ [CanSin04].

$(p, X) \in D^{2,-} u(t, x)$ means that there exists $\psi \in C^2(\mathbb{R}^d)$ such that $p = D\psi(x)$ and $X = D^2 \psi(x)$ and

$$\psi(y) - \psi(x) \leq u(t, y) - u(t, x)$$

for $y \in B_r(x)$. We can further assume that the minimum of $u(t, \cdot) - \psi$ is strict. We then consider the minimum of $u(s, x) - \psi(x) + \varepsilon^{-1}(s - t)^2$ in $(t - r, t + r) \times B_r(x)$. For ε small enough, this minimum is reached in an interior point $(t_\varepsilon, x_\varepsilon)$ and $(t_\varepsilon, x_\varepsilon) \to (t, x)$ as $\varepsilon \to 0$. Then

$$(\varepsilon^{-1}(s_\varepsilon - t), D\psi(x_\varepsilon), D^2\psi(x_\varepsilon)) \in \mathcal{P}^- u(t_\varepsilon, x_\varepsilon).$$

Hence, $D^2\psi(x_\varepsilon) \leq CI$. Letting $\varepsilon \to 0$ yields $X \leq CI$. This achieves Step 2.

The proof of the lemma is now complete.   □

Lemma 2.5.9. *Consider a convex set Ω of $\mathbb{R}^d$ and $v : (a, b) \times \Omega \to \mathbb{R}$ which is non-increasing with respect to $t \in (a, b)$ and convex with respect to $x \in \Omega$. Then for all $(\alpha, p, X) \in \mathcal{P}^- v(t, x)$, that there exists (α_n, p_n, X_n) such that*

$$(\alpha_n, p_n, X_n) \in \mathcal{P}^- v(t_n, x_n)$$

$$(t_n, x_n, \alpha_n, p_n) \to (t, x, \alpha, p)$$

$$X \leq X_n + o_n(1), X_n \geq 0.$$

The proof of this lemma relies on Alexandroff theorem in its classical form. A statement and a proof of this classical theorem can be found for instance in [EG92]. We will only use the following consequence of this theorem.

Theorem 2.5.10. *Consider a convex set Ω of $\mathbb{R}^d$ and a function $v : (a, b) \times \Omega \to \mathbb{R}$ which is convex with respect to $(t, x) \in (a, b) \times \Omega$. Then for almost $(t, x) \in (a, b) \times \Omega$, there exists $(\alpha, p, X) \in \mathcal{P}^- \cap \mathcal{P}^+ v(t, x)$, that is to say such that,*

$$v(s, y) = v(t, x) + \alpha(s - t) + p \cdot (y - x) + \frac{1}{2} X(y - x) \cdot (y - x) + o(|s - t| + |y - x|^2).$$

$$(2.62)$$

Jensen's lemma is also needed (stated here in a "parabolic" version for the sake of clarity).

Lemma 2.5.11 (Jensen). *Consider a convex set Ω of $\mathbb{R}^d$ and a function $v : (a, b) \times \Omega \to \mathbb{R}$ such that there exists $(\tau, C) \in \mathbb{R}^2$ such that $u(t, x) + \tau t^2 + C|x|^2$ is convex* *with respect to $(t, x) \in (a, b) \times \Omega$. If u reaches a strict local maximum at (t_0, x_0), then for $r > 0$ and $\delta > 0$ small enough, the set*

$$K = \{(t, x) \in (t_0 - r, t_0 + r) \times B_r(x_0) : \exists (\tau, p) \in (-\delta, \delta) \times B_\delta,$$

$$(s, y) \mapsto u(s, y) - \tau s - p \cdot y \text{ reaches a local maximum at } (t, x)\}$$

has a positive measure.

See [CIL92] for a proof. We can now turn to the proof of Lemma 2.5.8. The proof of Lemma 2.5.9 below mimics the proof of [ALL97, Lemma 3] in which there is no time dependence.

Proof of Lemma 2.5.9. Consider a test function ϕ such that $u - \phi$ reaches a local maximum at (t, x) and

$$(\alpha, p, X) = (\partial_t \phi, D\phi, D^2 \phi)(t, x).$$

Without loss of generality, we can assume that this maximum is strict; indeed, replace ϕ with $\phi(s, y) - |y - x|^2 - (s - t)^2$ for instance. Then consider the function

$$v_\varepsilon(t, x) = \inf_{y \in \mathbb{R}^d, s \geq 0} \left\{ v(s, y) + \frac{1}{\varepsilon} |y - x|^2 + \frac{1}{\varepsilon} (s - t)^2 \right\}.$$

One can check that v_ε is still convex with respect to x and non-increasing with respect to t and that

$$(t, x) \mapsto v_\varepsilon(t, x) + \frac{1}{\varepsilon} |x|^2 + \frac{1}{\varepsilon} t^2$$

is concave with respect to (t, x). Moreover, $v_\varepsilon \leq v$ and

$$\lim_{\varepsilon \to 0} v_\varepsilon(t, x) = v(t, x).$$

This implies that there exists $(t_\varepsilon, x_\varepsilon) \to 0$ as $\varepsilon \to 0$ such that $v_\varepsilon - \phi$ reaches a local maximum at $(t_\varepsilon, x_\varepsilon)$. Remarking that $v_\varepsilon - \phi$ satisfies the assumptions of Jensen's lemma, Lemma 2.5.11 above, we combine it with Theorem 2.5.10 and we conclude that we can find slopes $(\tau_n, p_n) \to (0, 0)$ and points $(t_n, x_n) \to (t_\varepsilon, x_\varepsilon)$ as $n \to \infty$ where $v_\varepsilon - \phi$ satisfies (2.62) and $v_\varepsilon - \phi - \tau_n s - p_n y$ reaches a local maximum at (t_n, x_n). In other words,

$$(\tau_n + \partial_t \phi(t_n, x_n), p_n + D\phi(t_n, x_n), D^2 v_\varepsilon(t_n, x_n)) \in \mathcal{P}^- v_\varepsilon(t_n, x_n)$$

with

$$D^2 v_\varepsilon(t_n, x_n) \geq 0$$

and

$$D^2\phi(t_n, x_n) \leq D^2 v_\varepsilon(t_n, x_n).$$

In order to conclude, we use the classical following result from viscosity solution theory (see [CIL92] for a proof):

Lemma 2.5.12. *Consider* (s_n, y_n) *such that*

$$v_\varepsilon(t_n, x_n) = v(s_n, y_n) + \varepsilon^{-1}|y_n - x_n|^2 + \varepsilon^{-1}(t_n - s_n)^2.$$

Then

$$|y_n - x_n|^2 + (t_n - s_n)^2 \leq \varepsilon |v^+|_{0,(a,b)\times\Omega}$$

and

$$\mathcal{P}^- u_\varepsilon(t_n, x_n) \subset \mathcal{P}^- u(s_n, y_n).$$

We used in the previous lemma that v is bounded from above since Ω is bounded. Putting all the previous pieces of information together yields the desired result. □

A.4   An Elementary Iteration Lemma

The following lemma is classical, see for instance [GT01, Lemma 8.23].

Lemma 2.5.13. *Consider a non-decreasing function* $h : (0, 1) \to \mathbb{R}^+$ *such that for all* $\rho \in (0, 1)$,

$$h(\gamma\rho) \leq \delta h(\rho) + C_0 \rho^\beta$$

for some $\delta, \gamma, \beta \in (0, 1)$. *Then for all* $\rho \in (0, 1)$,

$$h(\rho) \leq C_\alpha \rho^\alpha$$

for all $\alpha = \frac{1}{2}\min(\frac{\ln\delta}{\ln\gamma}, \beta) \in (0, 1)$.

Proof. Consider $k \in \mathbb{N}$, $k \geq 1$, and get by induction that for all $\rho_0, \rho_1 \in (0, 1)$ with $\rho_1 \leq \rho_0$,

$$h(\gamma^k \rho_1) \leq \delta^k h(\rho_1) + C_0 \rho_1^\beta \sum_{j=0}^{k-1} \gamma^{\beta j}.$$

Then write

$$h(\gamma^k \rho_1) \leq \delta^k h(\rho_0) + C_0 \frac{\rho_1^\beta}{1 - \gamma^\beta}$$

$$\leq (\gamma^k)^{\tilde{\beta}} h(\rho_0) + C_0 \frac{\rho_1^\beta}{1 - \gamma^\beta}$$

$$\leq (\gamma^k)^{2\alpha} h(\rho_0) + C_0 \frac{\rho_1^{2\alpha}}{1 - \gamma^\beta}$$

where $\tilde{\beta} = \frac{\ln \delta}{\ln \gamma}$. Now pick $\rho \in [\gamma^{k+1}\rho_1, \gamma^k \rho_1)$ and choose $\rho_1 = \sqrt{\rho_0 \rho}$ and get from the previous inequality the desired result for $\rho \in (0, \rho_0)$. Choose next $\rho_0 = \frac{1}{2}$ and conclude for $\rho \in (0, 1)$. $\square$

References

[ALL97] O. Alvarez, J.-M. Lasry, P.-L. Lions, Convex viscosity solutions and state constraints. J. Math. Pures Appl. (9) **76**(3), 265–288 (1997)

[Bar194] G. Barles, in *Solutions de viscosité des équations de Hamilton-Jacobi*. Mathématiques & Applications (Berlin), vol. 17 (Springer, Paris, 1994), x+194 pp.

[CafCab] L.A. Caffarelli, X. Cabre, in *Fully Nonlinear Elliptic Equations*. American Mathematical Society Colloquium Publications, vol. 43 (American Mathematical Society, Providence, 1995), vi+104 pp.

[CanSin04] P. Cannarsa, C. Sinestrari, Semiconcave functions, in *Hamilton-Jacobi Equations, and Optimal Control*. Progress in Nonlinear Differential Equations and Their Applications, vol. 58 (Birkhäuser, Boston, 2004), xiv+304 pp.

[CL81] M.G. Crandall, P.-L. Lions, Condition d'unicité pour les solutions généralisées des équations de Hamilton-Jacobi du premier ordre. C. R. Acad. Sci. Paris Sér. I Math. **292**(3), 183–186 (1981)

[CIL92] M.G. Crandall, H. Ishii, P.-L. Lions, User's guide to viscosity solutions of second order partial differential equations. Bull. Am. Math. Soc. (N.S.) **27**(1), 1–67 (1992)

[Dong91] G.C. Dong, in *Nonlinear Partial Differential Equations of Second Order*. Translated from the Chinese by Kai Seng Chou [Kaising Tso]. Translations of Mathematical Monographs, vol. 95 (American Mathematical Society, Providence, 1991), viii+251 pp.

[EG92] L.C. Evans, R.F. Gariepy, in *Measure Theory and Fine Properties of Functions*. Studies in Advanced Mathematics (CRC Press, Boca Raton, 1992), viii+268 pp.

[GT01] D. Gilbarg, N.S. Trudinger, in *Elliptic Partial Differential Equations of Second Order*, Reprint of the 1998 edn. Classics in Mathematics (Springer, Berlin, 2001), xiv+517 pp.

[HUL] J.-B. Hirriart-Urruty, C. Lemaréchal, Fundamentals of convex analysis. Abridged version of Convex analysis and minimization algorithms. I [Springer, Berlin, 1993; MR1261420 (95m:90001)] and II [Springer, Berlin, 1993; MR1295240 (95m:90002)]. Grundlehren Text Editions (Springer, Berlin, 2001), x+259 pp.

[Imb06] C. Imbert, Convexity of solutions and $C^{1,1}$ estimates for fully nonlinear elliptic equations. J. Math. Pure Appl. (9) **85**(6), 791–807 (2006)

[Ish87] H. Ishii, Perron's method for Hamilton-Jacobi equations. Duke Math. J. **55**(2), 369–384 (1987)

[Ish89] H.Ishii, On uniqueness and existence of viscosity solutions of fully nonlinear second-order elliptic PDEs. Comm. Pure Appl. Math. **42**(1), 15–45 (1989)

[IL90] H. Ishii, P.-L. Lions, Viscosity solutions of fully nonlinear second-order elliptic partial differential equations. J. Differ. Equat. **83**(1), 26–78 (1990)

[Jens88] R. Jensen, The maximum principle for viscosity solutions of fully nonlinear second order partial differential equations. Arch. Ration. Mech. Anal. **101**(1), 1–27 (1988)

[Kryl76] N.V. Krylov, Sequences of convex functions, and estimates of the maximum of the solution of a parabolic equation. Sibirsk. Mat. Z. **17**(2), 290–303, 478 (1976)

[Kryl87] N.V. Krylov, in *Nonlinear Elliptic and Parabolic Equations of the Second Order.* Translated from the Russian by P.L. Buzytsky. Mathematics and Its Applications (Soviet Series), vol. 7 (D. Reidel Publishing Co., Dordrecht, 1987), xiv+462 pp.

[Kryl96] N.V. Krylov, in *Lectures on Elliptic and Parabolic Equations in Hölder Spaces.* Graduate Studies in Mathematics, vol. 12 (American Mathematical Society, Providence, 1996), xii+164 pp.

[Kryl97] N.V. Krylov, Fully nonlinear second order elliptic equations: recent development. Dedicated to Ennio De Giorgi. Ann. Scuola Norm. Sup. Pisa Cl. Sci. (4) **25**(3–4), 569–595 (1997)

[LSU67] O.A. Ladyzenskaja, V.A. Solonnikov, N.N. Uralceva, in *Linear and Quasilinear Equations of Parabolic Type* (Russian). Translated from the Russian by S. Smith. Translations of Mathematical Monographs, vol. 23 (American Mathematical Society, Providence, 1967), xi+648 pp.

[Lieb96] G.M. Lieberman, *Second Order Parabolic Differential Equations* (World Scientific, River Edge, 1996)

[Lions83] P.-L. Lions, Optimal control of diffusion processes and Hamilton-Jacobi-Bellman equations, II. Viscosity solutions and uniqueness. Comm. Partial Differ. Equat. **8**(11), 1229–1276 (1983)

[Saf84] M.V. Safonov, The classical solution of the elliptic Bellman equation (Russian). Dokl. Akad. Nauk SSSR **278**(4), 810–813 (1984)

[Tso85] K. Tso, On an Aleksandrov-Bakelman type maximum principle for second-order parabolic equations. Comm. Partial Differ. Equat. **10**(5), 543–553 (1985)

[Wang92a] L. Wang, On the regularity theory of fully nonlinear parabolic equations, I. Comm. Pure Appl. Math. **45**(1), 27–76 (1992)

Chapter 3
An Introduction to the Kähler–Ricci Flow

Jian Song and Ben Weinkove

Abstract These notes give an introduction to the Kähler–Ricci flow. We give an exposition of a number of well-known results including: maximal existence time for the flow, convergence on manifolds with negative and zero first Chern class, and behavior of the flow in the case when the canonical bundle is big and nef. We also discuss the collapsing of the Kähler–Ricci flow on the product of a torus and a Riemann surface of genus greater than one. Finally, we discuss the connection between the flow and the minimal model program with scaling, the behavior of the flow on general Kähler surfaces and some other recent results and conjectures.

Introduction

The Ricci flow, first introduced by Hamilton [Ham82] three decades ago, is the equation

$$\frac{\partial}{\partial t} g_{ij} = -2 R_{ij}, \qquad (3.1)$$

evolving a Riemannian metric by its Ricci curvature. It now occupies a central position as one of the key tools of geometry. It was used in [Ham82, Ham86] to classify three-manifolds with positive Ricci curvature and four-manifolds with positive curvature operator. Hamilton later introduced the notion of *Ricci flow with surgery* [Ham95a] and laid out an ambitious program to prove the Poincaré and Geometrization conjectures. In a spectacular demonstration of the power of the Ricci flow, Perelman [Per02, Per03q, Per03b] developed new techniques which

J. Song (✉)
Department of Mathematics, Rutgers University, Piscataway, NJ 08854, USA
e-mail: jiansong@math.rutgers.edu

B. Weinkove
Department of Mathematics, University of California San Diego, La Jolla, CA 92093, USA

S. Boucksom et al. (eds.), *An Introduction to the Kähler–Ricci Flow*,
Lecture Notes in Mathematics 2086, DOI 10.1007/978-3-319-00819-6_3,
© Springer International Publishing Switzerland 2013

enabled him to complete Hamilton's program and settle these celebrated conjectures (see also [CZ06, KL08, MT07, MT08]). More recently, the Ricci flow was used to prove the Brendle–Schoen Differentiable Sphere Theorem [BS08] and other geometric classification results [BW08, NiW10].

In addition to these successes has been the development of the Kähler–Ricci flow. If the Ricci flow starts from a Kähler metric on a complex manifold, the evolving metrics will remain Kähler, and the resulting PDE is called the Kähler–Ricci flow. Cao [Cao85] used this flow, together with parabolic versions of the estimates of Yau [Yau78b] and Aubin [Aub78], to reprove the existence of Kähler–Einstein metrics on manifolds with negative and zero first Chern class. Since then, the study of the Kähler–Ricci flow has developed into a vast field in its own right. There have been several different avenues of research involving this flow, including: existence of Kähler–Einstein metrics on manifolds with positive first Chern class and notions of algebraic stability [Bando87, Cal82, ChW09, Don02, MSz09, PSS07, PSSW09, PSSW11, PS05, PS06, PS10, Rub09, SeT08, Sz10, Tian97, TZ07, Tos10a, Yau93, Zhu07] (Perelman, unpublished work on the Kähler–Ricci flow); the classification of Kähler manifolds with positive curvature in both the compact and non-compact cases [Bando84, Cao92, CZ09, ChauT06, CST09, CheT06, Gu09, Mok88, Ni04, PSSW08b]; and extensions of the flow to non-Kähler settings [Gill11, StT10]. (These lists of references are far from exhaustive.) In these notes we will not even manage to touch on these areas.

Our main goal is to give an introduction to the Kähler–Ricci flow. In the last two sections of the notes, we will also discuss some results related to the *analytic minimal model program* of the first-named author and Tian [ST07, ST12, ST09, Tian02, Tian08]. The field has been developing at a fast pace in the last several years, and we mention briefly now some of the ideas.

Ultimately, the goal is to see whether the Kähler–Ricci flow will give a geometric classification of algebraic varieties. In the case of real three-manifolds, the work of Perelman and Hamilton shows that the Ricci flow with surgery, starting at any Riemannian metric, can be used to break up the manifold into pieces, each of which has a particular geometric structure. We can ask the same question for the Kähler–Ricci flow on a projective algebraic variety: starting with any Kähler metric, will the Kähler–Ricci flow "with surgery" break up the variety into simpler pieces, each equipped with some canonical geometric structure?

A process of "simplifying" algebraic varieties through surgeries already exists and is known as the Minimal Model Program. In the case of complex dimension two, the idea is relatively simple. Start with a variety and find "(-1)-curves"— these are special holomorphic spheres embedded in the variety—and remove them using an algebraic procedure known as "blowing down". It can be shown that after a finite number of these algebraic surgeries, the final variety either has a "ruled" structure, or has *nef canonical bundle*, a condition that can be interpreted as being "nonpositively curved" in some appropriate sense. This last type of variety is known as a "minimal model". In higher dimensions, a similar, though more complicated, process also exists. It turns out that there are many different ways to arrive at the minimal model by algebraic procedures such as blow-downs. However, in [BCHM10] Birkar–Cascini–Hacon–McKernan introduced the notion of the *Minimal Model Program with scaling* (or MMP with scaling), which, ignoring some technical assumptions, takes a variety with a "polarization" and describes a particular sequence of algebraic operations which take it to a minimal model or a ruled surface (or its higher dimensional analogue). This process seems to be closely related to the Kähler–Ricci flow, with the polarization corresponding to a choice of initial Kähler metric.

Starting in 2007, Song–Tian [ST07, ST12, ST09] and Tian [Tian08] proposed the analytic MMP using the Kähler–Ricci flow with a series of conjectures, and showed [ST09] that, in a weak sense, the flow can be continued through singularities related to the MMP with scaling. In the case of complex dimension two, it was shown by the authors [SW10] that the algebraic procedure of "blowing down" a holomorphic sphere corresponds to a geometric "canonical surgical contraction" for the Kähler–Ricci flow.

Moreover, the minimal model is endowed with an analytic structure. Eyssidieux–Guedj–Zeriahi [EGZ11] generalized an estimate of Kolodziej [Kol98] (see also the work of Zhang [Zha06]) to construct singular Kähler–Einstein metrics on minimal models of general type. In the case of smooth minimal models, convergence of the Kähler–Ricci flow to this metric was already known by the work of Tsuji in the 1980s [Tsu88], results which were clarified and extended by Tian–Zhang [Tzha06]. On Iitaka fibrations, the Kähler–Ricci flow was shown by Song–Tian to converge to a "generalized Kähler–Einstein metric" [ST07, ST12].

These are very recent developments in a field which we expect is only just beginning. In these lecture notes we have decided to focus on describing the main tools and techniques which are now well-established, rather than give expositions of the most recent advances. In particular, we do not in any serious way address "surgery" for the Kähler–Ricci flow and we only give a sketchy outline of the Minimal Model Program and its relation to the Kähler–Ricci flow. On the other hand, we have taken the opportunity to include two new results in these notes: a detailed description of collapsing along the Kähler–Ricci flow in the case of a product elliptic surface (Sect. 3.6) and a description of the Kähler–Ricci flow on Kähler surfaces (Sect. 3.8), extending our previous work on algebraic surfaces [SW10].

We have aimed these notes at the non-expert and have tried to make them as self-contained and complete as possible. We do not expect the reader to be either a geometric analyst or an algebraic geometer. We assume only a basic knowledge of complex manifolds. We hope that these notes will provide enough background material for the non-expert reader to go on to begin research in this area.

We give now a brief outline of the contents of these notes. In Sect. 3.1, we give some preliminaries and background material on Kähler manifolds and curvature, describe some analytic tools such as the maximum principle, and provide some definitions and results from algebraic geometry. Readers may wish to skip this section at first and refer back to it if necessary. In Sect. 3.2, we describe a number of well-known basic analytic results for the Kähler–Ricci flow. Many of these results have their origin in the work of Calabi, Yau, Cheng, Aubin and others [Aub78, Cal58, Cao85, ChengYau75, Yau78b, Yau78]. We include a more recent argument, due to Phong–Šešum–Sturm [PSS07], for the "Calabi third-order" estimate in the setting of the Kähler–Ricci flow.

In Sect. 3.3 we prove one of the basic results for the Kähler–Ricci flow: the flow admits a smooth solution as long as the class of the metric remains Kähler. The result in this generality is due to Tian–Zhang [Tzha06], extending earlier results of Cao and Tsuji [Cao85, Tsu88, Tsu96]. In Sect. 3.4, we give an exposition of Cao's work [Cao85]—the first paper on the Kähler–Ricci flow. Namely, we describe the behavior of the flow on manifolds with negative or zero first Chern class. We include in this section the crucial C^0 estimate of Yau [Yau78b]. We give a different proof of convergence in this case, following Phong–Sturm [PS06]. In Sect. 3.5, we consider the Kähler–Ricci flow on manifolds with nef and big canonical bundle. This was first studied by Tsuji [Tsu88] and demonstrates how one can study the singular behavior of the Kähler–Ricci flow.

In Sect. 3.6, we address the case of *collapsing* along the Kähler–Ricci flow with the example of a product of an elliptic curve and a curve of higher genus. In Sect. 3.7, we describe some basic results in the case where a singularity for the flow occurs at a finite time, including the recent result of Zhang [Zha09] on the behavior of the scalar curvature. We also describe without proof some of the results of [SSW11, SW10].

In Sect. 3.8, we discuss the Kähler–Ricci flow and the Minimal Model Program. We give a brief sketch of some of the ideas of the MMP and how the Kähler–Ricci flow relates to it. We also describe some results from [SW10] and extend them to the case of Kähler surfaces.

The authors would like to mention that these notes arose from lectures given at the conference *Analytic aspects of complex algebraic geometry*, held at the Centre International de Rencontres Mathèmatiques in Luminy, in February 2011. The authors would like to thank S. Boucksom, P. Eyssidieux, and V. Guedj for organizing this wonderful conference. Additional thanks go to V. Guedj for his encouragement to write these notes.

We would also like to express our gratitude to the following people for providing helpful suggestions and corrections: Huai-Dong Cao, John Lott, Morgan Sherman, Gang Tian, Valentino Tosatti and Zhenlei Zhang. Finally the authors thank their former Ph.D. advisor D.H. Phong for his valuable advice, encouragement and support over the last several years. In addition, his teaching and ideas have had a strong influence on the style and point of view of these notes.

3.1   Preliminaries

In this section we describe some definitions and results which will be used throughout the text.

3.1.1   Kähler Manifolds

Let M be a compact complex manifold of complex dimension n. We will often work in a holomorphic coordinate chart U with coordinates $(z^1, \ldots, z^n)$ and write a tensor in terms of its components in such a coordinate system. We refer the reader to [GH78, KodMor71] for an introduction to complex manifolds etc.

A *Hermitian metric* on M is a smooth tensor $g = g_{i\bar{j}}$ such that $(g_{i\bar{j}})$ is a positive definite Hermitian matrix at each point of M. Associated to g is a $(1,1)$-form ω given by

$$\omega = \frac{\sqrt{-1}}{2\pi} g_{i\bar{j}} dz^i \wedge d\bar{z}^j, \qquad (3.2)$$

where here and henceforth we are summing over repeated indices from 1 to n. If $d\omega = 0$ then we say that g is a *Kähler* metric and that ω is the *Kähler form* associated to g. Henceforth, whenever, for example, $g(t), \hat{g}, g_0, \ldots$ are Kähler metrics we will use the obvious notation $\omega(t), \hat{\omega}, \omega_0, \ldots$ for the associated Kähler forms, and vice versa. Abusing terminology slightly, we will often refer to a Kähler form ω as a Kähler metric.

The Kähler condition $d\omega = 0$ is equivalent to:

$$\partial_k g_{i\bar{j}} = \partial_i g_{k\bar{j}}, \quad \text{for all } i, j, k, \qquad (3.3)$$

where we are writing $\partial_i = \partial/\partial z^i$. The condition (3.3) is independent of choice of holomorphic coordinate system.

For examples of Kähler manifolds, consider complex projective space $\mathbb{P}^n = (\mathbb{C}^{n+1} \setminus \{0\})/ \sim$ where $(z_0, \ldots, z_n) \sim (z_0', \ldots, z_n')$ if there exists $\lambda \in \mathbb{C}^*$ with $z_i = \lambda z_i'$ for all i. We denote by $[Z_0, \ldots, Z_n] \in \mathbb{P}^n$ the equivalence class of $(Z_0, \ldots, Z_n) \in \mathbb{C}^{n+1} \setminus \{0\}$. Define the *Fubini–Study metric* ω_{FS} by

$$\omega_{FS} = \frac{\sqrt{-1}}{2\pi} \partial\bar{\partial} \log(|Z_0|^2 + \cdots + |Z_n|^2). \qquad (3.4)$$

Note that although $|Z_0|^2 + \cdots + |Z_n|^2$ is not a well-defined function on $\mathbb{P}^n$, ω_{FS} is a well-defined $(1,1)$-form. We leave it as an exercise for the reader to check that ω_{FS} is Kähler. Moreover, since the restriction of a Kähler metric to a complex submanifold is Kähler, we can produce a large class of Kähler manifolds by considering complex submanifolds of $\mathbb{P}^n$. These are known as *smooth projective varieties*.

Let $X = X^i \partial_i$ and $Y = Y^{\bar{i}} \partial_{\bar{i}}$ be $T^{1,0}$ and $T^{0,1}$ vector fields respectively, and let $a = a_i dz^i$ and $b = b_{\bar{i}} d\bar{z}^i$ be $(1,0)$ and $(0,1)$ forms respectively. By definition this means that if $(\tilde{z}^1, \ldots, \tilde{z}^n)$ is another holomorphic coordinate system then on their overlap,

$$\tilde{X}^j = X^i \frac{\partial \tilde{z}^j}{\partial z^i}, \quad \tilde{Y}^{\bar{j}} = Y^{\bar{i}} \overline{\frac{\partial \tilde{z}^j}{\partial z^i}}, \quad \tilde{a}_j = a_i \frac{\partial z^i}{\partial \tilde{z}^j}, \quad \tilde{b}_{\bar{j}} = b_{\bar{i}} \overline{\frac{\partial z^i}{\partial \tilde{z}^j}}. \qquad (3.5)$$

Associated to a Kähler metric g are *covariant derivatives* ∇_k and $\nabla_{\bar{k}}$ which act on the tensors X, Y, a, b in the following way:

$$\nabla_k X^i = \partial_k X^i + \Gamma^i_{jk} X^j, \ \nabla_{\bar{k}} X^i = \partial_{\bar{k}} X^i, \ \nabla_k Y^{\bar{i}} = \partial_k Y^{\bar{i}},$$

$$\nabla_{\bar{k}} Y^{\bar{i}} = \partial_{\bar{k}} Y^{\bar{i}} + \overline{\Gamma^i_{jk}} Y^{\bar{j}}, \qquad (3.6)$$

$$\nabla_k a_i = \partial_k a_i - \Gamma^j_{ik} a_j, \ \nabla_{\bar{k}} a_i = \partial_{\bar{k}} a_i, \ \nabla_k b_{\bar{i}} = \partial_k b_{\bar{i}}, \ \nabla_{\bar{k}} b_{\bar{i}} = \partial_{\bar{k}} b_{\bar{i}} - \overline{\Gamma^j_{ik}} b_{\bar{j}}, \qquad (3.7)$$

where Γ^i_{jk} are the *Christoffel symbols* given by

$$\Gamma^i_{jk} = g^{\bar{\ell}i} \partial_j g_{k\bar{\ell}}, \tag{3.8}$$

for $(g^{\bar{\ell}i})$ the inverse of the matrix $(g_{i\bar{\ell}})$. Observe that $\Gamma^i_{jk} = \Gamma^i_{kj}$ from (3.3). The Christoffel symbols are not the components of a tensor, but $\nabla_k X^i, \nabla_{\bar{k}} X^i, \ldots$ do define tensors, as the reader can verify. Also, if g and $\hat{g}$ are Kähler metrics with Christoffel symbols Γ^i_{jk} and $\hat{\Gamma}^i_{jk}$ then the difference $\Gamma^i_{jk} - \hat{\Gamma}^i_{jk}$ is a tensor.

We extend covariant derivatives to act naturally on any type of tensor. For example, if W is a tensor with components $W^{i\bar{j}}_k$ then define

$$\nabla_m W^{i\bar{j}}_k = \partial_m W^{i\bar{j}}_k + \Gamma^i_{\ell m} W^{\ell\bar{j}}_k - \Gamma^\ell_{km} W^{i\bar{j}}_\ell, \quad \nabla_{\bar{m}} W^{i\bar{j}}_k = \partial_{\bar{m}} W^{i\bar{j}}_k + \overline{\Gamma^j_{\ell m}} W^{i\bar{\ell}}_k. \tag{3.9}$$

Note also that the Christoffel symbols are chosen so that $\nabla_k g_{i\bar{j}} = 0$.

The metric g defines a pointwise norm $|\cdot|_g$ on any tensor. For example, with X, Y, a, b as above, we define

$$|X|^2_g = g_{i\bar{j}} X^i \overline{X^j}, \quad |Y|^2_g = g_{i\bar{j}} Y^{\bar{j}} \overline{Y^i}, \quad |a|^2_g = g^{\bar{j}i} a_i \overline{a_j}, \quad |b|^2_g = g^{\bar{j}i} b_{\bar{j}} \overline{b_{\bar{i}}}. \tag{3.10}$$

This is extended to any type of tensor. For example, if W is a tensor with components $W^{i\bar{j}}_k$ then define $|W|^2_g = g^{\bar{\ell}k} g_{i\bar{j}} g_{p\bar{q}} W^{i\bar{q}}_k \overline{W^{j\bar{p}}_\ell}$.

Finally, note that a Kähler metric g defines a Riemannian metric $g_{\mathbb{R}}$. In local coordinates, write $z^i = x^i + \sqrt{-1} y^i$, so that $\partial_{z^i} = \frac{1}{2}(\partial_{x^i} - \sqrt{-1}\partial_{y^i})$ and $\partial_{\bar{z}^i} = \frac{1}{2}(\partial_{x^i} + \sqrt{-1}\partial_{y^i})$. Then

$$g_{\mathbb{R}}(\partial_{x^i}, \partial_{x^j}) = 2\text{Re}(g_{i\bar{j}}) = g_{\mathbb{R}}(\partial_{y^i}, \partial_{y^j}), \quad g_{\mathbb{R}}(\partial_{x^i}, \partial_{y^j}) = 2\text{Im}(g_{i\bar{j}}). \tag{3.11}$$

We will typically write g instead of $g_{\mathbb{R}}$.

3.1.2 Normal Coordinates

The following proposition is very useful in computations.

Proposition 3.1.1. *Let g be a Kähler metric on M and let $S = S_{i\bar{j}}$ be a tensor which is Hermitian (that is $\overline{S_{i\bar{j}}} = S_{\bar{j}i}$.) Then at each point p on M there exists a holomorphic coordinate system centered at p such that,*

$$g_{i\bar{j}}(p) = \delta_{ij}, \quad S_{i\bar{j}}(p) = \lambda_i \delta_{ij}, \quad \partial_k g_{i\bar{j}}(p) = 0, \quad \forall i, j, k = 1, \ldots, n, \tag{3.12}$$

for some $\lambda_1, \ldots, \lambda_n \in \mathbb{R}$, where δ_{ij} is the Kronecker delta.

Proof. It is an exercise in linear algebra to check that we can find a coordinate system $(z^1, \ldots, z^n)$ centered at p (so that $p \mapsto 0$) satisfying the first two conditions: g is the identity at p and S is diagonal at p. Indeed this amounts to the fact that a Hermitian matrix can be diagonalized by a unitary transformation.

For the last condition we make a change of coordinates. Define coordinates $(\tilde{z}^1, \ldots, \tilde{z}^n)$ in a neighborhood of p by

$$z^i = \tilde{z}^i - \frac{1}{2}\Gamma^i_{jk}(0)\tilde{z}^j\tilde{z}^k, \quad \text{for } i = 1, \ldots, n. \tag{3.13}$$

Writing $\tilde{g}_{i\bar{j}}$, $\tilde{S}_{i\bar{j}}$ for the components of g, S with respect to $(\tilde{z}^1, \ldots, \tilde{z}^n)$ we see that $\tilde{g}_{i\bar{j}}(0) = g_{i\bar{j}}(0)$ and $\tilde{S}_{i\bar{j}}(0) = S_{i\bar{j}}(0)$ since $\partial z^i / \partial \tilde{z}^j (0) = \delta_{ij}$. It remains to check that the first derivatives of $\tilde{g}_{i\bar{j}}$ vanish at 0. Compute at 0,

$$\begin{aligned}
\frac{\partial}{\partial \tilde{z}^k}\tilde{g}_{i\bar{j}} &= \frac{\partial}{\partial \tilde{z}^k}\left(\frac{\partial z^a}{\partial \tilde{z}^i}\frac{\overline{\partial z^b}}{\partial \tilde{z}^j}g_{a\bar{b}}\right) \\
&= \frac{\partial^2 z^a}{\partial \tilde{z}^k \partial \tilde{z}^i}\frac{\overline{\partial z^b}}{\partial \tilde{z}^j}g_{a\bar{b}} + \frac{\partial z^a}{\partial \tilde{z}^i}\frac{\overline{\partial z^b}}{\partial \tilde{z}^j}\frac{\partial z^m}{\partial \tilde{z}^k}\frac{\partial}{\partial z^m}g_{a\bar{b}} \\
&= -\Gamma^j_{ik} + \frac{\partial}{\partial z^k}g_{i\bar{j}} = 0, \tag{3.14}
\end{aligned}$$

as required. $\qquad\square$

We call a holomorphic coordinate system centered at p satisfying $g_{i\bar{j}}(p) = \delta_{ij}$ and $\partial_k g_{i\bar{j}}(p) = 0$ a *normal coordinate system* for g centered at p. It implies in particular that the Christoffel symbols of g vanish at p. Proposition 3.1.1 states that we can find a normal coordinate system for g at any point p, and that moreover we can simultaneously diagonalize any other Hermitian tensor (such as another Kähler metric) at that point.

3.1.3 Curvature

Define the *curvature tensor* of the Kähler metric g to be the tensor

$$R_i{}^m{}_{k\bar{\ell}} = -\partial_{\bar{\ell}}\Gamma^m_{ik}. \tag{3.15}$$

The reader can verify that this does indeed define a tensor on M. We often lower the second index using the metric g and define

$$R_{i\bar{j}k\bar{\ell}} = g_{m\bar{j}}R_i{}^m{}_{k\bar{\ell}}, \tag{3.16}$$

an object which we also refer to as the curvature tensor. In addition, we can lower or raise any index of curvature using the metric g. For example, $R_{i\bar{j}}^{\ \ \bar{q}p} := g^{\bar{q}k}g^{\bar{\ell}p}R_{i\bar{j}k\bar{\ell}}$.

Using the formula for the Christoffel symbols and (3.3), calculate

$$R_{i\bar{j}k\bar{\ell}} = -\partial_i\partial_{\bar{j}}g_{k\bar{\ell}} + g^{\bar{q}p}(\partial_i g_{k\bar{q}})(\partial_{\bar{j}}g_{p\bar{\ell}}). \tag{3.17}$$

The curvature tensor has a number of symmetries:

Proposition 3.1.2. *We have*

(i) $\overline{R_{i\bar{j}k\bar{\ell}}} = R_{j\bar{i}\ell\bar{k}}$.
(ii) $R_{i\bar{j}k\bar{\ell}} = R_{k\bar{j}i\bar{\ell}} = R_{i\bar{\ell}k\bar{j}}$.
(iii) $\nabla_m R_{i\bar{j}k\bar{\ell}} = \nabla_i R_{m\bar{j}k\bar{\ell}}$.

Proof. (i) and (ii) follow immediately from the formula (3.17) together with the Kähler condition (3.3). For (iii) we compute at a point p in normal coordinates for g,

$$\nabla_m R_{i\bar{j}k\bar{\ell}} = -\partial_m\partial_i\partial_{\bar{j}}g_{k\bar{\ell}} = -\partial_i\partial_m\partial_{\bar{j}}g_{k\bar{\ell}} = \nabla_i R_{m\bar{j}k\bar{\ell}}, \tag{3.18}$$

as required.   □

Parts (ii) and (iii) of Proposition 3.1.2 are often referred to as the *first and second Bianchi identities*, respectively. Define the *Ricci curvature* of g to be the tensor

$$R_{i\bar{j}} := g^{\bar{\ell}k}R_{i\bar{j}k\bar{\ell}} = g^{\bar{\ell}k}R_{k\bar{\ell}i\bar{j}} = R_{k\ \ i\bar{j}}^{\ k}, \tag{3.19}$$

and the *scalar curvature* $R = g^{\bar{j}i}R_{i\bar{j}}$ to be the trace of the Ricci curvature. For Kähler manifolds, the Ricci curvature takes on a simple form:

Proposition 3.1.3. *We have*

$$R_{i\bar{j}} = -\partial_i\partial_{\bar{j}}\log\det g. \tag{3.20}$$

Proof. First, recall the well-known formula for the derivative of the determinant of a Hermitian matrix. Let $A = (A_{i\bar{j}})$ be an invertible Hermitian matrix with inverse $(A^{\bar{j}i})$. If the entries of A depend on a variable s then an application of Cramer's rule shows that

$$\frac{d}{ds}\det A = A^{\bar{j}i}\left(\frac{d}{ds}A_{i\bar{j}}\right)\det A. \tag{3.21}$$

Using this, calculate

$$R_{i\bar{j}} = -\partial_{\bar{j}}\Gamma_{ki}^k = -\partial_{\bar{j}}(g^{\bar{q}k}\partial_i g_{k\bar{q}}) = -\partial_{\bar{j}}\partial_i\log\det g, \tag{3.22}$$

which gives the desired formula.   □

Associated to the tensor $R_{i\bar{j}}$ is a $(1, 1)$ form $\mathrm{Ric}(\omega)$ given by

$$\mathrm{Ric}(\omega) = \frac{\sqrt{-1}}{2\pi} R_{i\bar{j}} dz^i \wedge d\bar{z}^j. \tag{3.23}$$

Proposition 3.1.3 implies that $\mathrm{Ric}(\omega)$ is closed.

We end this subsection by showing that the curvature tensor arises when commuting covariant derivatives ∇_k and $\nabla_{\bar{\ell}}$. Indeed, the curvature tensor is often defined by this property.

Proposition 3.1.4. *Let* $X = X^i \partial_i$, $Y = Y^{\bar{i}} \partial_{\bar{i}}$ *be* $T^{1,0}$ *and* $T^{0,1}$ *vector fields respectively, and let* $a = a_i dz^i$ *and* $b = b_{\bar{i}} d\bar{z}^i$ *be* $(1, 0)$ *and* $(0, 1)$ *forms respectively. Then*

$$[\nabla_k, \nabla_{\bar{\ell}}] X^m = R_i{}^m{}_{k\bar{\ell}} X^i \tag{3.24}$$

$$[\nabla_k, \nabla_{\bar{\ell}}] Y^{\bar{m}} = -R^{\bar{m}}{}_{\bar{j}k\bar{\ell}} Y^{\bar{j}} \tag{3.25}$$

$$[\nabla_k, \nabla_{\bar{\ell}}] a_i = -R_i{}^m{}_{k\bar{\ell}} a_m \tag{3.26}$$

$$[\nabla_k, \nabla_{\bar{\ell}}] b_{\bar{j}} = R^{\bar{m}}{}_{\bar{j}k\bar{\ell}} b_{\bar{m}}, \tag{3.27}$$

where we are writing $[\nabla_k, \nabla_{\bar{\ell}}] = \nabla_k \nabla_{\bar{\ell}} - \nabla_{\bar{\ell}} \nabla_k$.

Proof. We prove the first and leave the other three as exercises. Compute at a point p in a normal coordinate system for g,

$$[\nabla_k, \nabla_{\bar{\ell}}] X^m = \partial_k \partial_{\bar{\ell}} X^m - \partial_{\bar{\ell}} (\partial_k X^m + \Gamma^m_{ik} X^i) = -(\partial_{\bar{\ell}} \Gamma^m_{ik}) X^i = R_i{}^m{}_{k\bar{\ell}} X^i, \tag{3.28}$$

as required. □

Note that the commutation formulae of Proposition 3.1.4 can naturally be extended to tensors of any type. Finally we remark that, when acting on any tensor, we have $[\nabla_i, \nabla_j] = 0 = [\nabla_{\bar{i}}, \nabla_{\bar{j}}]$, as the reader can verify.

3.1.4 The Maximum Principle

There are various notions of "maximum principle". In the setting of the Ricci flow, Hamilton introduced his *maximum principle for tensors* [CLN06, Ham82, Ham95a] which has been exploited in quite sophisticated ways to investigate the positivity of curvature tensors along the flow (see for example [Bando84, BW08, BS08, Ham86, Mok88, NiW10]). For our purposes however, we need only a simple version of the maximum principle.

We begin with an elementary lemma. As above, (M, ω) will be a compact Kähler manifold.

Proposition 3.1.5. *Let f be a smooth real-valued function on M which achieves its maximum (minimum) at a point x_0 in M. Then at x_0,*

$$df = 0 \quad \text{and} \quad \sqrt{-1}\partial\bar{\partial} f \leq 0 \, (\geq 0). \tag{3.29}$$

Here, if $\alpha = \sqrt{-1} a_{i\bar{j}} dz^i \wedge d\bar{z}^j$ is a real $(1, 1)$-form, we write $\alpha \leq 0 \, (\geq 0)$ to mean that the Hermitian matrix $(a_{i\bar{j}})$ is nonpositive (nonnegative). Proposition 3.1.5 is a simple consequence of the fact from calculus that a smooth function has nonpositive Hessian (and hence nonpositive complex Hessian) and zero first derivative at its maximum.

Next we introduce the Laplace operator Δ on functions. Define

$$\Delta f = g^{\bar{j}i} \partial_i \partial_{\bar{j}} f \tag{3.30}$$

for a function f.

In these lecture notes, we will often make use of the trace notation "tr". If $\alpha = \frac{\sqrt{-1}}{2\pi} a_{i\bar{j}} dz^i \wedge d\bar{z}^j$ is a real $(1, 1)$-form then we write

$$\text{tr}_\omega \alpha = g^{\bar{j}i} a_{i\bar{j}} = \frac{n\,\omega^{n-1} \wedge \alpha}{\omega^n}. \tag{3.31}$$

In this notation, we can write $\Delta f = \text{tr}_\omega \frac{\sqrt{-1}}{2\pi} \partial\bar{\partial} f$.

It follows immediately from this definition that Proposition 3.1.5 still holds if we replace $\sqrt{-1}\partial\bar{\partial} f \leq 0 \, (\geq 0)$ in (3.29) by $\Delta f \leq 0 \, (\geq 0)$.

For the *parabolic maximum principle* (which we still call the *maximum principle*) we introduce a time parameter t. The following proposition will be used many times in these lecture notes.

Proposition 3.1.6. *Fix $T > 0$. Let $f = f(x, t)$ be a smooth function on $M \times [0, T]$. If f achieves its maximum (minimum) at $(x_0, t_0) \in M \times [0, T]$ then either $t_0 = 0$ or at (x_0, t_0),*

$$\frac{\partial f}{\partial t} \geq 0 \, (\leq 0) \quad \text{and} \quad df = 0 \quad \text{and} \quad \sqrt{-1}\partial\bar{\partial} f \leq 0 \, (\geq 0). \tag{3.32}$$

Proof. Exercise for the reader.   □

We remark that, in practice, one is usually given a function f defined on a half-open time interval $[0, T)$ say, rather than a compact interval. To apply this proposition it may be necessary to fix an arbitrary $T_0 \in (0, T)$ and work on $[0, T_0]$. Since we use this procedure many times in the notes, we will often omit to mention the fact that we are restricting to such a compact interval. Note also that

Propositions 3.1.5 and 3.1.6 still hold with M replaced by an open set $U \subseteq M$ as long as the maximum (or minimum) of f is achieved in the interior of the set U.

We end this section with a useful application of the maximum principle in the case where f satisfies a heat-type differential inequality.

Proposition 3.1.7. *Fix T with $0 < T \leq \infty$. Suppose that $f = f(x,t)$ is a smooth function on $M \times [0, T)$ satisfying the differential inequality*

$$\left(\frac{\partial}{\partial t} - \Delta \right) f \leq 0. \tag{3.33}$$

Then $\sup_{(x,t)\in M\times[0,T)} f(x,t) \leq \sup_{x\in M} f(x,0)$.

Proof. Fix $T_0 \in (0, T)$. For $\varepsilon > 0$, define $f_\varepsilon = f - \varepsilon t$. Suppose that f_ε on $M \times [0, T_0]$ achieves its maximum at (x_0, t_0). If $t_0 > 0$ then by Proposition 3.1.6,

$$0 \leq \left(\frac{\partial}{\partial t} - \Delta \right) f_\varepsilon (x_0, t_0) \leq -\varepsilon, \tag{3.34}$$

a contradiction. Hence the maximum of f_ε is achieved at $t_0 = 0$ and

$$\sup_{(x,t)\in M\times[0,T_0]} f(x,t) \leq \sup_{(x,t)\in M\times[0,T_0]} f_\varepsilon(x,t) + \varepsilon T_0 \leq \sup_{x\in M} f(x,0) + \varepsilon T_0. \tag{3.35}$$

Let $\varepsilon \to 0$. Since T_0 is arbitrary, this proves the result. □

We remark that a similar result of course holds for the infimum of f if we replace $\left(\frac{\partial}{\partial t} - \Delta \right) f \leq 0$ by $\left(\frac{\partial}{\partial t} - \Delta \right) f \geq 0$. Finally, note that Proposition 3.1.7 holds, with the same proof, if the Laplace operator Δ in (3.33) is defined with respect to a metric $g = g(t)$ that depends on t.

3.1.5 Other Analytic Results and Definitions

In this subsection, we list a number of other results and definitions from analysis, besides the maximum principle, which we will need later. For a good reference, see [Aub82]. Let (M, ω) be a compact Kähler manifold of complex dimension n. In these lecture notes, we will be concerned only with smooth functions and tensors so for the rest of this section assume that all functions and tensors on M are smooth. The following is known as the *Poincaré inequality*.

Theorem 3.1.8. *There exists a constant C_P such that for any real-valued function f on M with $\int_M f\omega^n = 0$, we have*

$$\int_M f^2\omega^n \leq C_P \int_M |\partial f|^2 \omega^n, \tag{3.36}$$

for $|\partial f|^2 = g^{\bar{j}i}\partial_i f \partial_{\bar{j}} f$.

We remark that the constant C_P is (up to scaling by some universal factor) equal to λ^{-1} where λ is the first nonzero eigenvalue of the operator $-\Delta$ associated to g.

Next, we have the *Sobolev inequality*.

Theorem 3.1.9. *Assume $n > 1$. There exists a uniform constant C_S such that for any real-valued function f on M, we have*

$$\left(\int_M |f|^{2\beta} \omega^n \right)^{1/\beta} \le C_S \left(\int_M |\partial f|^2 \omega^n + \int_M |f|^2 \omega^n \right), \tag{3.37}$$

for $\beta = n/(n-1) > 1$.

We give now some definitions for later use. Given a function f, define the C^0 norm on M to be $\|f\|_{C^0(M)} = \sup_M |f|$. We give a similar definition for any subset $U \subset M$. Given a (real) tensor W and a Riemannian metric g, we define $|W|_g^2$ by contracting with g, in the obvious way (cf. Sect. 3.1.1). Define $\|W\|_{C^0(M,g)}$ to be the $C^0(M)$ norm of $|W|_g$. If no confusion arises, we will often drop the M and g in denoting norms.

Given a function f on M, we define for $p \ge 1$ the $L^p(M, \omega)$ norm with respect to a Kähler metric ω by

$$\|f\|_{L^p(M,\omega)} = \left(\int_M |f|^p \omega^n \right)^{1/p}. \tag{3.38}$$

Note that $\|f\|_{L^p(M,\omega)} \to \|f\|_{C^0(M)}$ as $p \to \infty$.

We use $\nabla_{\mathbb{R}}$ to denote the (real) covariant derivative of g. Given a function f, write $\nabla_{\mathbb{R}}^m f$ for the tensor with components (in real coordinates) $(\nabla_{\mathbb{R}})_{i_1} \cdots (\nabla_{\mathbb{R}})_{i_m} f$ and similarly for ∇ acting on tensors.

For a function f and a subset $U \subseteq M$, define

$$\|f\|_{C^k(U,g)} = \sum_{m=0}^{k} \|\nabla_{\mathbb{R}}^m f\|_{C^0(U,g)}, \tag{3.39}$$

and similarly for tensors.

We say that a tensor T has *uniform $C^\infty(M, g)$ bounds* if for each $k = 0, 1, 2, \ldots$ there exists a uniform constant C_k such that $\|T\|_{C^k(M,g)} \le C_k$. Given an open subset $U \subseteq M$ we say that T has *uniform $C^\infty_{\mathrm{loc}}(U, g)$ bounds* if for any compact subset $K \subseteq U$ there exist constants $C_{k,K}$ such that $\|T\|_{C^k(K,g)} \le C_{k,K}$. We say that a family of tensors T_t *converges in $C^\infty_{\mathrm{loc}}(U, g)$* to a tensor T_∞ if for every compact $K \subseteq U$, and each $k = 0, 1, 2, \ldots$, the tensors T_t converge to T_∞ in $C^k(K, g)$.

Given $\beta \in (0, 1)$, define the Hölder norm $C^\beta(M, g)$ of a function f by

$$\|f\|_{C^\beta(M,g)} = \|f\|_{C^0(M)} + \sup_{p \ne q} \frac{|f(p) - f(q)|}{d(p, q)^\beta}, \tag{3.40}$$

for d the distance function of g. The $C^\beta(M, g)$ norm for tensors T is defined similarly, except that we must use parallel transport with respect to g construct the difference $T(p) - T(q)$. For a positive integer k, define $\|f\|_{C^{k+\beta}(M,g)} = \|f\|_{C^k(M,g)} + \|\nabla_\mathbb{R}^k f\|_{C^\beta(M,g)}$, and similarly for tensors.

Finally, we define what is meant by *Gromov–Hausdorff convergence*. This is a notion of convergence for metric spaces. Given two subsets A and B of a metric space (X, d), we define the *Hausdorff distance* between A and B to be

$$d_\mathrm{H}(A, B) = \inf\{\varepsilon > 0 \mid A \subseteq B_\varepsilon \text{ and } B \subseteq A_\varepsilon\} \tag{3.41}$$

where $A_\varepsilon = \cup_{a \in A}\{x \in X \mid d(a, x) \leq \varepsilon\}$. We then define the *Gromov–Hausdorff distance* between two compact metric spaces X and Y to be

$$d_\mathrm{GH}(X, Y) = \inf_{f,g} d_\mathrm{H}(f(X), g(Y)), \tag{3.42}$$

where the infimum is taken over all isometric embeddings $f : X \to Z$, $g : Y \to Z$ into a metric space Z (for all possible Z). We then say that a family X_t of compact metric spaces *converges in the Gromov–Hausdorff sense* to a compact metric space X_∞ if the X_t converge to X_∞ with respect to d_GH.

3.1.6 Dolbeault Cohomology, Line Bundles and Divisors

In this section we introduce cohomology classes, line bundles, divisors, Hermitian metrics etc. Good references for this and the next subsection are [GH78,KodMor71]. Let M be a compact complex manifold. We say that a form α is $\bar\partial$-*closed* if $\bar\partial\alpha = 0$ and $\bar\partial$-*exact* if $\alpha = \bar\partial\eta$ for some form η. Define the *Dolbeault cohomology group* $H_{\bar\partial}^{1,1}(M, \mathbb{R})$ by

$$H_{\bar\partial}^{1,1}(M, \mathbb{R}) = \frac{\{\bar\partial\text{--closed real}(1, 1) - \text{forms}\}}{\{\bar\partial\text{--exact real}(1, 1) - \text{forms}\}}. \tag{3.43}$$

A Kähler metric ω on M defines a nonzero element $[\omega]$ of $H_{\bar\partial}^{1,1}(M, \mathbb{R})$. If a cohomology class $\alpha \in H_{\bar\partial}^{1,1}(M, \mathbb{R})$ can be written $\alpha = [\omega]$ for some Kähler metric ω then we say that α is a *Kähler class* and write $\alpha > 0$.

A basic result of Kähler geometry is the $\partial\bar\partial$-Lemma.

Theorem 3.1.10. *Let (M, ω) be a compact Kähler manifold. Suppose that $0 = [\alpha] \in H_{\bar\partial}^{1,1}(M, \mathbb{R})$ for a real smooth $\bar\partial$-closed $(1, 1)$-form α. Then there exists a real-valued smooth function φ with $\alpha = \frac{\sqrt{-1}}{2\pi}\partial\bar\partial\varphi$, which is uniquely determined up to the addition of a constant.*

In other words, a real $(1, 1)$-form α is $\bar{\partial}$-exact if and only if it is $\partial\bar{\partial}$-exact. It is an immediate consequence of the $\partial\bar{\partial}$-Lemma that if ω and ω' are Kähler metrics in the same Kähler class then $\omega' = \omega + \frac{\sqrt{-1}}{2\pi}\partial\bar{\partial}\varphi$ for some smooth function φ, which is uniquely determined up to a constant, and sometimes referred to as a *(Kähler) potential*.

A *line bundle* L over M is given by an open cover $\{U_\alpha\}$ of M together with collection of *transition functions* $\{t_{\alpha\beta}\}$ which are holomorphic maps $t_{\alpha\beta} : U_\alpha \cap U_\beta \to \mathbb{C}^*$ satisfying

$$t_{\alpha\beta}t_{\beta\alpha} = 1, \quad t_{\alpha\beta}t_{\beta\gamma} = t_{\alpha\gamma}. \tag{3.44}$$

We identify two such collections of transition functions $\{t_{\alpha\beta}\}$ and $\{t'_{\alpha\beta}\}$ if we can find holomorphic functions $f_\alpha : U_\alpha \to \mathbb{C}^*$ with $t'_{\alpha\beta} = \frac{f_\alpha}{f_\beta}t_{\alpha\beta}$. (In addition, we also need to identify $(\{U_\alpha\}, \{t_{\alpha\beta}\})$, $(\{U'_\gamma\}, \{t'_{\gamma\delta}\})$ whenever $\{U'_\gamma\}$ is a refinement of $\{U_\alpha\}$ and the $t'_{\gamma\delta}$ are restrictions of the $t_{\alpha\beta}$. We will not dwell on technical details about refinements etc. and instead refer the reader to [GH78] or [KodMor71].) Given line bundles L, L' with transition functions $\{t_{\alpha\beta}\}$, $\{t'_{\alpha\beta}\}$ write LL' for the new line bundle with transition functions $\{t_{\alpha\beta}t'_{\alpha\beta}\}$. Similarly, for any $m \in \mathbb{Z}$, we define line bundles L^m by $\{t^m_{\alpha\beta}\}$. We call L^{-1} the inverse of L. Sometimes we use the additive notation for line bundles, writing $L + L'$ for LL' and mL for L^m.

A *holomorphic section* s of L is a collection $\{s_\alpha\}$ of holomorphic maps $s_\alpha : U_\alpha \to \mathbb{C}$ satisfying the transformation rule $s_\alpha = t_{\alpha\beta}s_\beta$ on $U_\alpha \cap U_\beta$. A *Hermitian metric* h on L is a collection $\{h_\alpha\}$ of smooth positive functions $h_\alpha : U_\alpha \to \mathbb{R}$ satisfying the transformation rule $h_\alpha = |t_{\beta\alpha}|^2 h_\beta$ on $U_\alpha \cap U_\beta$. Given a holomorphic section s and a Hermitian metric h, we can define the *pointwise norm squared* of s with respect to h by $|s|^2_h = h_\alpha s_\alpha \overline{s_\alpha}$ on U_α. The reader can check that $|s|^2_h$ is a well-defined function on M.

We define the *curvature* R_h of a Hermitian metric h on L to be the closed $(1, 1)$ form on M given by $R_h = -\frac{\sqrt{-1}}{2\pi}\partial\bar{\partial}\log h_\alpha$ on U_α. Again, we let the reader check that this is well-defined. Define the *first Chern class* $c_1(L)$ of L to be the cohomology class $[R_h] \in H^{1,1}_{\bar{\partial}}(M, \mathbb{R})$. Since any two Hermitian metrics h, h' on L are related by $h' = he^{-\varphi}$ for some smooth function φ, we see that $R_{h'} = R_h + \frac{\sqrt{-1}}{2\pi}\partial\bar{\partial}\varphi$ and hence $c_1(L)$ is well-defined independent of choice of Hermitian metric. Note that if h is a Hermitian metric on L then h^m is a Hermitian metric on L^m and $c_1(L^m) = mc_1(L)$.

Every complex manifold M is equipped with a line bundle K_M, known as the *canonical bundle*, whose transition functions are given by $t_{\alpha\beta} = \det\left(\partial z^i_\beta / \partial z^j_\alpha\right)$ on $U_\alpha \cap U_\beta$, where U_α are coordinate charts for M with coordinates $z^1_\alpha, \ldots, z^n_\alpha$. If g is a Kähler metric (or more generally, a Hermitian metric) on M then $h_\alpha = \det(g^\alpha_{i\bar{j}})^{-1}$ on U_α defines a Hermitian metric on K_M. The inverse K_M^{-1} of K_M is sometimes called the *anti-canonical bundle*. Its first Chern class $c_1(K_M^{-1})$ is called the *first Chern class of M* and is often denoted by $c_1(M)$. It follows from Proposition 3.1.3 and the above definitions that $c_1(M) = [\text{Ric}(\omega)]$ for any Kähler metric ω on M.

We now discuss *divisors* on M. First, we say that a subset V of M is an *analytic hypersurface* if V is locally given as the zero set $\{f = 0\}$ of a locally defined holomorphic function f. In general, V may not be a submanifold. Denote by V^{reg} the set of points $p \in V$ for which V is a submanifold of M near p. We say that V is *irreducible* if V^{reg} is connected. A *divisor* D on M is a formal finite sum $\sum_i a_i V_i$ where $a_i \in \mathbb{Z}$ and each V_i is an irreducible analytic hypersurface of M. We say that D is *effective* if the a_i are all nonnegative. The *support* of D is the union of the V_i for each i with $a_i \neq 0$.

Given a divisor D we define an *associated line bundle* as follows. Suppose that D is given by local defining functions f_α (vanishing on D to order 1) over an open cover U_α. Define transition functions $t_{\alpha\beta} = f_\alpha/f_\beta$ on $U_\alpha \cap U_\beta$. These are holomorphic and nonvanishing in $U_\alpha \cap U_\beta$, and satisfy (3.44). Write $[D]$ for the associated line bundle, which is well-defined independent of choice of local defining functions. Note that the map $D \mapsto [D]$ is not injective. Indeed if $D \neq 0$ is defined by a meromorphic function f on M then $[D]$ is trivial.

As an example: associated to a hyperplane $\{Z_i = 0\}$ in $\mathbb{P}^n$ is the line bundle H, called the *hyperplane bundle*. Taking the open cover $U_\alpha = \{Z_\alpha \neq 0\}$, the hyperplane is given by $Z_i/Z_\alpha = 0$ in U_α. Thus we can define H by the transition functions $t_{\alpha\beta} = Z_\beta/Z_\alpha$. Define a Hermitian metric h_{FS} on H by

$$(h_{\mathrm{FS}})_\alpha = \frac{|Z_\alpha|^2}{|Z_0|^2 + \cdots + |Z_n|^2} \quad \text{on} \quad U_\alpha. \tag{3.45}$$

Notice that $R_{h_{\mathrm{FS}}} = \omega_{\mathrm{FS}}$. The canonical bundle of $\mathbb{P}^n$ is given by $K_{\mathbb{P}^n} = -(n+1)H$ and $c_1(\mathbb{P}^n) = (n+1)[\omega_{\mathrm{FS}}] > 0$. The line bundle H is sometimes written $\mathcal{O}(1)$.

3.1.7 Notions of Positivity of Line Bundles

Let L be a line bundle over a compact Kähler manifold (M, ω). We say that L is *positive* if $c_1(L) > 0$. This is equivalent to saying that there exists a Hermitian metric h on L for which R_h is a Kähler form.

The Kodaira Embedding Theorem relates the positivity of L with embeddings of M into projective space via sections of L. More precisely, write $H^0(M, L)$ for the vector space of holomorphic sections of L. This is finite dimensional if not empty. We say that L is *very ample* if for any ordered basis $\underline{s} = (s_0, \ldots, s_N)$ of $H^0(M, L)$, the map $\iota_{\underline{s}} : M \to \mathbb{P}^N$ given by

$$\iota_{\underline{s}}(x) = [s_0(x), \ldots, s_N(x)], \tag{3.46}$$

is well-defined and an embedding. Note that $s_0(x), \ldots, s_N(x)$ are not well-defined as elements of $\mathbb{C}$, but $[s_0(x), \ldots, s_N(x)]$ is a well-defined element of $\mathbb{P}^N$ as long as not all the $s_i(x)$ vanish. We say that L is *ample* if there exists a positive integer m_0 such that L^m is very ample for all $m \geq m_0$. The Kodaira Embedding Theorem states:

Theorem 3.1.11. *L is ample if and only if L is positive.*

The hard part of this theorem is the "if" direction. For the other direction, assume that L^m is very ample, with $(s_0, \ldots, s_N)$ a basis of $H^0(M, L^m)$. Since M is a submanifold of $\mathbb{P}^n$, we see that $\iota_{\underline{s}}^* \omega_{FS}$ is a Kähler form on M and if h is any Hermitian metric on L^m then by definition of $\iota_{\underline{s}}$,

$$\iota_{\underline{s}}^* \omega_{FS} = -\frac{\sqrt{-1}}{2\pi} \partial \bar{\partial} \log h + \frac{\sqrt{-1}}{2\pi} \partial \bar{\partial} \log(|s_0|_h^2 + \cdots + |s_N|_h^2) = R_h + \frac{\sqrt{-1}}{2\pi} \partial \bar{\partial} f, \tag{3.47}$$

for a globally defined function f. This implies that $\frac{1}{m} \iota_{\underline{s}}^* \omega_{FS} \in c_1(L)$ and hence $c_1(L) > 0$.

We say that a line bundle L is *globally generated* if for each $x \in M$ there exists a holomorphic section s of L such that $s(x) \neq 0$. If L is globally generated then given an ordered basis $\underline{s} = (s_0, \ldots, s_N)$ of holomorphic sections of L, we have a well-defined holomorphic map $\iota_{\underline{s}} : M \to \mathbb{P}^N$ given by (3.46) (although it is not necessarily an embedding). We say that a line bundle L is *semi-ample* if there exists a positive integer m_0 such that L^{m_0} is globally generated. Observe that if L is semi-ample then, by considering again the pull-back of ω_{FS} to M by an appropriate map $\iota_{\underline{s}}$, there exists a Hermitian metric h on L such that R_h is a nonnegative $(1,1)$-form. That is, $c_1(L)$ contains a nonnegative representative.

We next discuss the pairing of line bundles with curves in M. By a *curve* in M we mean an analytic subvariety of dimension 1. If C is smooth, then we define

$$L \cdot C = \int_C R_h, \tag{3.48}$$

where h is any Hermitian metric on L. By Stokes' Theorem, $L \cdot C$ is independent of choice of h. If C is not smooth then we integrate over C^{reg}, the smooth part of C (Stokes' Theorem still holds—see for example [GH78], p. 33). We can also pair a divisor D with a curve by setting $D \cdot C = [D] \cdot C$, and we may pair a general element $\alpha \in H^{1,1}(M, \mathbb{R})$ with a curve C by setting $\alpha \cdot C = \int_C \eta$ for $\eta \in \alpha$.

We say that a line bundle L is *nef* if $L \cdot C \geq 0$ for all curves C in M ("nef" is an abbreviation of either "numerically eventually free" or "numerically effective", depending on whom you ask). It follows immediately from the definitions that:

$$L \text{ ample} \quad \Rightarrow \quad L \text{ semi} - \text{ample} \quad \Rightarrow \quad L \text{ nef.} \tag{3.49}$$

We may also pair a line bundle with itself n times, where n is the complex dimension of M. Define

$$c_1(L)^n := \int_M (R_h)^n. \tag{3.50}$$

Moreover, given any $\alpha \in H^{1,1}(M, \mathbb{R})$ we define $\alpha^n = \int_M \eta^n$ for $\eta \in \alpha$.

Assume now that M is a smooth projective variety. We say that a line bundle L on M is *big* if there exist constants m_0 and $c > 0$ such that $\dim H^0(M, L^m) \geq c\, m^n$ for all $m \geq m_0$. It follows from the Riemann–Roch Theorem (see [Ha77, Laz04], for example) that a nef line bundle is big if and only if $c_1(L)^n > 0$. It follows that an ample line bundle is both nef and big. If M has K_M big then we say that M is of *general type*. If M has K_M nef then we say that M is a *smooth minimal model*.

We define the *Kodaira dimension* of M to be the infimum of $\kappa \in [-\infty, \infty)$ such that there exists a constant C with $\dim H^0(M, K_M^m) \leq C m^\kappa$ for all positive m. In the special case that all $H^0(M, K_M^m)$ are empty, we have $\kappa = -\infty$. The largest possible value of κ is n. We write $\mathrm{kod}(M)$ for the Kodaira dimension κ of M. Thus if M is of general type then $\mathrm{kod}(M) = n$. If M is *Fano*, which means that $c_1(M) > 0$ then $\mathrm{kod}(M) = -\infty$.

If K_M is semi-ample then for m sufficiently large, the map $\iota_{\underline{s}} : M \to \mathbb{P}^N$ given by sections of K_M^m has image a subvariety Y, which is uniquely determined up to isomorphism. Y is called the *canonical model* of M and $\dim Y = \mathrm{kod}(M)$ [Laz04].

We now quote some results from algebraic geometry:

Theorem 3.1.12. *Let M be a projective algebraic manifold.*

(i) *Let α be a Kähler class and let L be a nef line bundle. Then $\alpha + s\, c_1(L)$ is Kähler for all $s > 0$.*

(ii) *(Kawamata's Base Point Free Theorem) If L is nef and $aL - K_X$ is nef and big for some $a > 0$ then L is semi-ample.*

(iii) *(Kodaira's Lemma) Let L be a nef and big line bundle on M. Then there exists an effective divisor E and $\delta > 0$ such that $c_1(L) - \varepsilon c_1([E]) > 0$ for all $\varepsilon \in (0, \delta]$.*

Proof. For part (i), see for example Proposition 6.2 in [Dem96] or Corollary 1.4.10 in [Laz04]. For part (ii), see [KMM87, Shok85]. For part (iii), see for example p. 43 of [Dem96]. □

It will be useful to gather here some results from complex surfaces which we will make use of later. First we have the *Adjunction Formula* for surfaces. See for example [GH78] or [BHPV].

Theorem 3.1.13. *Let M be a Kähler surface, with C an irreducible smooth curve in M. Then if $g(C)$ is the genus of C, we have*

$$1 + \frac{K_M \cdot C + C \cdot C}{2} = g(C). \tag{3.51}$$

Moreover, if C is an irreducible, possibly singular, curve in M, we have

$$1 + \frac{K_M \cdot C + C \cdot C}{2} \geq 0, \tag{3.52}$$

with equality if and only if C is smooth and isomorphic to $\mathbb{P}^1$.

Note that $C \cdot C$ is well-defined, since M has complex dimension 2 and so C is both a curve and a divisor. We may write C^2 instead of $C \cdot C$. Generalizing the intersection pairing, we have the *cup product form* on $H^{1,1}(M, \mathbb{R})$ given by $\alpha \cdot \beta = \int_M \alpha \wedge \beta$. Again, we write α^2 instead of $\alpha \cdot \alpha$. A divisor D in M defines an element of $H^{1,1}(M, \mathbb{R})$ by $D \mapsto [R_h] \in H^{1,1}(M, \mathbb{R})$ for h a Hermitian metric on the line bundle $[D]$, and this is consistent with our previous definitions.

We have the *Hodge Index Theorem* for Kähler surfaces (see for example Theorem IV.2.14 of [BHPV] or p. 470 of [GH78]).

Theorem 3.1.14. *The cup product form on $H^{1,1}(M, \mathbb{R})$ is non-degenerate of type $(1, k - 1)$, where k is the dimension of $H^{1,1}(M, \mathbb{R})$. In particular, if $\alpha \in H^{1,1}(M, \mathbb{R})$ satisfies $\alpha^2 > 0$ then for any $\beta \in H^{1,1}(M, \mathbb{R})$,*

$$\alpha \cdot \beta = 0 \quad \Rightarrow \quad \beta^2 < 0 \text{ or } \beta = 0. \tag{3.53}$$

Finally, we state the Nakai–Moishezon criterion for Kähler surfaces, due to Buchdahl and Lamari [Buch99, Lam99].

Theorem 3.1.15. *Let M be a Kähler surface and β be a Kähler class on M. If $\alpha \in H^{1,1}(M, \mathbb{R})$ is a class satisfying*

$$\alpha^2 > 0, \quad \alpha \cdot \beta > 0, \quad \alpha \cdot C > 0$$

for every irreducible curve C on M, then α is a Kähler class on M.

A generalization of this to Kähler manifolds of any dimension was established by Demailly–Paun [DemPaun04].

3.2 General Estimates for the Kähler–Ricci Flow

In this section we introduce the Kähler–Ricci flow equation. We derive a number of fundamental evolution equations and estimates for the flow which will be used extensively throughout these notes. In addition, we discuss higher order estimates for the flow.

3.2.1 The Kähler–Ricci Flow

Let (M, ω_0) be a compact Kähler manifold of complex dimension n. A solution of the *Kähler–Ricci flow* on M starting at ω_0 is a family of Kähler metrics $\omega = \omega(t)$ solving

$$\frac{\partial}{\partial t} \omega = -\text{Ric}(\omega), \quad \omega|_{t=0} = \omega_0. \tag{3.54}$$

Note that this differs from Hamilton's equation (3.1) by a factor of 2: see Remark 3.2.11.

For later use it will be convenient to consider a more general equation than (3.54), namely

$$\frac{\partial}{\partial t}\omega = -\text{Ric}(\omega) - \nu\omega, \qquad \omega|_{t=0} = \omega_0, \qquad (3.55)$$

where ν is a fixed real number which we take to be either $\nu = 0$ or $\nu = 1$. As we will discuss later in Sect. 3.4, the case $\nu = 1$ corresponds to a rescaling of (3.54). When $\nu = 1$ we call (3.55) the *normalized Kähler–Ricci flow*.

We have the following existence and uniqueness result.

Theorem 3.2.1. *There exists a unique solution $\omega = \omega(t)$ to (3.55) on some maximal time interval $[0, T)$ for some T with $0 < T \le \infty$.*

Since the case $\nu = 1$ is a rescaling of (3.54), it suffices to consider (3.54). We will provide a proof of this in Sect. 3.3, and show in addition that T can be prescribed in terms of the cohomology class of $[\omega_0]$ and the manifold M. Theorem 3.2.1 also follows from the well-known results of Hamilton. Indeed we can use the short time existence result of Hamilton [Ham82] (see also [Det83]) to obtain a maximal solution to the Ricci flow $\frac{\partial}{\partial t}g_{ij} = -R_{ij}$ on $[0, T)$ starting at g_0 for some $T > 0$. Since the Ricci flow preserves the Kähler condition (see e.g. [Ham95a]), $g(t)$ solves (3.54) on $[0, T)$. Note that this argument does not explicitly give us the value of T.

A remark about notation. When we write tensorial objects such as curvature tensors $R_{i\bar{j}k\bar{\ell}}$, covariant derivatives ∇_i, Laplace operators Δ, we refer to the objects corresponding to the evolving metric $\omega = \omega(t)$, unless otherwise indicated.

3.2.2 Evolution of Scalar Curvature

Let $\omega = \omega(t)$ be a solution to the Kähler–Ricci flow (3.55) on $[0, T)$ for T with $0 < T \le \infty$. We compute the well-known evolution of the scalar curvature.

Theorem 3.2.2. *The scalar curvature R of $\omega = \omega(t)$ evolves by*

$$\frac{\partial}{\partial t}R = \Delta R + |\text{Ric}(\omega)|^2 + \nu R, \qquad (3.56)$$

where $|\text{Ric}(\omega)|^2 = g^{\bar{\ell}i}g^{\bar{j}k}R_{i\bar{j}}R_{k\bar{\ell}}$. Hence the scalar curvature has a lower bound

$$R(t) \ge -\nu n - C_0 e^{-\nu t}, \qquad (3.57)$$

for $C_0 = -\inf_M R(0) - \nu n$.

Proof. Taking the trace of the evolution equation (3.55) gives

$$g^{\bar{\ell}k}\frac{\partial}{\partial t}g_{k\bar{\ell}} = -R - vn. \tag{3.58}$$

Since $R = -g^{\bar{j}i}\partial_i\partial_{\bar{j}}\log\det g$ we have

$$\frac{\partial}{\partial t}R = -g^{\bar{j}i}\partial_i\partial_{\bar{j}}\left(g^{\bar{\ell}k}\frac{\partial}{\partial t}g_{k\bar{\ell}}\right) - \left(\frac{\partial}{\partial t}g^{\bar{j}i}\right)\partial_i\partial_{\bar{j}}\log\det g \tag{3.59}$$

$$= \Delta R + g^{\bar{\ell}i}g^{\bar{j}k}R_{k\bar{\ell}}R_{i\bar{j}} + vR, \tag{3.60}$$

as required. For (3.57), we use the elementary fact that $n|\mathrm{Ric}(\omega)|^2 \geq R^2$ to obtain

$$\left(\frac{\partial}{\partial t} - \Delta\right)R \geq \frac{1}{n}R(R + vn) = \frac{1}{n}(R + vn)^2 - v(R + vn). \tag{3.61}$$

Hence

$$\left(\frac{\partial}{\partial t} - \Delta\right)(e^{vt}(R + vn)) \geq 0. \tag{3.62}$$

By the maximum principle (see Proposition 3.1.7 and the remark following it), the quantity $e^{vt}(R + vn)$ is bounded below by $\inf_M R(0) + vn$, its value at time $t = 0$. $\qquad\square$

We remark that although we used the Kähler condition to prove Theorem 3.2.2, in fact it holds in full generality for the Riemannian Ricci flow [Ham82] (see also [ChowKnopf]).

Theorem 3.2.2 implies a bound on the volume form of the metric.

Corollary 3.2.3. *Let $\omega = \omega(t)$ be a solution of (3.55) on $[0, T)$ and C_0 as in Theorem 3.2.2.*

(i) If $v = 0$ then

$$\omega^n(t) \leq e^{C_0 t}\omega^n(0). \tag{3.63}$$

In particular, if T is finite then the volume form $\omega^n(t)$ is uniformly bounded from above for $t \in [0, T)$.

(ii) If $v = 1$ there exists a uniform constant C such that

$$\omega^n(t) \leq e^{C_0(1-e^{-t})}\omega^n(0). \tag{3.64}$$

In particular, the volume form $\omega^n(t)$ is uniformly bounded from above for $t \in [0, T)$.

Proof. We have

$$\frac{\partial}{\partial t} \log \frac{\omega^n(t)}{\omega^n(0)} = g^{\bar{j}i} \frac{\partial}{\partial t} g_{i\bar{j}} = -R - \nu n \le C_0 e^{-\nu t}. \tag{3.65}$$

Integrating in time, we obtain (3.63) and (3.64). □

3.2.3 Evolution of the Trace of the Metric

We now prove an estimate for the trace of the metric along the Kähler–Ricci flow. This is originally due to Cao [Cao85] and is the parabolic version of an estimate for the complex Monge–Ampère equation due to Yau and Aubin [Aub78, Yau78]. We give the estimate in the form of an evolution inequality. We begin by computing the evolution of $\text{tr}_{\hat{\omega}} \omega$, the trace of ω with respect to a fixed metric $\hat{\omega}$, using the notation of Sect. 3.1.4.

Proposition 3.2.4. *Let $\hat{\omega}$ be a fixed Kähler metric on M, and let $\omega = \omega(t)$ be a solution to the Kähler–Ricci flow (3.55). Then*

$$\left(\frac{\partial}{\partial t} - \Delta\right) \text{tr}_{\hat{\omega}} \omega = -\nu \, \text{tr}_{\hat{\omega}} \omega - g^{\bar{\ell}k} \hat{R}_{k\bar{\ell}}{}^{\bar{j}i} g_{i\bar{j}} - \hat{g}^{\bar{j}i} g^{\bar{q}p} g^{\bar{\ell}k} \hat{\nabla}_i g_{p\bar{\ell}} \hat{\nabla}_{\bar{j}} g_{k\bar{q}}, \tag{3.66}$$

where $\hat{R}_{k\bar{\ell}}{}^{\bar{j}i}$, $\hat{\nabla}$ denote the curvature and covariant derivative with respect to $\hat{g}$.

Proof. Compute using normal coordinates for $\hat{g}$ and the formula (3.17),

$$\begin{aligned}
\Delta \text{tr}_{\hat{\omega}} \omega &= g^{\bar{\ell}k} \partial_k \partial_{\bar{\ell}} (\hat{g}^{\bar{j}i} g_{i\bar{j}}) \\
&= g^{\bar{\ell}k} (\partial_k \partial_{\bar{\ell}} \hat{g}^{\bar{j}i}) g_{i\bar{j}} + g^{\bar{\ell}k} \hat{g}^{\bar{j}i} \partial_k \partial_{\bar{\ell}} g_{i\bar{j}} \\
&= g^{\bar{\ell}k} \hat{R}_{k\bar{\ell}}{}^{\bar{j}i} g_{i\bar{j}} - \hat{g}^{\bar{j}i} R_{i\bar{j}} + \hat{g}^{\bar{j}i} g^{\bar{q}p} g^{\bar{\ell}k} \partial_i g_{p\bar{\ell}} \partial_{\bar{j}} g_{k\bar{q}},
\end{aligned} \tag{3.67}$$

and

$$\frac{\partial}{\partial t} \text{tr}_{\hat{\omega}} \omega = -\hat{g}^{\bar{j}i} R_{i\bar{j}} - \nu \, \text{tr}_{\hat{\omega}} \omega, \tag{3.68}$$

and combining these gives (3.66). □

We use Proposition 3.2.4 to prove the following estimate, which will be used frequently in the sequel:

Proposition 3.2.5. *Let $\hat{\omega}$ be a fixed Kähler metric on M, and let $\omega = \omega(t)$ be a solution to (3.55). Then there exists a constant $\hat{C}$ depending only on the lower bound of the bisectional curvature for $\hat{g}$ such that*

$$\left(\frac{\partial}{\partial t} - \Delta\right) \log \operatorname{tr}_{\hat{\omega}} \omega \le \hat{C} \operatorname{tr}_{\omega} \hat{\omega} - \nu. \tag{3.69}$$

Proof. First observe that for a positive function f,

$$\Delta \log f = \frac{\Delta f}{f} - \frac{|\partial f|_g^2}{f^2}. \tag{3.70}$$

It follows immediately from Proposition 3.2.4 that

$$\left(\frac{\partial}{\partial t} - \Delta\right) \log \operatorname{tr}_{\hat{\omega}} \omega$$

$$= \frac{1}{\operatorname{tr}_{\hat{\omega}} \omega} \left(-\nu \operatorname{tr}_{\hat{\omega}} \omega - g^{\bar{\ell}k} \hat{R}_{k\bar{\ell}}{}^{\bar{j}i} g_{i\bar{j}} + \frac{|\partial \operatorname{tr}_{\hat{\omega}} \omega|_g^2}{\operatorname{tr}_{\hat{\omega}} \omega} - \hat{g}^{\bar{j}i} g^{\bar{q}p} g^{\bar{\ell}k} \hat{\nabla}_i g_{p\bar{\ell}} \hat{\nabla}_{\bar{j}} g_{k\bar{q}} \right). \tag{3.71}$$

We claim that

$$\frac{|\partial \operatorname{tr}_{\hat{\omega}} \omega|_g^2}{\operatorname{tr}_{\hat{\omega}} \omega} - \hat{g}^{\bar{j}i} g^{\bar{q}p} g^{\bar{\ell}k} \hat{\nabla}_i g_{p\bar{\ell}} \hat{\nabla}_{\bar{j}} g_{k\bar{q}} \le 0. \tag{3.72}$$

To prove this we choose normal coordinates for $\hat{g}$ for which g is diagonal. Compute using the Cauchy–Schwarz inequality

$$|\partial \operatorname{tr}_{\hat{\omega}} \omega|_g^2 = \sum_i g^{\bar{i}i} \partial_i \left(\sum_j g_{j\bar{j}}\right) \partial_{\bar{i}} \left(\sum_k g_{k\bar{k}}\right)$$

$$= \sum_{j,k} \sum_i g^{\bar{i}i} (\partial_i g_{j\bar{j}})(\partial_{\bar{i}} g_{k\bar{k}})$$

$$\le \sum_{j,k} \left(\sum_i g^{\bar{i}i} |\partial_i g_{j\bar{j}}|^2\right)^{1/2} \left(\sum_i g^{\bar{i}i} |\partial_i g_{k\bar{k}}|^2\right)^{1/2}$$

$$= \left(\sum_j \left(\sum_i g^{\bar{i}i} |\partial_i g_{j\bar{j}}|^2\right)^{1/2}\right)^2$$

$$= \left(\sum_j \sqrt{g_{j\bar{j}}} \left(\sum_i g^{\bar{i}i} g^{\bar{j}j} |\partial_i g_{j\bar{j}}|^2\right)^{1/2}\right)^2$$

$$\le \sum_\ell g_{\ell\bar{\ell}} \sum_{i,j} g^{\bar{i}i} g^{\bar{j}j} |\partial_j g_{i\bar{j}}|^2$$

$$\le (\operatorname{tr}_{\hat{\omega}} \omega) \sum_{i,j,k} g^{\bar{i}i} g^{\bar{j}j} \partial_k g_{i\bar{j}} \partial_{\bar{k}} g_{j\bar{i}}, \tag{3.73}$$

where in the second-to-last line we used the Kähler condition to give $\partial_i g_{j\bar{j}} = \partial_j g_{i\bar{j}}$. The inequality (3.73) gives exactly (3.72)

We can now complete the proof of the proposition. Define a constant $\hat{C}$ by

$$\hat{C} = - \inf_{x \in M} \{ \hat{R}_{i\bar{i}j\bar{j}}(x) \mid \{\partial_{z^1}, \ldots, \partial_{z^n}\} \text{is orthonormal w.r.t. } \hat{g} \text{ at } x, \ i, j = 1, \ldots, n\}, \tag{3.74}$$

which is finite since we are taking the infimum of a continuous function over a compact set.

Then computing at a point using normal coordinates for $\hat{g}$ for which the metric g is diagonal we have

$$g^{\bar{\ell}k} \hat{R}_{k\bar{\ell}}{}^{\bar{j}i} g_{i\bar{j}} = \sum_{k,i} g^{\bar{k}k} \hat{R}_{k\bar{k}i\bar{i}} g_{i\bar{i}} \geq -\hat{C} \sum_k g^{\bar{k}k} \sum_i g_{i\bar{i}} = -\hat{C}\,(\operatorname{tr}_{\hat{\omega}}\omega)(\operatorname{tr}_{\omega}\hat{\omega}). \tag{3.75}$$

Combining (3.71), (3.72) and (3.75) yields (3.69).   □

3.2.4 The Parabolic Schwarz Lemma

In this section we prove the parabolic Schwarz lemma of [ST07]. This is a parabolic version of Yau's Schwarz lemma [Yau78b]. We state it here in the form of an evolution inequality.

Theorem 3.2.6. *Let $f : M \to N$ be a holomorphic map between compact complex manifolds M and N of complex dimension n and κ respectively. Let ω_0 and ω_N be Kähler metrics on M and N respectively and let $\omega = \omega(t)$ be a solution of (3.55) on $M \times [0, T)$, namely*

$$\frac{\partial}{\partial t}\omega = -\operatorname{Ric}(\omega) - v\omega, \qquad \omega|_{t=0} = \omega_0, \tag{3.76}$$

for $t \in [0, T)$, with either $v = 0$ or $v = 1$. Then for all points of $M \times [0, T)$ with $\operatorname{tr}_{\omega}(f^*\omega_N)$ positive we have

$$\left(\frac{\partial}{\partial t} - \Delta \right) \log \operatorname{tr}_{\omega}(f^*\omega_N) \leq C_N \operatorname{tr}_{\omega}(f^*\omega_N) + v, \tag{3.77}$$

where C_N is an upper bound for the bisectional curvature of ω_N.

Observe that a simple maximum principle argument immediately gives the following consequence which the reader will recognize as similar to the conclusion of Yau's Schwarz lemma.

Corollary 3.2.7. *If the bisectional curvature of ω_N has a negative upper bound $C_N < 0$ on N then there exists a constant $C > 0$ depending only on C_N, ω_0, ω_N and v such that* $\text{tr}_\omega(f^*\omega_N) \leq C$ *on $M \times [0, T)$ and hence*

$$\omega \geq \frac{1}{C} f^*\omega_N, \quad \text{on } M \times [0, T). \tag{3.78}$$

In practice, we will find the inequality (3.77) more useful than this corollary, since the assumption of negative bisectional curvature is rather strong. For the proof of Theorem 3.2.6, we will follow quite closely the notation and calculations given in [ST07].

Proof of Theorem 3.2.6. Fix x in M with $f(x) = y \in N$, and choose normal coordinate systems $(z^i)_{i=1,\ldots,n}$ for g centered at x and $(w^\alpha)_{\alpha=1,\ldots,\kappa}$ for g_N centered at y. The map f is given locally as $(f^1, \ldots, f^\kappa)$ for holomorphic functions $f^\alpha = f^\alpha(z^1, \ldots, z^n)$. Write f_i^α for $\frac{\partial}{\partial z^i} f^\alpha$. To simplify notation we write the components of g_N as $h_{\alpha\bar\beta}$ instead of $(g_N)_{\alpha\bar\beta}$. The components of the tensor f^*g_N are then $f_i^\alpha \overline{f_j^\beta} h_{\alpha\bar\beta}$ and hence $\text{tr}_\omega(f^*\omega_N) = g^{\bar{j}i} f_i^\alpha \overline{f_j^\beta} h_{\alpha\bar\beta}$. Writing $u = \text{tr}_\omega(f^*\omega_N) > 0$, we compute at x,

$$\Delta u = g^{\bar{\ell}k} \partial_k \partial_{\bar\ell} \left(g^{\bar{j}i} f_i^\alpha \overline{f_j^\beta} h_{\alpha\bar\beta} \right)$$

$$= R^{\bar{j}i} f_i^\alpha \overline{f_j^\beta} h_{\alpha\bar\beta} + g^{\bar{\ell}k} g^{\bar{j}i} (\partial_k f_i^\alpha)(\overline{\partial_\ell f_j^\beta}) h_{\alpha\bar\beta} - g^{\bar{\ell}k} g^{\bar{j}i} S_{\alpha\bar\beta\gamma\bar\delta} f_i^\alpha \overline{f_j^\beta} f_k^\gamma \overline{f_\ell^\delta}, \tag{3.79}$$

for $S_{\alpha\bar\beta\gamma\bar\delta}$ the curvature tensor of g_N on N. Next,

$$\frac{\partial}{\partial t} u = -g^{\bar{\ell}i} g^{\bar{j}k} \left(\frac{\partial}{\partial t} g_{k\bar\ell} \right) f_i^\alpha \overline{f_j^\beta} h_{\alpha\bar\beta} = R^{\bar{j}i} f_i^\alpha \overline{f_j^\beta} h_{\alpha\bar\beta} + vu. \tag{3.80}$$

Combining (3.79) and (3.80) with (3.70), we obtain

$$\left(\frac{\partial}{\partial t} - \Delta \right) \log u = \frac{1}{u} g^{\bar{\ell}k} g^{\bar{j}i} S_{\alpha\bar\beta\gamma\bar\delta} f_i^\alpha \overline{f_j^\beta} f_k^\gamma \overline{f_\ell^\delta}$$

$$+ \frac{1}{u} \left(\frac{|\partial u|_g^2}{u} - g^{\bar{\ell}k} g^{\bar{j}i} (\partial_k f_i^\alpha)(\overline{\partial_\ell f_j^\beta}) h_{\alpha\bar\beta} \right) + v. \tag{3.81}$$

If C_N is an upper bound for the bisectional curvature of g_N we see that

$$g^{\bar{\ell}k} g^{\bar{j}i} S_{\alpha\bar\beta\gamma\bar\delta} f_i^\alpha \overline{f_j^\beta} f_k^\gamma \overline{f_\ell^\delta} \leq C_N u^2, \tag{3.82}$$

and hence (3.77) will follow from the inequality

$$\frac{|\partial u|_g^2}{u} - g^{\bar{\ell}k} g^{\bar{j}i} (\partial_k f_i^\alpha)(\overline{\partial_\ell f_j^\beta}) h_{\alpha\bar{\beta}} \le 0. \tag{3.83}$$

The inequality (3.83) is analogous to (3.72) and the proof is almost identical. Indeed, at the point x,

$$|\partial u|_g^2 = \sum_{i,j,k,\alpha,\beta} \overline{f_i^\alpha} f_j^\beta \partial_k f_i^\alpha \overline{\partial_k f_j^\beta}$$

$$\le \sum_{i,j,\alpha,\beta} |f_i^\alpha||f_j^\beta| \left(\sum_k |\partial_k f_i^\alpha|^2 \right)^{1/2} \left(\sum_\ell |\partial_\ell f_j^\beta|^2 \right)^{1/2}$$

$$= \left(\sum_{i,\alpha} |f_i^\alpha| \left(\sum_k |\partial_k f_i^\alpha|^2 \right)^{1/2} \right)^2$$

$$\le \left(\sum_{j,\beta} |f_j^\beta|^2 \right) \left(\sum_{i,k,\alpha} |\partial_k f_i^\alpha|^2 \right) = u\, g^{\bar{\ell}k} g^{\bar{j}i} (\partial_k f_i^\alpha)(\overline{\partial_\ell f_j^\beta}) h_{\alpha\bar{\beta}}, \tag{3.84}$$

which gives (3.83).   □

3.2.5 The Third Order Estimate

In this section we prove the so-called "third order" estimate for the Kähler–Ricci flow assuming that the metric is uniformly bounded. By third order estimate we mean an estimate on the first derivative of the Kähler metric, which is of order 3 in terms of the potential function. Since the work of Yau [Yau78] on the elliptic Monge–Ampère equation, such estimates have often been referred to as *Calabi estimates* in reference to a well-known calculation of Calabi [Cal58]. There are now many generalizations of the Calabi estimate [Cher87, ShW11, Tos10b, TWY08, ZhaZha11]. A parabolic Calabi estimate was applied to the Kähler–Ricci flow in [Cao85]. Phong–Šešum–Sturm [PSS07] later gave a succinct and explicit formula, which we will describe here.

Let $\omega = \omega(t)$ be a solution of the normalized Kähler–Ricci flow (3.55) on $[0, T)$ for $0 < T \le \infty$ and let $\hat{\omega}$ be a fixed Kähler metric on M. We wish to estimate the quantity $S = |\hat{\nabla} g|^2$ where $\hat{\nabla}$ is the covariant derivative with respect to $\hat{g}$ and the norm $| \cdot |$ is taken with respect to the evolving metric g. Namely

$$S = g^{\bar{j}i} g^{\bar{\ell}k} g^{\bar{q}p} \hat{\nabla}_i g_{k\bar{q}} \overline{\hat{\nabla}_j g_{\ell\bar{p}}}. \tag{3.85}$$

Define a tensor Ψ_{ij}^k by

$$\Psi_{ij}^k := \Gamma_{ij}^k - \hat{\Gamma}_{ij}^k = g^{\bar{\ell}k}\hat{\nabla}_i g_{j\bar{\ell}}. \tag{3.86}$$

We may rewrite S as

$$S = |\Psi|^2 = g^{\bar{j}i} g^{\bar{q}p} g_{k\bar{\ell}} \Psi_{ip}^k \overline{\Psi_{jq}^\ell}. \tag{3.87}$$

We have the following key equality of Phong–Šešum–Sturm [PSS07].

Proposition 3.2.8. *With the notation above, S evolves by*

$$\left(\frac{\partial}{\partial t} - \Delta\right) S = -|\overline{\nabla}\Psi|^2 - |\nabla\Psi|^2 + v|\Psi|^2 - 2\mathrm{Re}\left(g^{\bar{j}i} g^{\bar{q}p} g_{k\bar{\ell}} \nabla^{\bar{b}} \hat{R}_{i\bar{b}p}^{\phantom{i\bar{b}p}k} \overline{\Psi_{jq}^\ell} \right), \tag{3.88}$$

where $\nabla^{\bar{b}} = g^{\bar{b}a}\nabla_a$ and $\hat{R}_{i\bar{b}p}^{\phantom{i\bar{b}p}k} := \hat{g}^{\bar{m}k} \hat{R}_{i\bar{b}p\bar{m}}$.

Proof. Compute

$$\Delta S = g^{\bar{j}i} g^{\bar{q}p} g_{k\bar{\ell}} \left((\Delta\Psi_{ip}^k)\overline{\Psi_{jq}^\ell} + \Psi_{ip}^k (\overline{\Delta\Psi_{jq}^\ell}) \right) + |\overline{\nabla}\Psi|^2 + |\nabla\Psi|^2, \tag{3.89}$$

where we are writing $\Delta = g^{\bar{b}a}\nabla_a\nabla_{\bar{b}}$ for the "rough" Laplacian and $\overline{\Delta} = g^{\bar{b}a}\nabla_{\bar{b}}\nabla_a$ for its conjugate. While Δ and $\overline{\Delta}$ agree when acting on functions, they differ in general when acting on tensors. In particular, using the commutation formulae (see Sect. 3.1.3),

$$\overline{\Delta}\Psi_{jq}^\ell = \Delta\Psi_{jq}^\ell + R_j^{b}\Psi_{bq}^\ell + R_q^{b}\Psi_{jb}^\ell - R_b^{\ell}\Psi_{jq}^b. \tag{3.90}$$

Combining (3.89) and (3.90),

$$\Delta S = 2\mathrm{Re}\left(g^{\bar{j}i} g^{\bar{q}p} g_{k\bar{\ell}} (\Delta\Psi_{ip}^k)\overline{\Psi_{jq}^\ell} \right) + |\overline{\nabla}\Psi|^2 + |\nabla\Psi|^2$$
$$+ R^{\bar{j}i} g^{\bar{q}p} g_{k\bar{\ell}} \Psi_{ip}^k \overline{\Psi_{jq}^\ell} + g^{\bar{j}i} R^{\bar{q}p} g_{k\bar{\ell}} \Psi_{ip}^k \overline{\Psi_{jq}^\ell} - g^{\bar{j}i} g^{\bar{q}p} R_{k\bar{\ell}} \Psi_{ip}^k \overline{\Psi_{jq}^\ell} \tag{3.91}$$

We now compute the time derivative of S given by (3.87). We claim that

$$\frac{\partial}{\partial t}\Psi_{ip}^k = \Delta\Psi_{ip}^k - \nabla^{\bar{b}} \hat{R}_{i\bar{b}p}^{\phantom{i\bar{b}p}k}. \tag{3.92}$$

Given this, together with

$$\frac{\partial}{\partial t} g^{\bar{j}i} = R^{\bar{j}i} + vg^{\bar{j}i}, \quad \frac{\partial}{\partial t} g_{k\bar{\ell}} = -R_{k\bar{\ell}} - vg_{k\bar{\ell}}, \tag{3.93}$$

we obtain

$$\frac{\partial}{\partial t} S = 2\mathrm{Re}\left(g^{\bar{j}i}g^{\bar{q}p}g_{k\bar{\ell}}\left(\Delta\Psi^k_{ip} - \nabla^{\bar{b}}\hat{R}_{i\bar{b}p}{}^k\right)\overline{\Psi^\ell_{jq}}\right) + R^{\bar{j}i}g^{\bar{q}p}g_{k\bar{\ell}}\Psi^k_{ip}\overline{\Psi^\ell_{jq}}$$
$$+ g^{\bar{j}i}R^{\bar{q}p}g_{k\bar{\ell}}\Psi^k_{ip}\overline{\Psi^\ell_{jq}} - g^{\bar{j}i}g^{\bar{q}p}R_{k\bar{\ell}}\Psi^k_{ip}\overline{\Psi^\ell_{jq}} + v|\Psi|^2. \tag{3.94}$$

Then (3.88) follows from (3.91) and (3.94).

To establish (3.92), compute

$$\frac{\partial}{\partial t}\Psi^k_{ip} = \frac{\partial}{\partial t}\Gamma^k_{ip} = -\nabla_i R_p{}^k. \tag{3.95}$$

On the other hand,

$$\nabla_{\bar{b}}\Psi^k_{ip} = \partial_{\bar{b}}(\Gamma^k_{ip} - \hat{\Gamma}^k_{ip}) = \hat{R}_{i\bar{b}p}{}^k - R_{i\bar{b}p}{}^k, \tag{3.96}$$

and hence

$$\Delta\Psi^k_{ip} = g^{\bar{b}a}\nabla_a\nabla_{\bar{b}}\Psi^k_{ip} = \nabla^{\bar{b}}\hat{R}_{i\bar{b}p}{}^k - \nabla_i R_p{}^k. \tag{3.97}$$

where for the last equality we have used the second Bianchi identity [part (iii) of Proposition 3.1.2]. Then (3.92) follows from (3.95) and (3.97). $\square$

Using this evolution equation together with Proposition 3.2.4, we obtain a third order estimate assuming a metric bound.

Theorem 3.2.9. *Let $\omega = \omega(t)$ solve (3.55) and assume that there exists a constant $C_0 > 0$ such that*

$$\frac{1}{C_0}\omega_0 \le \omega \le C_0\omega_0. \tag{3.98}$$

Then there exists a constant C depending only on C_0 and ω_0 such that

$$S := |\nabla_{g_0}g|^2 \le C. \tag{3.99}$$

In addition, there exists a constant C' depending only on C_0 and ω_0 such that

$$\left(\frac{\partial}{\partial t} - \Delta\right)S \le -\frac{1}{2}|\mathrm{Rm}|^2 + C', \tag{3.100}$$

where $|\mathrm{Rm}|^2$ denotes the norm squared of the curvature tensor $R_{i\bar{j}k\bar{\ell}}$.

Proof. We apply (3.88). First note that

$$\nabla^{\bar{b}}\hat{R}_{i\bar{b}p}{}^k = g^{\bar{b}r}\hat{\nabla}_r\hat{R}_{i\bar{b}p}{}^k - g^{\bar{b}r}\Psi^a_{ir}\hat{R}_{a\bar{b}p}{}^k - g^{\bar{b}r}\Psi^a_{pr}\hat{R}_{i\bar{b}a}{}^k + g^{\bar{b}r}\Psi^k_{ar}\hat{R}_{i\bar{b}p}{}^a. \tag{3.101}$$

Then with $\hat{g} = g_0$, we have, using (3.98),

$$\left| 2\mathrm{Re}\left(g^{\bar{j}i} g^{\bar{q}p} g_{k\bar{\ell}} \nabla^{\bar{b}} \hat{R}_{i\bar{b}p}{}^k \overline{\Psi^{\ell}_{jq}} \right) \right| \le C_1(S + \sqrt{S}) \le 2C_1(S + 1), \qquad (3.102)$$

for some uniform constant C_1. Hence for a uniform C_2,

$$\left(\frac{\partial}{\partial t} - \Delta \right) S \le -|\overline{\nabla}\Psi|^2 - |\nabla\Psi|^2 + C_2 S + C_2. \qquad (3.103)$$

On the other hand, from Proposition 3.2.4 and the assumption (3.98) again,

$$\left(\frac{\partial}{\partial t} - \Delta \right) \mathrm{tr}_{\hat{\omega}}\omega \le C_3 - \frac{1}{C_3}S, \qquad (3.104)$$

for a uniform $C_3 > 0$. Define $Q = S + C_3(1 + C_2)\mathrm{tr}_{\hat{\omega}}\omega$ and compute

$$\left(\frac{\partial}{\partial t} - \Delta \right) Q \le -S + C_4, \qquad (3.105)$$

for a uniform constant C_4. It follows that S is bounded from above at a point at which Q achieves a maximum, and (3.99) follows.

For (3.100), observe from (3.96) that

$$|\overline{\nabla}\Psi|^2 = |\hat{R}_{i\bar{b}p}{}^k - R_{i\bar{b}p}{}^k|^2 \ge \frac{1}{2}|\mathrm{Rm}|^2 - C_5. \qquad (3.106)$$

Then (3.100) follows from (3.103), (3.106) and (3.99). $\square$

3.2.6 Curvature and Higher Derivative Bounds

In this section we assume that we have a solution $\omega = \omega(t)$ of (3.55) on $[0, T)$ with $0 < T \le \infty$ which satisfies the estimates

$$\frac{1}{C_0}\omega_0 \le \omega \le C_0\omega_0, \qquad (3.107)$$

for some uniform constant C_0. We show that the curvature and all derivatives of the curvature of ω are uniformly bounded, and that we have uniform C^{∞} estimates of g with respect to the fixed metric ω_0. We first compute the evolution of the curvature tensor.

Lemma 3.2.10. *Along the flow (3.55), the curvature tensor evolves by*

$$
\frac{\partial}{\partial t} R_{i\bar{j}k\bar{\ell}} = \frac{1}{2} \Delta_{\mathbb{R}} R_{i\bar{j}k\bar{\ell}} - \nu R_{i\bar{j}k\bar{\ell}} + R_{i\bar{j}a\bar{b}} R^{\bar{b}a}{}_{k\bar{\ell}} + R_{i\bar{b}a\bar{\ell}} R^{\bar{b}}{}_{\bar{j}k}{}^{a} - R_{i\bar{a}k\bar{b}} R^{\bar{a}}{}_{\bar{j}}{}^{\bar{b}}{}_{\bar{\ell}}
$$
$$
- \frac{1}{2} \left(R_{i}{}^{a} R_{a\bar{j}k\bar{\ell}} + R^{\bar{a}}{}_{\bar{j}} R_{i\bar{a}k\bar{\ell}} + R_{k}{}^{a} R_{i\bar{j}a\bar{\ell}} + R^{\bar{a}}{}_{\bar{\ell}} R_{i\bar{j}k\bar{a}} \right) \tag{3.108}
$$

where we write $\Delta_{\mathbb{R}} = \Delta + \overline{\Delta}$ *and* $\Delta = g^{\bar{q}p} \nabla_p \nabla_{\bar{q}}$.

Proof. Using the formula $\frac{\partial}{\partial t} \Gamma^p_{ik} = -\nabla_i R_k{}^p$ and the Bianchi identity, compute

$$
\frac{\partial}{\partial t} R_{i\bar{j}k\bar{\ell}} = -\left(\frac{\partial}{\partial t} g_{p\bar{j}} \right) \partial_{\bar{\ell}} \Gamma^p_{ik} - g_{p\bar{j}} \partial_{\bar{\ell}} \left(\frac{\partial}{\partial t} \Gamma^p_{ik} \right) = -R^{\bar{a}}{}_{\bar{j}} R_{i\bar{a}k\bar{\ell}} - \nu R_{i\bar{j}k\bar{\ell}} + \nabla_{\bar{\ell}} \nabla_k R_{i\bar{j}}.
$$
$$
\tag{3.109}
$$

Using the Bianchi identity again and the commutation formulae, we obtain

$$
\Delta R_{i\bar{j}k\bar{\ell}} = g^{\bar{b}a} \nabla_a \nabla_{\bar{\ell}} R_{i\bar{j}k\bar{b}}
$$
$$
= g^{\bar{b}a} \nabla_{\bar{\ell}} \nabla_a R_{i\bar{j}k\bar{b}} + g^{\bar{b}a} [\nabla_a, \nabla_{\bar{\ell}}] R_{i\bar{j}k\bar{b}}
$$
$$
= \nabla_{\bar{\ell}} \nabla_k R_{i\bar{j}} - R^{\bar{b}}{}_{\bar{\ell}k}{}^{a} R_{a\bar{b}i\bar{j}} + R^{\bar{b}}{}_{\bar{\ell}} R_{k\bar{b}i\bar{j}} - R^{\bar{b}}{}_{\bar{\ell}i}{}^{a} R_{k\bar{b}a\bar{j}} + R^{\bar{b}}{}_{\bar{\ell}}{}^{\bar{a}}{}_{\bar{j}} R_{k\bar{b}i\bar{a}}.
$$
$$
\tag{3.110}
$$

And

$$
\overline{\Delta} R_{i\bar{j}k\bar{\ell}} = g^{\bar{b}a} \nabla_{\bar{b}} \nabla_k R_{i\bar{j}a\bar{\ell}}
$$
$$
= g^{\bar{b}a} \nabla_k \nabla_{\bar{b}} R_{i\bar{j}a\bar{\ell}} + g^{\bar{b}a} [\nabla_{\bar{b}}, \nabla_k] R_{i\bar{j}a\bar{\ell}}
$$
$$
= \nabla_{\bar{\ell}} \nabla_k R_{i\bar{j}} + [\nabla_k, \nabla_{\bar{\ell}}] R_{i\bar{j}} + g^{\bar{b}a} [\nabla_{\bar{b}}, \nabla_k] R_{i\bar{j}a\bar{\ell}}
$$
$$
= \nabla_{\bar{\ell}} \nabla_k R_{i\bar{j}} - R_{k\bar{\ell}i}{}^{a} R_{a\bar{j}} + R_{k\bar{\ell}}{}^{\bar{b}}{}_{\bar{j}} R_{i\bar{b}}
$$
$$
+ R_k{}^{a}{}_i{}^{b} R_{b\bar{j}a\bar{\ell}} - R_k{}^{a\bar{b}}{}_{\bar{j}} R_{i\bar{b}a\bar{\ell}} + R_k{}^{b} R_{i\bar{j}b\bar{\ell}} - R_k{}^{a\bar{b}}{}_{\bar{\ell}} R_{i\bar{j}a\bar{b}}. \tag{3.111}
$$

Combining (3.109), (3.110) and (3.111) gives (3.108). $\qquad\square$

In fact we do not need the precise formula (3.108) in what follows, but merely the fact that it has the general form

$$
\frac{\partial}{\partial t} \mathrm{Rm} = \frac{1}{2} \Delta_{\mathbb{R}} \mathrm{Rm} - \nu \mathrm{Rm} + \mathrm{Rm} * \mathrm{Rm} + \mathrm{Rc} * \mathrm{Rm}. \tag{3.112}
$$

To clarify notation: if A and B are tensors, we write $A * B$ for any linear combination of products of the tensors A and B formed by contractions on $A_{i_1 \cdots i_k}$ and $B_{j_1 \cdots j_\ell}$ using the metric g. We are writing Rc for the Ricci tensor.

Remark 3.2.11. A word about notation. The operator $\Delta_{\mathbb{R}}$ is the usual "rough" Laplace operator associated to the Riemannian metric $g_{\mathbb{R}}$ defined in (3.11). Hamilton defined his Ricci flow as $\frac{\partial}{\partial t} g_{ij} = -2R_{ij}$ precisely to remove the factor of $\frac{1}{2}$ appearing in evolution equations such as (3.112). In real coordinates, the Kähler–Ricci flow we consider in these notes is $\frac{\partial}{\partial t} g_{ij} = -R_{ij}$.

Lemma 3.2.12. *There exists a universal constant C such that*

$$\left(\frac{\partial}{\partial t} - \Delta\right) |\mathrm{Rm}|^2 \leq -|\nabla \mathrm{Rm}|^2 - |\overline{\nabla} \mathrm{Rm}|^2 + C|\mathrm{Rm}|^3 - \nu |\mathrm{Rm}|^2, \qquad (3.113)$$

and, for all points of $M \times [0, T)$ where $|\mathrm{Rm}|$ is not zero,

$$\left(\frac{\partial}{\partial t} - \Delta\right) |\mathrm{Rm}| \leq \frac{C}{2} |\mathrm{Rm}|^2 - \frac{\nu}{2} |\mathrm{Rm}|. \qquad (3.114)$$

Proof. The inequality (3.113) follows from (3.112). Next, note that

$$\left(\frac{\partial}{\partial t} - \Delta\right) |\mathrm{Rm}| = \frac{1}{2|\mathrm{Rm}|} \left(\frac{\partial}{\partial t} - \Delta\right) |\mathrm{Rm}|^2 + \frac{1}{4|\mathrm{Rm}|^3} g^{\overline{j}i} \nabla_i |\mathrm{Rm}|^2 \nabla_{\overline{j}} |\mathrm{Rm}|^2, \qquad (3.115)$$

and

$$g^{\overline{j}i} \nabla_i |\mathrm{Rm}|^2 \nabla_{\overline{j}} |\mathrm{Rm}|^2 \leq 2|\mathrm{Rm}|^2 (|\nabla \mathrm{Rm}|^2 + |\overline{\nabla} \mathrm{Rm}|^2). \qquad (3.116)$$

Then (3.114) follows from (3.113) and (3.116). $\qquad \square$

We combine this result with the third order estimate from Sect. 3.2.5 to obtain:

Theorem 3.2.13. *Let $\omega = \omega(t)$ solve (3.55) and assume that there exists a constant $C_0 > 0$ such that*

$$\frac{1}{C_0} \omega_0 \leq \omega \leq C_0 \omega_0. \qquad (3.117)$$

Then there exists a constant C depending only on C_0 and ω_0 such that

$$|\mathrm{Rm}|^2 \leq C. \qquad (3.118)$$

In addition, there exists a constant C' depending only on C_0 and ω_0 such that

$$\left(\frac{\partial}{\partial t} - \Delta\right) |\mathrm{Rm}|^2 \leq -|\nabla \mathrm{Rm}|^2 - |\overline{\nabla} \mathrm{Rm}|^2 + C', \qquad (3.119)$$

Proof. From Theorem 3.2.9, the quantity $S = |\nabla_{g_0} g|^2$ is uniformly bounded from above. We compute the evolution of $Q = |\mathrm{Rm}| + AS$ for a constant A. From (3.100) and (3.114), if A is chosen to be sufficiently large, we obtain

$$\left(\frac{\partial}{\partial t} - \Delta\right) Q \leq -|\mathrm{Rm}|^2 + C', \tag{3.120}$$

for a uniform constant C'. Then the upper bound of $|\mathrm{Rm}|^2$ follows from the maximum principle. Finally, (3.119) follows from (3.113).   □

Moreover, once we have bounded curvature, it is a result of Hamilton [Ham82] that bounds on all derivatives of curvature follow. For convenience we change to a real coordinate system. Writing $\nabla_{\mathbb{R}}$ for the covariant derivative with respect to g as a Riemannian metric, we have:

Theorem 3.2.14. *Let $\omega = \omega(t)$ solve (3.55) on $[0, T)$ with $0 < T \leq \infty$ and assume that there exists a constant $C > 0$ such that*

$$|\mathrm{Rm}|^2 \leq C. \tag{3.121}$$

Then there exist uniform constants C_m for $m = 1, 2, \ldots$ such that

$$|\nabla_{\mathbb{R}}^m \mathrm{Rm}|^2 \leq C_m. \tag{3.122}$$

Proof. We give a sketch of the proof and leave the details as an exercise to the reader. We use a maximum principle argument due to Shi [Shi89] (see [CLN06] for a good exposition). In fact we do not need the full force of Shi's results, which are local, since we are assuming a global curvature bound.

From Lemma 3.2.10 and an induction argument (see Theorem 13.2 of [Ham82])

$$\left(\frac{\partial}{\partial t} - \frac{1}{2}\Delta_{\mathbb{R}}\right) \nabla_{\mathbb{R}}^m \mathrm{Rm} = \sum_{p+q=m} \nabla_{\mathbb{R}}^p \mathrm{Rm} * \nabla_{\mathbb{R}}^q \mathrm{Rm}. \tag{3.123}$$

It follows that

$$\left(\frac{\partial}{\partial t} - \frac{1}{2}\Delta_{\mathbb{R}}\right) |\nabla_{\mathbb{R}}^m \mathrm{Rm}|^2 = -|\nabla_{\mathbb{R}}^{m+1} \mathrm{Rm}|^2 + \sum_{p+q=m} \nabla_{\mathbb{R}}^p \mathrm{Rm} * \nabla_{\mathbb{R}}^q \mathrm{Rm} * \nabla_{\mathbb{R}}^m \mathrm{Rm}. \tag{3.124}$$

Moreover, since $|\mathrm{Rm}|^2$ is bounded we have from Lemma 3.2.12 that

$$\left(\frac{\partial}{\partial t} - \frac{1}{2}\Delta_{\mathbb{R}}\right) |\mathrm{Rm}|^2 \leq -|\nabla_{\mathbb{R}} \mathrm{Rm}|^2 + C', \tag{3.125}$$

for some uniform constant C'. For the case $m = 1$, if we set $Q = |\nabla_{\mathbb{R}} \mathrm{Rm}|^2 + A|\mathrm{Rm}|^2$ for $A > 0$ sufficiently large then from (3.123),

$$\left(\frac{\partial}{\partial t} - \frac{1}{2}\Delta_{\mathbb{R}}\right) Q \leq -|\nabla_{\mathbb{R}}\text{Rm}|^2 + C'', \tag{3.126}$$

and it follows from the maximum principle that $|\nabla_{\mathbb{R}}\text{Rm}|^2$ is uniformly bounded from above. In addition,

$$\left(\frac{\partial}{\partial t} - \frac{1}{2}\Delta_{\mathbb{R}}\right) |\nabla_{\mathbb{R}}\text{Rm}|^2 \leq -|\nabla_{\mathbb{R}}^2\text{Rm}|^2 + C''', \tag{3.127}$$

and an induction completes the proof.   □

Next, we show that once we have a uniform bound on a metric evolving by the Kähler–Ricci flow, together with bounds on derivatives of curvature, then we have C^∞ bounds for the metric. Moreover, this result is local:

Theorem 3.2.15. *Let $\omega = \omega(t)$ solve (3.55) on $U \times [0, T)$ with $0 \leq T \leq \infty$, where U is an open subset of M. Assume that there exist constants C_m for $m = 0, 1, 2 \ldots$ such that*

$$\frac{1}{C_0}\omega_0 \leq \omega \leq C_0\omega_0, \quad S \leq C_0 \quad \text{and} \quad |\nabla_{\mathbb{R}}^m\text{Rm}|^2 \leq C_m. \tag{3.128}$$

Then for any compact subset $K \subset U$ and for $m = 1, 2, \ldots$, there exist constants C_m' depending only on ω_0, K, U and C_m such that

$$\|\omega(t)\|_{C^m(K,g_0)} \leq C_m'. \tag{3.129}$$

Proof. This is a well-known result. See [ChowKnopf], for example, or the discussion in [PSSW11]. We give just a sketch of the proof following quite closely the arguments in [ShW11, SW10]. It suffices to prove the result on the ball B say, in a fixed holomorphic coordinate chart. We will obtain the C^∞ estimates for $\omega(t)$ on a slightly smaller ball. Fix a time $t \in (0, T]$. Consider the equations

$$\Delta_{\text{E}}g_{i\bar{j}} = -\sum_k R_{k\bar{k}i\bar{j}} + \sum_{k,p,q} g^{q\bar{p}}\partial_k g_{i\bar{q}}\partial_{\bar{k}}g_{p\bar{j}} =: Q_{i\bar{j}}. \tag{3.130}$$

where $\Delta_{\text{E}} = \sum_k \partial_k\partial_{\bar{k}}$. For each fixed i, j, we can regard (3.130) as Poisson's equation $\Delta_{\text{E}}g_{i\bar{j}} = Q_{i\bar{j}}$.

Fix $p > 2n$. From our assumptions, each $\|Q_{i\bar{j}}\|_{L^p(B)}$ is uniformly bounded. Applying the standard elliptic estimates (see Theorem 9.11 of [GT01] for example) to (3.130) we see that the Sobolev norm $\|g_{i\bar{j}}\|_{L_2^p}$ is uniformly bounded on a slightly smaller ball. From now on, the estimates that we state will always be modulo shrinking the ball slightly. Morrey's embedding theorem (Theorem 7.17 of [GT01]) gives that $\|g_{i\bar{j}}\|_{C^{1+\beta}}$ is uniformly bounded for some $0 < \beta < 1$.

The key observation we now need is as follows: the mth derivative of $Q_{i\bar{j}}$ can be written in the form $A * B$ where each A or B represents either a covariant derivative of Rm or a quantity involving derivatives of g up to order at most $m + 1$. Hence if g is uniformly bounded in $C^{m+1+\beta}$ then each $Q_{i\bar{j}}$ is uniformly bounded in $C^{m+\beta}$.

Applying this observation with $m = 0$ we see that each $\|Q_{i\bar{j}}\|_{C^\beta}$ is uniformly bounded. The standard Schauder estimates (see Theorem 4.8 of [GT01]) give that $\|g_{i\bar{j}}\|_{C^{2+\beta}}$ is uniformly bounded.

We can now apply a bootstrapping argument. Applying the observation with $m = 1$ we see that $Q_{i\bar{j}}$ is uniformly bounded in $C^{1+\beta}$, and so on. This completes the proof. $\qquad\square$

Combining Theorems 3.2.13–3.2.15, we obtain:

Corollary 3.2.16. *Let* $\omega = \omega(t)$ *solve (3.55) on* $M \times [0, T)$ *with* $0 \le T \le \infty$. *Assume that there exists a constant* C_0 *such that*

$$\frac{1}{C_0}\omega_0 \le \omega \le C_0\omega_0. \tag{3.131}$$

Then for $m = 1, 2, \ldots,$ *there exist uniform constants* C_m *such that*

$$\|\omega(t)\|_{C^m(g_0)} \le C_m. \tag{3.132}$$

In fact, there is a local version of Corollary 3.2.16. Although we will not actual make use of it in these lecture notes, we state here the result:

Theorem 3.2.17. *Let* $\omega = \omega(t)$ *solve (3.55) on* $U \times [0, T)$ *with* $0 \le T \le \infty$, *where* U *is an open subset of* M. *Assume that there exists a constant* C_0 *for such that*

$$\frac{1}{C_0}\omega_0 \le \omega \le C_0\omega_0. \tag{3.133}$$

Then for any compact subset $K \subset U$ *and for* $m = 1, 2, \ldots,$ *there exist constants* C_m' *depending only on* ω_0, K *and* U *such that*

$$\|\omega(t)\|_{C^m(K, g_0)} \le C_m'. \tag{3.134}$$

Proof. This can either be proved using the Schauder estimates of Evans–Krylov [Eva82, Kryl82] (see also [CLN06, Gill11]) or using local maximum principle arguments [ShW11]. We omit the proof. $\qquad\square$

3.3 Maximal Existence Time for the Kähler–Ricci Flow

In this section we identify the maximal existence time for a smooth solution of the Kähler–Ricci flow. To do this, we rewrite the Kähler–Ricci flow as a parabolic complex Monge–Ampère equation.

3.3.1 The Parabolic Monge–Ampère Equation

Let $\omega = \omega(t)$ be a solution of the Kähler–Ricci flow

$$\frac{\partial}{\partial t}\omega = -\mathrm{Ric}(\omega), \qquad \omega|_{t=0} = \omega_0. \tag{3.135}$$

As long as the solution exists, the cohomology class $[\omega(t)]$ evolves by

$$\frac{d}{dt}[\omega(t)] = -c_1(M), \qquad [\omega(0)] = [\omega_0], \tag{3.136}$$

and solving this ordinary differential equation gives $[\omega(t)] = [\omega_0] - tc_1(M)$. Immediately we see that a necessary condition for the Kähler–Ricci flow to exist for $t \in [0, t')$ is that $[\omega_0] - tc_1(M) > 0$ for $t \in [0, t')$. This necessary condition is in fact sufficient. If we define

$$T = \sup\{t > 0 \mid [\omega_0] - tc_1(M) > 0\}, \tag{3.137}$$

then we have:

Theorem 3.3.1. *There exists a unique maximal solution $g(t)$ of the Kähler–Ricci flow (3.135) for $t \in [0, T)$.*

This theorem was proved by Cao [Cao85] in the special case when $c_1(M)$ is zero or definite. In this generality, the result is due to Tian–Zhang [Tzha06]. Weaker versions appeared earlier in the work of Tsuji (see [Tsu88] and Theorem 8 of [Tsu96]).

We now begin the proof of Theorem 3.3.1. Fix $T' < T$. We will show that there exists a solution to (3.135) on $[0, T')$. First we observe that (3.135) can be rewritten as a parabolic complex Monge–Ampère equation.

To do this, we need to choose reference metrics $\hat{\omega}_t$ in the cohomology classes $[\omega_0] - tc_1(M)$. Since $[\omega_0] - T'c_1(M)$ is a Kähler class, there exists a Kähler form η in $[\omega_0] - T'c_1(M)$. We choose our family of reference metrics $\hat{\omega}_t$ to be the linear path of metrics between ω_0 and η. Namely, define

$$\chi = \frac{1}{T'}(\eta - \omega_0) \in -c_1(M), \tag{3.138}$$

and

$$\hat{\omega}_t = \omega_0 + t\chi = \frac{1}{T'}((T' - t)\omega_0 + t\eta) \in [\omega_0] - tc_1(M). \tag{3.139}$$

Fix a volume form Ω on M with

$$\frac{\sqrt{-1}}{2\pi}\partial\bar\partial \log \Omega = \chi = \frac{\partial}{\partial t}\hat\omega_t \in -c_1(M), \tag{3.140}$$

which exists by the discussion in Sect. 3.1.6. Notice that here we are abusing notation somewhat by writing $\frac{\sqrt{-1}}{2\pi}\partial\bar\partial \log \Omega$. To clarify, we mean that if the volume form Ω is written in local coordinates z^i as

$$\Omega = a(z^1,\dots,z^n)(\sqrt{-1})^n dz^1 \wedge d\overline{z^1} \wedge \cdots \wedge dz^n \wedge d\overline{z^n},$$

for a locally defined smooth positive function a then we define $\frac{\sqrt{-1}}{2\pi}\partial\bar\partial \log \Omega = \frac{\sqrt{-1}}{2\pi}\partial\bar\partial \log a$. Although the function a depends on the choice of holomorphic coordinates, the $(1,1)$-form $\frac{\sqrt{-1}}{2\pi}\partial\bar\partial \log a$ does not, as the reader can easily verify.

We now consider the *parabolic complex Monge–Ampère equation*, for $\varphi = \varphi(t)$ a real-valued function on M,

$$\frac{\partial}{\partial t}\varphi = \log \frac{(\hat\omega_t + \frac{\sqrt{-1}}{2\pi}\partial\bar\partial\varphi)^n}{\Omega}, \qquad \hat\omega_t + \frac{\sqrt{-1}}{2\pi}\partial\bar\partial\varphi > 0, \qquad \varphi|_{t=0} = 0. \tag{3.141}$$

This equation is equivalent to the Kähler–Ricci flow (3.135). Indeed, given a smooth solution φ of (3.141) on $[0, T')$, we can obtain a solution $\omega = \omega(t)$ of (3.135) on $[0, T')$ as follows. Define $\omega(t) = \hat\omega_t + \frac{\sqrt{-1}}{2\pi}\partial\bar\partial\varphi$ and observe that $\omega(0) = \hat\omega_0 = \omega_0$ and

$$\frac{\partial}{\partial t}\omega = \frac{\partial}{\partial t}\hat\omega_t + \frac{\sqrt{-1}}{2\pi}\partial\bar\partial\left(\frac{\partial}{\partial t}\varphi\right) = -\mathrm{Ric}(\omega), \tag{3.142}$$

as required. Conversely, suppose that $\omega = \omega(t)$ solves (3.135) on $[0, T')$. Then since $\hat\omega_t \in [\omega(t)]$, we can apply the $\partial\bar\partial$-Lemma to find a family of potential functions $\tilde\varphi(t)$ such that $\omega(t) = \hat\omega_t + \frac{\sqrt{-1}}{2\pi}\partial\bar\partial\tilde\varphi(t)$ and $\int_M \tilde\varphi(t)\omega_0^n = 0$. By standard elliptic regularity theory the family $\tilde\varphi(t)$ is smooth on $M \times [0, T')$. Then

$$\frac{\sqrt{-1}}{2\pi}\partial\bar\partial \log \omega^n = \frac{\partial}{\partial t}\omega = \frac{\sqrt{-1}}{2\pi}\partial\bar\partial \log \Omega + \frac{\sqrt{-1}}{2\pi}\partial\bar\partial\left(\frac{\partial}{\partial t}\tilde\varphi\right), \tag{3.143}$$

and since the only pluriharmonic functions on M are the constants, we see that

$$\frac{\partial}{\partial t}\tilde\varphi = \log \frac{\omega^n}{\Omega} + c(t),$$

for some smooth function $c : [0, T') \to \mathbb{R}$. Now set $\varphi(t) = \tilde\varphi(t) - \int_0^t c(s)ds - \tilde\varphi(0)$, noting that since $\omega(0) = \omega_0$ the function $\tilde\varphi(0)$ is constant. It follows that $\varphi = \varphi(t)$ solves the parabolic complex Monge–Ampère equation (3.141).

To prove Theorem 3.3.1 then, it suffices to study (3.141). Since the linearization of the right hand side of (3.141) is the Laplace operator $\Delta_{g(t)}$, which is elliptic, it follows that (3.141) is a strictly parabolic (nonlinear) partial differential equation for φ. The standard parabolic theory [Lieb96] gives a unique maximal solution of (3.141) for some time interval $[0, T_{max})$ with $0 < T_{max} \leq \infty$. We may assume without loss of generality that $T_{max} < T'$. We will then obtain a contradiction by showing that a solution of (3.141) exists beyond T_{max}. This will be done in the next two subsections.

3.3.2 Estimates for the Potential and the Volume Form

We assume now that we have a solution $\varphi = \varphi(t)$ to the parabolic complex Monge–Ampère equation (3.141) on $[0, T_{max})$, for $0 < T_{max} < T' < T$. Our goal is to establish uniform estimates for φ on $[0, T_{max})$. In this subsection we will prove a C^0 estimate for φ and a lower bound for the volume form.

Note that $\hat{\omega}_t$ is a family of smooth Kähler forms on the closed interval $[0, T_{max}]$. Hence by compactness we have uniform bounds on $\hat{\omega}_t$ from above and below (away from zero).

Lemma 3.3.2. *There exists a uniform C such that for all $t \in [0, T_{max})$,*

$$\|\varphi(t)\|_{C^0(M)} \leq C. \tag{3.144}$$

Proof. For the upper bound of φ, we will apply the maximum principle to $\theta := \varphi - At$ for $A > 0$ a uniform constant to be determined later. From (3.141) we have

$$\frac{\partial}{\partial t}\theta = \log \frac{(\hat{\omega}_t + \frac{\sqrt{-1}}{2\pi}\partial\bar{\partial}\theta)^n}{\Omega} - A. \tag{3.145}$$

Fix $t' \in (0, T_{max})$. Since $M \times [0, t']$ is compact, θ attains a maximum at some point $(x_0, t_0) \in M \times [0, t']$. We claim that if A is sufficiently large we have $t_0 = 0$.

Otherwise $t_0 > 0$. Then by Proposition 3.1.6, at (x_0, t_0),

$$0 \leq \frac{\partial}{\partial t}\theta = \log \frac{(\hat{\omega}_{t_0} + \frac{\sqrt{-1}}{2\pi}\partial\bar{\partial}\theta)^n}{\Omega} - A \leq \log \frac{\hat{\omega}_{t_0}^n}{\Omega} - A \leq -1, \tag{3.146}$$

a contradiction, where we have chosen $A \geq 1 + \sup_{M \times [0, T_{max}]} \log(\hat{\omega}_t^n / \Omega)$. Hence we have proved the claim that $t_0 = 0$, giving $\sup_{M \times [0, t']} \theta \leq \sup_M \theta|_{t=0} = 0$ and thus

$$\varphi(x, t) \leq At \leq AT_{max}, \qquad \text{for } (x, t) \in M \times [0, t']. \tag{3.147}$$

Since $t' \in (0, T_{\max})$ was arbitrary, this gives a uniform upper bound for φ on $[0, T_{\max})$.

We apply a similar argument to $\psi = \varphi + Bt$ for B a positive constant with $B \geq 1 - \inf_{M \times [0, T_{\max}]} \log(\hat{\omega}_t^n / \Omega)$ and obtain

$$\varphi(x, t) \geq -B T_{\max}, \qquad \text{for } (x, t) \in M \times [0, t'], \tag{3.148}$$

giving the lower bound. $\qquad\qquad\square$

Next we prove a lower bound for the volume form along the flow, or equivalently a lower bound for $\dot{\varphi} = \partial \varphi / \partial t$. This argument is due to Tian–Zhang [Tzha06].

Lemma 3.3.3. *There exists a uniform $C > 0$ such that on $M \times [0, T_{\max})$,*

$$\frac{1}{C} \Omega \leq \omega^n(t) \leq C\Omega, \tag{3.149}$$

or equivalently, $\|\dot{\varphi}\|_{C^0}$ is uniformly bounded.

Proof. The upper bound of ω^n follows from part (i) of Corollary 3.2.3. Note that since this is equivalent to an upper bound of $\dot{\varphi}$, we have given an alternative proof of the upper bound part of Lemma 3.3.2.

For the lower bound of ω^n, differentiate (3.141):

$$\frac{\partial}{\partial t} \dot{\varphi} = \Delta \dot{\varphi} + \text{tr}_\omega \chi, \tag{3.150}$$

where we recall that $\chi = \partial \hat{\omega}_t / \partial t$ is defined by (3.138). Define a quantity $Q = (T' - t)\dot{\varphi} + \varphi + nt$ and compute using (3.150),

$$\left(\frac{\partial}{\partial t} - \Delta \right) Q = (T' - t)\text{tr}_\omega \chi + n - \Delta\varphi = \text{tr}_\omega(\hat{\omega}_t + (T' - t)\chi) = \text{tr}_\omega \hat{\omega}_{T'} > 0, \tag{3.151}$$

where we have used the fact that

$$\Delta\varphi = \text{tr}_\omega(\omega - \hat{\omega}_t) = n - \text{tr}_\omega \hat{\omega}_t. \tag{3.152}$$

Then by the maximum principle (Proposition 3.1.7), Q is uniformly bounded from below on $M \times [0, T_{\max})$ by its infimum at the initial time. Thus

$$(T' - t)\dot{\varphi} + \varphi + nt \geq T' \inf_M \log \frac{\omega_0^n}{\Omega}, \qquad \text{on } M \times [0, T_{\max}), \tag{3.153}$$

and since φ is uniformly bounded from Lemma 3.3.2 and $T' - t \geq T' - T_{\max} > 0$, this gives the desired lower bound of $\dot{\varphi}$. $\qquad\square$

3.3.3 A Uniform Bound for the Evolving Metric

Again we assume that we have a solution $\varphi = \varphi(t)$ to (3.141) on $[0, T_{\max})$, for $0 < T_{\max} < T' < T$. From Lemma 3.3.2, we have a uniform bound for $\|\varphi\|_{C^0(M)}$ and we will use this together with Proposition 3.2.5 to obtain an upper bound for the quantity $\operatorname{tr}_{\omega_0} \omega$ on $[0, T_{\max})$. This argument is similar to those in [Aub78, Yau78] (see also [Cao85] and Lemmas 3.4.3 and 3.4.8 below). We will then complete the proof of Theorem 3.3.1.

Lemma 3.3.4. *There exists a uniform C such that on $M \times [0, T_{\max})$,*

$$\operatorname{tr}_{\omega_0} \omega \leq C. \tag{3.154}$$

Proof. We consider the quantity

$$Q = \log \operatorname{tr}_{\omega_0} \omega - A\varphi, \tag{3.155}$$

for $A > 0$ a uniform constant to be determined later. For a fixed $t' \in (0, T_{\max})$, assume that Q on $M \times [0, t']$ attains a maximum at a point (x_0, t_0). Without loss of generality, we may suppose that $t_0 > 0$. Then at (x_0, t_0), applying Proposition 3.2.5 with $\hat{\omega} = \omega_0$,

$$0 \leq \left(\frac{\partial}{\partial t} - \Delta \right) Q \leq C_0 \operatorname{tr}_{\omega} \omega_0 - A\dot{\varphi} + A\Delta\varphi$$

$$= \operatorname{tr}_{\omega}(C_0\omega_0 - A\hat{\omega}_{t_0}) - A \log \frac{\omega^n}{\Omega} + An, \tag{3.156}$$

for C_0 depending only on the lower bound of the bisectional curvature of g_0. Choose A sufficiently large so that $A\hat{\omega}_{t_0} - (C_0 + 1)\omega_0$ is Kähler on M. Then

$$\operatorname{tr}_{\omega}(C_0\omega_0 - A\hat{\omega}_{t_0}) \leq -\operatorname{tr}_{\omega}\omega_0, \tag{3.157}$$

and so at (x_0, t_0),

$$\operatorname{tr}_{\omega}\omega_0 + A \log \frac{\omega^n}{\Omega} \leq An, \tag{3.158}$$

and hence

$$\operatorname{tr}_{\omega}\omega_0 + A \log \frac{\omega^n}{\omega_0^n} \leq C, \tag{3.159}$$

for some uniform constant C. At (x_0, t_0), choose coordinates so that

$$(g_0)_{i\bar{j}} = \delta_{ij} \quad \text{and} \quad g_{i\bar{j}} = \lambda_i \delta_{ij}, \quad \text{for } i, j = 1, \ldots, n, \tag{3.160}$$

for positive $\lambda_1, \ldots, \lambda_n$. Then (3.159) is precisely

$$\sum_{i=1}^{n} \left(\frac{1}{\lambda_i} + A \log \lambda_i \right) \le C. \tag{3.161}$$

Since the function $x \mapsto \frac{1}{x} + A \log x$ for $x > 0$ is uniformly bounded from below, we have (for a different C),

$$\left(\frac{1}{\lambda_i} + A \log \lambda_i \right) \le C, \quad \text{for } i = 1, \ldots, n. \tag{3.162}$$

Then $A \log \lambda_i \le C$, giving a uniform upper bound for λ_i and hence $(\mathrm{tr}_{\omega_0} \omega)(x_0, t_0)$. Since φ is uniformly bounded on $M \times [0, T_{\max})$ we see that $Q(x_0, t_0)$ is uniformly bounded from above. Hence Q is bounded from above on $M \times [0, t']$ for any $t' < T_{\max}$. Using again that φ is uniformly bounded we obtain the required estimate (3.154). $\qquad\square$

Note that we did not make use of the bound on $\dot{\varphi}$ in the above argument. By doing so we could have simplified the proof slightly. However, it turns out that the argument of Lemma 3.3.4 will be useful later (see Lemma 3.5.5 and Sect. 3.7 below) where we do not have a uniform lower bound of $\dot{\varphi}$.

As a consequence of Lemma 3.3.4, we have:

Corollary 3.3.5. *There exists a uniform $C > 0$ such that on $M \times [0, T_{\max})$,*

$$\frac{1}{C} \omega_0 \le \omega \le C \omega_0. \tag{3.163}$$

Proof. The upper bound follows from Lemma 3.3.4. For the lower bound,

$$\mathrm{tr}_{\omega} \omega_0 \le \frac{1}{(n-1)!} (\mathrm{tr}_{\omega_0} \omega)^{n-1} \frac{\omega_0^n}{\omega^n} \le C, \tag{3.164}$$

using Lemma 3.3.3. To verify the first inequality of (3.164), choose coordinates as in (3.160) and observe that

$$\frac{1}{\lambda_1} + \cdots + \frac{1}{\lambda_n} \le \frac{1}{(n-1)!} \frac{(\lambda_1 + \cdots + \lambda_n)^{n-1}}{\lambda_1 \cdots \lambda_n}, \tag{3.165}$$

for positive λ_i. $\qquad\square$

We can now finish the proof of Theorem 3.3.1.

Proof of Theorem 3.3.1. Combining Corollary 3.3.5 with Corollary 3.2.16, we obtain uniform C^∞ estimates for $g(t)$ on $[0, T_{\max})$. Hence as $t \to T_{\max}$, the metrics $g(t)$ converge in C^∞ to a smooth Kähler metric $g(T_{\max})$ and thus we obtain a smooth solution to the Kähler–Ricci flow on $[0, T_{max}]$. But we have already seen from Theorem 3.2.1 (or by the discussion at the end of Sect. 3.3.1) that we can always find a smooth solution of the Kähler–Ricci flow on some, possibly short, time interval with any initial Kähler metric. Applying this to $g(T_{max})$, we obtain a solution of the Kähler–Ricci flow $g(t)$ on $[0, T_{max} + \varepsilon)$ for $\varepsilon > 0$. But this contradicts the definition of T_{max}, and completes the proof of Theorem 3.3.1. $\square$

3.4 Convergence of the Flow

In this section we show that the Kähler–Ricci flow converges, after appropriate normalization, to a Kähler–Einstein metric in the cases $c_1(M) < 0$ and $c_1(M) = 0$. This was originally proved by Cao [Cao85] and makes use of parabolic versions of estimates due to Yau and Aubin [Aub78, Yau78] and also Yau's well-known C^0 estimate for the complex Monge–Ampère equation [Yau78].

3.4.1 The Normalized Kähler–Ricci Flow When $c_1(M) < 0$

We first consider the case of a manifold M with $c_1(M) < 0$. We restrict to the case when $[\omega_0] = -c_1(M)$. By Theorem 3.3.1 we have a solution to the Kähler–Ricci flow (3.135) for $t \in [0, \infty)$. The Kähler class $[\omega(t)]$ is given by $(1 + t)[\omega_0]$ which diverges as $t \to \infty$. To avoid this we consider instead the *normalized Kähler–Ricci flow*

$$\frac{\partial}{\partial t}\omega = -\mathrm{Ric}(\omega) - \omega, \qquad \omega|_{t=0} = \omega_0. \tag{3.166}$$

This is just a rescaling of (3.135) and we have a solution $\omega(t)$ to (3.166) for all time. Indeed if $\tilde{\omega}(s)$ solves $\frac{\partial}{\partial s}\tilde{\omega}(s) = -\mathrm{Ric}(\tilde{\omega}(s))$ for $s \in [0, \infty)$ then $\omega(t) = \tilde{\omega}(s)/(s + 1)$ with $t = \log(s + 1)$ solves (3.166). Conversely, given a solution to (3.166) we can rescale to obtain a solution to (3.135).

Since we have chosen $[\omega_0] = -c_1(M)$, we immediately see that $[\omega(t)] = [\omega_0]$ for all t. The following result is due to Cao [Cao85].

Theorem 3.4.1. *The solution* $\omega = \omega(t)$ *to (3.166) converges in* C^∞ *to the unique Kähler–Einstein metric* $\omega_{KE} \in -c_1(M)$.

We recall that a *Kähler–Einstein metric* is a Kähler metric ω_{KE} with $\mathrm{Ric}(\omega_{KE}) = \mu\omega_{KE}$ for some constant $\mu \in \mathbb{R}$. If $\omega_{KE} \in -c_1(M)$ then we necessarily have $\mu = -1$. The existence of a Kähler–Einstein metric on M with $c_1(M) < 0$ is due to Yau [Yau78] and Aubin [Aub78] independently.

The uniqueness of $\omega_{KE} \in -c_1(M)$ is due to Calabi [Cal57] and follows from the maximum principle. Indeed, suppose $\omega'_{KE}, \omega_{KE} \in -c_1(M)$ are both

Kähler–Einstein. Writing $\omega'_{KE} = \omega_{KE} + \frac{\sqrt{-1}}{2\pi}\partial\bar{\partial}\varphi$, we have $\mathrm{Ric}(\omega'_{KE}) = -\omega'_{KE} = \mathrm{Ric}(\omega_{KE}) - \frac{\sqrt{-1}}{2\pi}\partial\bar{\partial}\varphi$ and hence

$$\log\frac{(\omega_{KE} + \frac{\sqrt{-1}}{2\pi}\partial\bar{\partial}\varphi)^n}{\omega_{KE}^n} = \varphi + C, \tag{3.167}$$

for some constant C. By considering the maximum and minimum values of $\varphi + C$ on M we see that $\varphi + C = 0$ and hence $\omega_{KE} = \omega'_{KE}$.

To prove Theorem 3.4.1, we reduce (3.166) to a parabolic complex Monge–Ampère equation as in the previous section. Let Ω be a volume form on M satisfying

$$\frac{\sqrt{-1}}{2\pi}\partial\bar{\partial}\log\Omega = \omega_0 \in -c_1(M), \qquad \int_M \Omega = \int_M \omega_0^n. \tag{3.168}$$

Then we consider the *normalized parabolic complex Monge–Ampère equation*,

$$\frac{\partial}{\partial t}\varphi = \log\frac{(\omega_0 + \frac{\sqrt{-1}}{2\pi}\partial\bar{\partial}\varphi)^n}{\Omega} - \varphi, \qquad \omega_0 + \frac{\sqrt{-1}}{2\pi}\partial\bar{\partial}\varphi > 0, \qquad \varphi|_{t=0} = 0. \tag{3.169}$$

Given a solution $\varphi = \varphi(t)$ of (3.169), the metrics $\omega = \omega_0 + \frac{\sqrt{-1}}{2\pi}\partial\bar{\partial}\varphi$ solve (3.166). Conversely, as in Sect. 3.3.1, given a solution $\omega = \omega(t)$ of (3.166) we can obtain via the $\partial\bar{\partial}$-Lemma a solution $\varphi = \varphi(t)$ of (3.169).

We wish to obtain estimates for φ solving (3.169). First:

Lemma 3.4.2. *We have*

(i) *There exists a uniform constant C such that for t in $[0, \infty)$,*

$$\|\dot{\varphi}(t)\|_{C^0(M)} \le Ce^{-t}. \tag{3.170}$$

(ii) *There exists a continuous real-valued function φ_∞ on M such that for t in $[0, \infty)$,*

$$\|\varphi(t) - \varphi_\infty\|_{C^0(M)} \le Ce^{-t} \tag{3.171}$$

(iii) $\|\varphi(t)\|_{C^0(M)}$ *is uniformly bounded for $t \in [0, \infty)$.*

(iv) *There exists a uniform constant C' such that on $M \times [0, \infty)$, the volume form of $\omega = \omega(t)$ satisfies*

$$\frac{1}{C'}\omega_0^n \le \omega^n \le C'\omega_0^n. \tag{3.172}$$

Proof. Compute

$$\frac{\partial}{\partial t}\dot{\varphi} = \Delta\dot{\varphi} - \dot{\varphi}, \tag{3.173}$$

and hence

$$\frac{\partial}{\partial t}(e^t \dot\varphi) = \Delta(e^t \dot\varphi). \tag{3.174}$$

Then (i) follows from the maximum principle (Proposition 3.1.7). For (ii), let $s, t \geq 0$ and x be in M. Then

$$|\varphi(x, s) - \varphi(x, t)| = \left| \int_t^s \dot\varphi(x, u) du \right| \leq \int_t^s |\dot\varphi(x, u)| du$$

$$\leq \int_t^s C e^{-u} du = C(e^{-t} - e^{-s}), \tag{3.175}$$

which shows that $\varphi(t)$ converges uniformly to some continuous function φ_∞ on M. Taking the limit in (3.175) as $s \to \infty$ gives (ii). (iii) follows immediately from (ii). (iv) follows from (3.169) together with (i) and (iii). □

We use the C^0 bound on φ to obtain an upper bound on the evolving metric.

Lemma 3.4.3. *There exists a uniform constant C such that on $M \times [0, \infty)$, $\omega = \omega(t)$ satisfies*

$$\frac{1}{C}\omega_0 \leq \omega \leq C\omega_0. \tag{3.176}$$

Proof. By part (iv) of Lemma 3.4.2 and the argument of Corollary 3.3.5, it suffices to obtain a uniform upper bound for $\mathrm{tr}\,_{\omega_0}\omega$.

Applying Proposition 3.2.5,

$$\left(\frac{\partial}{\partial t} - \Delta\right) \log \mathrm{tr}\,_{\omega_0}\omega \leq C_0 \mathrm{tr}\,_\omega \omega_0 - 1, \tag{3.177}$$

for C_0 depending only on g_0. We apply the maximum principle to the quantity $Q = \log \mathrm{tr}\,_{\omega_0}\omega - A\varphi$ as in the proof of Lemma 3.3.4, where A is to be chosen later. We have

$$\left(\frac{\partial}{\partial t} - \Delta\right) Q \leq C_0 \mathrm{tr}\,_\omega \omega_0 - 1 - A\dot\varphi + An - A\mathrm{tr}\,_\omega \omega_0. \tag{3.178}$$

Assume that Q achieves a maximum at a point (x_0, t_0) with $t_0 > 0$. Choosing $A = C_0 + 1$ and using the fact that $\dot\varphi$ is uniformly bounded, we see that $\mathrm{tr}\,_\omega \omega_0$ is uniformly bounded at (x_0, t_0). Arguing as in (3.164), we have,

$$(\mathrm{tr}\,_{\omega_0}\omega)(x_0, t_0) \leq \frac{1}{(n-1)!} (\mathrm{tr}\,_\omega \omega_0)^{n-1} (x_0, t_0)\frac{\omega^n}{\omega_0^n}(x_0, t_0) \leq C, \tag{3.179}$$

using part (iv) of Lemma 3.4.2. Since φ is uniformly bounded, this shows that Q is bounded from above at (x_0, t_0). Hence tr$_{\omega_0}\omega$ is uniformly bounded from above. □

We can now complete the proof of Theorem 3.4.1. By Corollary 3.2.16 we have uniform C^∞ estimates on $\omega(t)$. Since $\varphi(t)$ is bounded in C^0 it follows that we have uniform C^∞ estimates on $\varphi(t)$. Recall that $\varphi(t)$ converges uniformly to a continuous function φ_∞ on M as $t \to \infty$. By the Arzela–Ascoli Theorem and the uniqueness of limits, it follows immediately that there exist times $t_k \to \infty$ such that the sequence of functions $\varphi(t_k)$ converges in C^∞ to φ_∞, which is smooth. In fact we have this convergence without passing to a subsequence. Indeed, suppose not. Then there exists an integer k, an $\varepsilon > 0$ and a sequence of times $t_i \to \infty$ such that

$$\|\varphi(t_i) - \varphi_\infty\|_{C^k(M)} \geq \varepsilon, \quad \text{for all } i. \tag{3.180}$$

But since $\varphi(t_i)$ is a sequence of functions with uniform C^{k+1} bounds we apply the Arzela–Ascoli Theorem to obtain a subsequence $\varphi(t_{i_j})$ which converges in C^k to φ'_∞, say, with

$$\|\varphi'_\infty - \varphi_\infty\|_{C^k(M)} \geq \varepsilon, \tag{3.181}$$

so that $\varphi'_\infty \neq \varphi_\infty$. But $\varphi(t_{i_j})$ converges uniformly to φ_∞, a contradiction. Hence $\varphi(t)$ converges to φ_∞ in C^∞ as $t \to \infty$.

It remains to show that the limit metric $\omega_\infty = \omega_0 + \frac{\sqrt{-1}}{2\pi}\partial\bar{\partial}\varphi_\infty$ is Kähler–Einstein. Since from Lemma 3.4.2, $\dot{\varphi}(t) \to 0$ as $t \to \infty$, we can take a limit as $t \to \infty$ of (3.169) to obtain

$$\log \frac{\omega_\infty^n}{\Omega} - \varphi_\infty = 0, \tag{3.182}$$

and applying $\frac{\sqrt{-1}}{2\pi}\partial\bar{\partial}$ to both sides of this equation gives that Ric$(\omega_\infty) = -\omega_\infty$ as required. This completes the proof of Theorem 3.4.1.

3.4.2 The Case of $c_1(M) = 0$: Yau's Zeroth Order Estimate

In this section we discuss the case of the Kähler–Ricci flow on a Kähler manifold (M, g_0) with vanishing first Chern class. Unlike the case of $c_1(M) < 0$ dealt with above, there will be no restriction on the Kähler class $[\omega_0]$.

By Theorem 3.3.1, there is a solution $\omega(t)$ of the Kähler–Ricci flow (3.135) for $t \in [0, \infty)$ and we have $[\omega(t)] = [\omega_0]$. The following result is due to Cao [Cao85] and makes use of Yau's celebrated zeroth order estimate, which we will describe in this subsection.

Theorem 3.4.4. *The solution $\omega(t)$ to (3.135) converges in C^∞ to the unique Kähler–Einstein metric $\omega_{KE} \in [\omega_0]$.*

Since $c_1(M) = 0$, the Kähler–Einstein metric ω_{KE} must be Kähler–Ricci flat (if $\mathrm{Ric}(\omega_{KE}) = \mu\omega_{KE}$ then $c_1(M) = [\mu\omega_{KE}] = 0$ implies $\mu = 0$). Note that, as Theorem 3.4.4 implies, there is a unique Kähler–Einstein metric in *every* Kähler class on M.

The uniqueness part of the argument is due to Calabi [Cal57]. Suppose $\omega'_{KE} = \omega_{KE} + \frac{\sqrt{-1}}{2\pi}\partial\bar{\partial}\varphi$ is another Kähler–Einstein metric in the same cohomology class. Then the equation $\mathrm{Ric}(\omega'_{KE}) = \mathrm{Ric}(\omega_{KE})$ gives

$$\log \frac{\omega'^n_{KE}}{\omega^n_{KE}} = C, \tag{3.183}$$

for some constant C. Exponentiating and then integrating gives $C = 1$ and hence $\omega'^n_{KE} = \omega^n_{KE}$. Then compute, using integration by parts,

$$
\begin{aligned}
0 = \int_M \varphi(\omega^n_{KE} - \omega'^n_{KE}) &= -\int_M \varphi \frac{\sqrt{-1}}{2\pi}\partial\bar{\partial}\varphi \wedge \left(\sum_{i=0}^{n-1} \omega^i_{KE} \wedge \omega'^{n-1-i}_{KE}\right) \\
&= \int_M \frac{\sqrt{-1}}{2\pi}\partial\varphi \wedge \bar{\partial}\varphi \wedge \left(\sum_{i=0}^{n-1} \omega^i_{KE} \wedge \omega'^{n-1-i}_{KE}\right) \\
&\geq \frac{1}{n}\int_M |\partial\varphi|^2_{\omega_{KE}}\omega^n_{KE}, \tag{3.184}
\end{aligned}
$$

which implies that φ is constant and hence $\omega_{KE} = \omega'_{KE}$.

As usual, we reduce (3.135) to a parabolic complex Monge–Ampère equation. Since $c_1(M) = 0$ there exists a unique volume form Ω satisfying

$$\frac{\sqrt{-1}}{2\pi}\partial\bar{\partial}\log\Omega = 0, \qquad \int_M \Omega = \int_M \omega^n_0. \tag{3.185}$$

Then solving (3.135) is equivalent to solving the parabolic complex Monge–Ampère equation

$$\frac{\partial}{\partial t}\varphi = \log\frac{(\omega_0 + \frac{\sqrt{-1}}{2\pi}\partial\bar{\partial}\varphi)^n}{\Omega}, \qquad \omega_0 + \frac{\sqrt{-1}}{2\pi}\partial\bar{\partial}\varphi > 0, \qquad \varphi|_{t=0} = 0. \tag{3.186}$$

We first observe:

Lemma 3.4.5. *We have*

(i) There exists a uniform constant C such that for $t \in [0, \infty)$

$$\|\dot{\varphi}(t)\|_{C^0(M)} \leq C. \tag{3.187}$$

(ii) *There exists a uniform constant C' such that on $M \times [0, \infty)$ the volume form of $\omega = \omega(t)$ satisfies*

$$\frac{1}{C'}\omega_0^n \le \omega^n \le C'\omega_0^n. \tag{3.188}$$

Proof. Differentiating (3.186) with respect to t we obtain

$$\frac{\partial}{\partial t}\dot{\varphi} = \Delta\dot{\varphi}, \tag{3.189}$$

and (i) follows immediately from the maximum principle. Part (ii) follows from (i).
□

We will obtain a bound on the oscillation of $\varphi(t)$ using Yau's zeroth order estimate for the elliptic complex Monge–Ampère equation. Note that Yau's estimate holds for any Kähler manifold (not just those with $c_1(M) = 0$):

Theorem 3.4.6. *Let (M, ω_0) be a compact Kähler manifold and let φ be a smooth function on M satisfying*

$$(\omega_0 + \frac{\sqrt{-1}}{2\pi}\partial\bar{\partial}\varphi)^n = e^F\omega_0^n, \quad \omega_0 + \frac{\sqrt{-1}}{2\pi}\partial\bar{\partial}\varphi > 0 \tag{3.190}$$

for some smooth function F. Then there exists a uniform C depending only on $\sup_M F$ and ω_0 such that

$$\operatorname{osc}_M \varphi := \sup_M \varphi - \inf_M \varphi \le C. \tag{3.191}$$

Proof. We will follow quite closely the exposition of Siu [Siu87]. We assume without loss of generality that $\int_M \varphi\, \omega_0^n = 0$. We also assume $n > 1$ (the case $n = 1$ is easier, and we leave it as an exercise for the reader).
Write $\omega = \omega_0 + \frac{\sqrt{-1}}{2\pi}\partial\bar{\partial}\varphi$. Then

$$C\int_M |\varphi|\omega_0^n \ge \int_M \varphi(\omega_0^n - \omega^n)$$

$$= -\int_M \varphi\frac{\sqrt{-1}}{2\pi}\partial\bar{\partial}\varphi \wedge \sum_{i=0}^{n-1}\omega_0^i \wedge \omega^{n-1-i}$$

$$= \int_M \frac{\sqrt{-1}}{2\pi}\partial\varphi \wedge \bar{\partial}\varphi \wedge \sum_{i=0}^{n-1}\omega_0^i \wedge \omega^{n-1-i}$$

$$\ge \frac{1}{n}\int_M |\partial\varphi|_{\omega_0}^2\omega_0^n. \tag{3.192}$$

By the Poincaré (Theorem 3.1.8) and Cauchy–Schwarz inequalities we have

$$\int_M |\varphi|^2 \omega_0^n \leq C \int_M |\partial\varphi|_{\omega_0}^2 \omega_0^n \leq C' \int_M |\varphi| \omega_0^n \leq C'' \left(\int_M |\varphi|^2 \omega_0^n\right)^{1/2}, \quad (3.193)$$

and hence $\|\varphi\|_{L^2(\omega_0)} \leq C$. We now repeat this argument with φ replaced by $\varphi|\varphi|^\alpha$ for $\alpha \geq 0$. Observe that the map of real numbers $x \mapsto x|x|^\alpha$ is differentiable with derivative $(\alpha + 1)|x|^\alpha$. Then

$$C \int_M |\varphi|^{\alpha+1} \omega_0^n \geq \int_M \varphi|\varphi|^\alpha (\omega_0^n - \omega^n)$$

$$= -\int_M \varphi|\varphi|^\alpha \frac{\sqrt{-1}}{2\pi} \partial\bar{\partial}\varphi \wedge \sum_{i=0}^{n-1} \omega_0^i \wedge \omega^{n-1-i}$$

$$= (\alpha + 1) \int_M |\varphi|^\alpha \sqrt{-1}\partial\varphi \wedge \bar{\partial}\varphi \wedge \sum_{i=0}^{n-1} \omega_0^i \wedge \omega^{n-1-i}$$

$$= \frac{(\alpha + 1)}{\left(\frac{\alpha}{2} + 1\right)^2} \int_M \sqrt{-1}\partial\left(\varphi|\varphi|^{\alpha/2}\right)$$

$$\wedge \bar{\partial}\left(\varphi|\varphi|^{\alpha/2}\right) \wedge \sum_{i=0}^{n-1} \omega_0^i \wedge \omega^{n-1-i}. \quad (3.194)$$

It then follows that for some uniform $C > 0$,

$$\int_M \left|\partial\left(\varphi|\varphi|^{\alpha/2}\right)\right|_{\omega_0}^2 \omega_0^n \leq C(\alpha + 1) \int_M |\varphi|^{\alpha+1} \omega_0^n. \quad (3.195)$$

Now apply the Sobolev inequality (Theorem 3.1.9) to $f = \varphi|\varphi|^{\alpha/2}$. Then for $\beta = n/(n-1)$ we have

$$\left(\int_M |\varphi|^{(\alpha+2)\beta} \omega_0^n\right)^{1/\beta} \leq C \left((\alpha + 1) \int_M |\varphi|^{\alpha+1} \omega_0^n + \int_M |\varphi|^{\alpha+2} \omega_0^n\right). \quad (3.196)$$

By Hölder's inequality we have for a uniform constant C,

$$\int_M |\varphi|^{\alpha+1} \omega_0^n \leq 1 + C \int_M |\varphi|^{\alpha+2} \omega_0^n. \quad (3.197)$$

Now substituting $p = \alpha + 2$ we have from (3.196),

$$\|\varphi\|_{L^{p\beta}(\omega_0)}^p \leq Cp \max\left(1, \|\varphi\|_{L^p(\omega_0)}^p\right). \quad (3.198)$$

Raising to the power $1/p$ we have for all $p \geq 2$,

$$\max(1, \|\varphi\|_{L^{p\beta}(\omega_0)}) \leq C^{1/p} p^{1/p} \max(1, \|\varphi\|_{L^p(\omega_0)}). \tag{3.199}$$

Fix an integer $k > 0$. Replace p in (3.199) by $p\beta^k$ and then $p\beta^{k-1}$ and so on, to obtain

$$\max(1, \|\varphi\|_{L^{p\beta^{k+1}}(\omega_0)}) \leq C^{\frac{1}{p\beta^k}} (p\beta^k)^{\frac{1}{p\beta^k}} \max(1, \|\varphi\|_{L^{p\beta^k}(\omega_0)}) \leq \cdots$$

$$\leq C^{\frac{1}{p\beta^k} + \frac{1}{p\beta^{k-1}} + \cdots + \frac{1}{p}} (p\beta^k)^{\frac{1}{p\beta^k}} (p\beta^{k-1})^{\frac{1}{p\beta^{k-1}}} \cdots$$

$$p^{\frac{1}{p}} \max(1, \|\varphi\|_{L^p(\omega_0)}) \tag{3.200}$$

$$= C_k \max(1, \|\varphi\|_{L^p(\omega_0)}) \tag{3.201}$$

for

$$C_k = C^{\frac{1}{p}\left(\frac{1}{\beta^k} + \frac{1}{\beta^{k-1}} + \cdots + 1\right)} p^{\frac{1}{p}\left(\frac{1}{\beta^k} + \frac{1}{\beta^{k-1}} + \cdots + 1\right)} \beta^{\frac{1}{p}\left(\frac{k}{\beta^k} + \frac{k-1}{\beta^{k-1}} + \cdots + \frac{1}{\beta}\right)}. \tag{3.202}$$

Since the infinite sums $\sum \frac{1}{\beta^i}$ and $\sum \frac{i}{\beta^i}$ converge for $\beta = n/(n-1) > 1$ we see that for any fixed p, the constants C_k are uniformly bounded from above, independent of k.

Setting $p = 2$ and letting $k \to \infty$ in (3.201) we finally obtain

$$\max(1, \|\varphi\|_{C^0}) \leq C \max(1, \|\varphi\|_{L^2(\omega_0)}) \leq C', \tag{3.203}$$

and hence (3.191). $\square$

Now the oscillation bound for $\varphi = \varphi(t)$ along the Kähler–Ricci flow (3.186) follows immediately:

Lemma 3.4.7. *There exists a uniform constant C such that for $t \in [0, \infty)$,*

$$\mathrm{osc}_M \, \varphi \leq C. \tag{3.204}$$

Proof. From Lemma 3.4.5 we have uniform bounds for $\dot{\varphi}$. Rewrite the parabolic complex Monge–Ampère equation (3.186) as

$$\left(\omega_0 + \frac{\sqrt{-1}}{2\pi} \partial\bar{\partial}\varphi(t)\right)^n = e^{F(t)} \omega_0^n \quad \text{with} \quad F(t) = \log \frac{\Omega}{\omega_0^n} + \dot{\varphi}(t) \tag{3.205}$$

and apply Theorem 3.4.6. $\square$

3.4.3 Higher Order Estimates and Convergence When $c_1(M) = 0$

In this subsection we complete the proof of Theorem 3.4.4. The proof for the higher order estimates follows along similar lines as in the case for $c_1(M) < 0$. As above, let $\varphi(t)$ solve the parabolic complex Monge–Ampère equation (3.186) on M with $c_1(M) = 0$ and write $\omega = \omega_0 + \frac{\sqrt{-1}}{2\pi}\partial\bar{\partial}\varphi$.

Lemma 3.4.8. *There exists a uniform constant C such that on $M \times [0, \infty)$, $\omega = \omega(t)$ satisfies*

$$\frac{1}{C}\omega_0 \leq \omega \leq C\omega_0. \tag{3.206}$$

Proof. By Lemma 3.4.5 and the argument of Corollary 3.3.5, it suffices to obtain a uniform upper bound for $\operatorname{tr}_{\omega_0}\omega$. As in the case of Lemma 3.4.3, define $Q = \log \operatorname{tr}_{\omega_0}\omega - A\varphi$ for A a constant to be determined later. Compute using Proposition 3.2.5,

$$\left(\frac{\partial}{\partial t} - \Delta\right)Q \leq C_0\operatorname{tr}_{\omega}\omega_0 - A\dot{\varphi} + An - A\operatorname{tr}_{\omega}\omega_0, \tag{3.207}$$

for C_0 depending only on g_0. Choosing $A = C_0 + 1$ we have, since $\dot{\varphi}$ is uniformly bounded,

$$\left(\frac{\partial}{\partial t} - \Delta\right)Q \leq -\operatorname{tr}_{\omega}\omega_0 + C. \tag{3.208}$$

We claim that for any $(x, t) \in M \times [0, \infty)$,

$$(\operatorname{tr}_{\omega_0}\omega)(x, t) \leq Ce^{A(\varphi(x,t) - \inf_{M \times [0,t]}\varphi)}. \tag{3.209}$$

To see this, suppose that Q achieves a maximum on $M \times [0, t]$ at the point (x_0, t_0). We assume without loss of generality that $t_0 > 0$. Applying the maximum principle to (3.208) we see that $(\operatorname{tr}_{\omega}\omega_0)(x_0, t_0) \leq C$ and, by the argument of Lemma 3.4.3, $(\operatorname{tr}_{\omega_0}\omega)(x_0, t_0) \leq C'$. Then for any $x \in M$,

$$(\log \operatorname{tr}_{\omega_0}\omega)(x, t) - A\varphi(x, t) = Q(x, t) \leq Q(x_0, t_0) \leq \log C' - A\varphi(x_0, t_0). \tag{3.210}$$

Exponentiating gives (3.209).
   Define

$$\tilde{\varphi} := \varphi - \frac{1}{V}\int_M \varphi\,\Omega, \quad \text{where} \quad V := \int_M \Omega = \int_M \omega^n. \tag{3.211}$$

From Lemma 3.4.7, $\|\tilde{\varphi}\|_{C^0(M)} \leq C$. The estimate (3.209) can be rewritten as:

$$(\operatorname{tr}_{\omega_0}\omega)(x,t) \leq Ce^{A\left(\tilde{\varphi}(x,t)+\frac{1}{V}\int_M \varphi(t)\,\Omega-\inf_{M\times[0,t]}\tilde{\varphi}-\inf_{[0,t]}\frac{1}{V}\int_M \varphi\,\Omega\right)}$$

$$\leq Ce^{C'+\frac{A}{V}\left(\int_M \varphi(t)\Omega-\inf_{[0,t]}\int_M \varphi\,\Omega\right)}. \tag{3.212}$$

Using Jensen's inequality,

$$\frac{d}{dt}\left(\frac{1}{V}\int_M \varphi\,\Omega\right) = \frac{1}{V}\int_M \dot{\varphi}\,\Omega = \frac{1}{V}\int_M \log\left(\frac{\omega^n}{\Omega}\right)\Omega \leq \log\left(\frac{1}{V}\int_M \omega^n\right) = 0, \tag{3.213}$$

and hence $\inf_{[0,t]}\int_M \varphi\,\Omega = \int_M \varphi(t)\Omega$. The required upper bound of $\operatorname{tr}_{\omega_0}\omega$ follows then from (3.212). □

It follows from Corollary 3.2.16 that we have uniform C^∞ estimates on $g(t)$ and the normalized potential function $\tilde{\varphi}(t) = \varphi(t) - V^{-1}\int_M \varphi(t)\Omega$. It remains to prove the C^∞ convergence part of Theorem 3.4.4. We follow the method of Phong–Sturm [PS06] (see also [MSz09, PSSW09]) and use a functional known as the Mabuchi energy [Mab86]. It is noted in [Cao, DT92] that the monotonicity of the Mabuchi energy along the Kähler–Ricci flow was established in unpublished work of H.-D. Cao in 1991.

We fix a metric ω_0 as above. The Mabuchi energy is a functional $\operatorname{Mab}_{\omega_0}$ on the space

$$\operatorname{PSH}(M,\omega_0) = \left\{\varphi \in C^\infty(M) \mid \omega_0 + \frac{\sqrt{-1}}{2\pi}\partial\bar{\partial}\varphi > 0\right\} \tag{3.214}$$

with the property that if φ_t is any smooth path in $\operatorname{PSH}(M,\omega_0)$ then

$$\frac{d}{dt}\operatorname{Mab}_{\omega_0}(\varphi_t) = -\int_M \dot{\varphi}_t R_{\varphi_t}\,\omega_{\varphi_t}^n, \tag{3.215}$$

where $\omega_{\varphi_t} = \omega_0 + \frac{\sqrt{-1}}{2\pi}\partial\bar{\partial}\varphi_t$, and R_{φ_t} is the scalar curvature of ω_{φ_t}. Observe that if φ_∞ is a critical point of $\operatorname{Mab}_{\omega_0}$ then $\omega_\infty = \omega_0 + \frac{\sqrt{-1}}{2\pi}\partial\bar{\partial}\varphi_\infty$ has zero scalar curvature and hence is Ricci flat (for that last statement: since $c_1(M) = 0$, then $\operatorname{Ric}(\omega_\infty) = \frac{\sqrt{-1}}{2\pi}\partial\bar{\partial}h_\infty$ for some function h_∞ and taking the trace gives $\Delta_{\omega_\infty}h_\infty = 0$ which implies h_∞ is constant and $\operatorname{Ric}(\omega_\infty) = 0$).

Typically, the Mabuchi energy is defined in terms of its derivative using the formula (3.215) but instead we will use the explicit formula as derived in [Tian]. Define

$$\operatorname{Mab}_{\omega_0}(\varphi) = \int_M \log\left(\frac{\omega_\varphi^n}{\omega_0^n}\right)\omega_\varphi^n - \int_M h_0(\omega_\varphi^n - \omega_0^n), \tag{3.216}$$

where $\omega_\varphi = \omega_0 + \frac{\sqrt{-1}}{2\pi}\partial\bar\partial\varphi$ and h_0 is the *Ricci potential* for ω_0 given by

$$\mathrm{Ric}(\omega_0) = \frac{\sqrt{-1}}{2\pi}\partial\bar\partial h_0, \qquad \int_M e^{h_0}\omega_0^n = \int_M \omega_0^n. \qquad (3.217)$$

Observe that Mab_{ω_0} depends only on the metric ω_φ and so can be regarded as a functional on the space of Kähler metrics cohomologous to ω_0. We now need to check that Mab_{ω_0} defined by (3.216) satisfies (3.215). Let φ_t be any smooth path in $\mathrm{PSH}(M, \omega_0)$. Using integration by parts, we compute

$$\begin{aligned}
\frac{d}{dt}\mathrm{Mab}_{\omega_0}(\varphi_t) &= \int_M \Delta\dot\varphi_t\, \omega_{\varphi_t}^n + \int_M \log\frac{\omega_{\varphi_t}^n}{\omega_0^n}\, \Delta\dot\varphi_t\, \omega_{\varphi_t}^n - \int_M h_0\Delta\dot\varphi_t\, \omega_{\varphi_t}^n \\
&= \int_M \dot\varphi_t(-R_{\varphi_t} + \mathrm{tr}_\omega\mathrm{Ric}(\omega_0))\omega_{\varphi_t}^n - \int_M \dot\varphi_t\Delta h_0\, \omega_{\varphi_t}^n \\
&= -\int_M \dot\varphi_t R_{\varphi_t}\omega_{\varphi_t}^n.
\end{aligned} \qquad (3.218)$$

The key fact we need is as follows:

Lemma 3.4.9. *Let $\varphi = \varphi(t)$ solve the Kähler–Ricci flow (3.186). Then*

$$\frac{d}{dt}\mathrm{Mab}_{\omega_0}(\varphi) = -\int_M |\partial\dot\varphi|_\omega^2\omega^n. \qquad (3.219)$$

In particular, the Mabuchi energy is decreasing along the Kähler–Ricci flow. Moreover, there exists a uniform constant C such that

$$\frac{d}{dt}\int_M |\partial\dot\varphi|_\omega^2\omega^n \le C\int_M |\partial\dot\varphi|_\omega^2\omega^n. \qquad (3.220)$$

Proof. Observe that from the Kähler–Ricci flow equation we have $\frac{\sqrt{-1}}{2\pi}\partial\bar\partial\dot\varphi = -\mathrm{Ric}(\omega)$ and taking the trace of this gives $\Delta\dot\varphi = -R$. Then

$$\frac{d}{dt}\mathrm{Mab}_{\omega_0}(\varphi) = -\int_M \dot\varphi R\,\omega^n = \int_M \dot\varphi\Delta\dot\varphi\,\omega^n = -\int_M |\partial\dot\varphi|_\omega^2\omega^n, \qquad (3.221)$$

giving (3.219). For (3.220), compute

$$\frac{d}{dt}\int_M |\partial\dot\varphi|_\omega^2\omega^n = \int_M (\frac{\partial}{\partial t}g^{\bar j i})\partial_i\dot\varphi\partial_{\bar j}\dot\varphi\,\omega^n + 2\mathrm{Re}\left(\int_M g^{\bar j i}\partial_i(\Delta\dot\varphi)\partial_{\bar j}\dot\varphi\,\omega^n\right)$$
$$+ \int_M |\partial\dot\varphi|^2\Delta\dot\varphi\,\omega^n \qquad (3.222)$$

$$= \int_M R^{\overline{j}i} \partial_i \varphi \partial_{\overline{j}} \dot{\varphi} \, \omega^n - 2 \int_M (\Delta \dot{\varphi})^2 \omega^n - \int_M |\partial \dot{\varphi}|^2 R \, \omega^n$$

$$\leq C \int_M |\partial \dot{\varphi}|^2_\omega \omega^n, \tag{3.223}$$

using (3.189), an integration by parts and the fact that, since we have C^∞ estimates for ω, we have uniform bounds of the Ricci and scalar curvatures of ω. □

It is now straightforward to complete the proof of the convergence of the Kähler–Ricci flow. Since we have uniform estimates for $\omega(t)$ along the flow, we see from the formula (3.216) that the Mabuchi energy is uniformly bounded. From (3.219) there is a sequence of times $t_i \in [i, i+1]$ for which

$$\left(\int_M \left| \partial \log \frac{\omega^n}{\Omega} \right|^2_\omega \omega^n \right)(t_i) = \left(\int_M |\partial \dot{\varphi}|^2_\omega \, \omega^n \right)(t_i) \to 0, \quad \text{as } i \to \infty. \tag{3.224}$$

By the differential inequality (3.220),

$$\left(\int_M \left| \partial \log \frac{\omega^n}{\Omega} \right|^2_\omega \omega^n \right)(t) \to 0, \quad \text{as } t \to \infty. \tag{3.225}$$

But since we have C^∞ estimates for $\varphi(t)$ we can apply the Arzela–Ascoli Theorem to obtain a sequence of times t_j such that $\varphi(t_j)$ converges in C^∞ to φ_∞, say. Writing $\omega_\infty = \omega_0 + \frac{\sqrt{-1}}{2\pi} \partial \overline{\partial} \varphi_\infty > 0$, we have from (3.225),

$$\left(\int_M \left| \partial \log \frac{\omega^n_\infty}{\Omega} \right|^2_{\omega_\infty} \omega^n_\infty \right) = 0, \tag{3.226}$$

and hence

$$\log \frac{\omega^n_\infty}{\Omega} = C, \tag{3.227}$$

for some constant C. Taking $\frac{\sqrt{-1}}{2\pi} \partial \overline{\partial}$ of (3.227) gives $\mathrm{Ric}(\omega_\infty) = 0$. Hence for a sequence of times $t_j \to \infty$ the Kähler–Ricci flow converges to ω_∞, the unique Kähler–Einstein metric in the cohomology class $[\omega_0]$.

To see that the convergence of the metrics $\omega(t)$ is in C^∞ without passing to a subsequence, we argue as follows. If not, then by the same argument as in the proof of Theorem 3.4.1 we can find a sequence of times $t_k \to \infty$ such that $\omega(t_k)$ converges in C^∞ to $\omega'_\infty \neq \omega_\infty$. But by (3.225), ω'_∞ is Kähler–Einstein, contradicting the uniqueness of Kähler–Einstein metrics in $[\omega_0]$. This completes the proof of Theorem 3.4.4.

Remark 3.4.10. It was pointed out to the authors by Zhenlei Zhang that one can equivalently consider the functional $\int_M h\, \omega^n$, where h is the Ricci potential of the evolving metric.

3.5 The Case When K_M Is Big and nef

In the previous section we considered the Kähler–Ricci flow on manifolds with $c_1(M) < 0$, which is equivalent to the condition that the canonical line bundle K_M is ample. In this section we consider the case where the line bundle K_M is not necessarily ample, but nef and big. Such a manifold is known as a smooth minimal model of general type.

3.5.1 Smooth Minimal Models of General Type

As in the case of $c_1(M) < 0$ we consider the normalized Kähler–Ricci flow

$$\frac{\partial}{\partial t}\omega = -\mathrm{Ric}(\omega) - \omega, \qquad \omega|_{t=0} = \omega_0, \tag{3.228}$$

but we impose no restrictions on the Kähler class of ω_0. We will prove:

Theorem 3.5.1. *Let M be a projective algebraic manifold which is a smooth minimal model of general type (that is, K_M is nef and big). Then*

(i) The solution $\omega = \omega(t)$ of the normalized Kähler–Ricci flow (3.228) starting at any Kähler metric ω_0 on M exists for all time.
(ii) There exists a codimension 1 analytic subvariety S of M such that $\omega(t)$ converges in $C^\infty_{\mathrm{loc}}(M \setminus S)$ to a Kähler metric ω_{KE} defined on $M \setminus S$ which satisfies the Kähler–Einstein equation

$$\mathrm{Ric}(\omega_{\mathrm{KE}}) = -\omega_{\mathrm{KE}}, \quad on\ M \setminus S. \tag{3.229}$$

We will see later in Sect. 3.5.3 that ω_{KE} is unique under some suitable conditions. Note that if K_M is not ample, then ω_{KE} cannot extend to be a smooth Kähler metric on M, and we call ω_{KE} a *singular Kähler–Einstein metric*. The first proof of Theorem 3.5.1 appeared in the work of Tsuji [Tsu88]. Later, Tian–Zhang [Tzha06] extended this result (see Sect. 3.5.4 below) and clarified some parts of Tsuji's proof. Our exposition will for the most part follow [Tzha06].

From part (i) of Theorem 3.1.12 we see that under the assumptions of Theorem 3.5.1, the cohomology class $[\omega_0] - t c_1(M)$ is Kähler for all $t \geq 0$ and hence by Theorem 3.3.1, the (unnormalized) Kähler–Ricci flow has a smooth solution $\omega(t)$ for all time t. Rescaling as in Sect. 3.4.1 we obtain a solution of the normalized

Kähler–Ricci flow (3.228) for all time. This establishes part (i) of Theorem 3.5.1. Observe that in fact we only need K_M to be nef to obtain a solution to the Kähler–Ricci flow for all time.

It is straightforward to calculate the Kähler class of the evolving metric along the flow. Indeed, $[\omega(t)]$ evolves according to the ordinary differential equation

$$\frac{d}{dt}[\omega(t)] = -c_1(M) - [\omega], \quad [\omega(0)] = [\omega_0], \tag{3.230}$$

and this has a solution

$$[\omega(t)] = -(1 - e^{-t})c_1(M) + e^{-t}[\omega_0]. \tag{3.231}$$

This shows that, in particular, $[\omega(t)] \to -c_1(M)$ as $t \to \infty$.

We now rewrite (3.228) as a parabolic complex Monge–Ampère equation. First, from the Base Point Free Theorem [part (ii) of Theorem 3.1.12], K_M is semi-ample. Hence there exists a smooth closed nonnegative $(1, 1)$-form $\hat{\omega}_\infty$ on M with $[\hat{\omega}_\infty] = -c_1(M)$. Indeed, we may take $\hat{\omega}_\infty = \frac{1}{m}\Phi^*\omega_{FS}$ where $\Phi : M \to \mathbb{P}^N$ is a holomorphic map defined by holomorphic sections of K_M^m for m large and ω_{FS} is the Fubini–Study metric (see Sect. 3.1.7).

Define reference metrics in $[\omega(t)]$ by

$$\hat{\omega}_t = e^{-t}\omega_0 + (1 - e^{-t})\hat{\omega}_\infty, \quad \text{for } t \in [0, \infty). \tag{3.232}$$

Let Ω be the smooth volume form on M satisfying

$$\frac{\sqrt{-1}}{2\pi}\partial\bar{\partial}\log\Omega = \hat{\omega}_\infty \in -c_1(M), \quad \int_M \Omega = \int_M \omega_0^n. \tag{3.233}$$

We then consider the parabolic complex Monge–Ampère equation

$$\frac{\partial}{\partial t}\varphi = \log\frac{(\hat{\omega}_t + \frac{\sqrt{-1}}{2\pi}\partial\bar{\partial}\varphi)^n}{\Omega} - \varphi, \quad \hat{\omega}_t + \frac{\sqrt{-1}}{2\pi}\partial\bar{\partial}\varphi > 0, \quad \varphi|_{t=0} = 0, \tag{3.234}$$

which is equivalent to (3.228). Hence a solution to (3.234) exists for all time.

3.5.2 Estimates

In this section we prove the estimates needed for the second part of Theorem 3.5.1. Assume that $\varphi = \varphi(t)$ solves (3.234). We have:

Lemma 3.5.2. *There exists a uniform constants C and $t' > 0$ such that on M,*

(i) $\varphi(t) \le C$ *for* $t \ge 0$.
(ii) $\dot{\varphi}(t) \le Cte^{-t}$ *for* $t \ge t'$. *In particular,* $\dot{\varphi}(t) \le C$ *for* $t \ge 0$.
(iii) $\omega^n(t) \le C\Omega$ *for* $t \ge 0$.

Proof. Part (i) follows immediately from the maximum principle. Indeed if φ achieves a maximum at a point (x_0, t_0) with $t_0 > 0$ then, directly from (3.234),

$$0 \le \frac{\partial}{\partial t}\varphi \le \log \frac{\hat{\omega}_t^n}{\Omega} - \varphi \quad \text{at } (x_0, t_0), \tag{3.235}$$

and hence $\varphi \le \log(\hat{\omega}_t^n / \Omega) \le C$.

Part (ii) is a result of [Tzha06]. Compute

$$\left(\frac{\partial}{\partial t} - \Delta\right)\varphi = \dot{\varphi} - n + \text{tr}_\omega \hat{\omega}_t \tag{3.236}$$

$$\left(\frac{\partial}{\partial t} - \Delta\right)\dot{\varphi} = -e^{-t}\text{tr}_\omega(\omega_0 - \hat{\omega}_\infty) - \dot{\varphi}, \tag{3.237}$$

using the fact that $\frac{\partial}{\partial t}\hat{\omega}_t = -e^{-t}(\omega_0 - \hat{\omega}_\infty)$. Hence

$$\left(\frac{\partial}{\partial t} - \Delta\right)(e^t\dot{\varphi}) = -\text{tr}_\omega(\omega_0 - \hat{\omega}_\infty) \tag{3.238}$$

$$\left(\frac{\partial}{\partial t} - \Delta\right)(\dot{\varphi} + \varphi + nt) = \text{tr}_\omega\hat{\omega}_\infty. \tag{3.239}$$

Subtracting (3.239) from (3.238) gives

$$\left(\frac{\partial}{\partial t} - \Delta\right)\left((e^t - 1)\dot{\varphi} - \varphi - nt\right) = -\text{tr}_\omega\omega_0 < 0, \tag{3.240}$$

which implies that the maximum of $(e^t - 1)\dot{\varphi} - \varphi - nt$ is decreasing in time, giving

$$(e^t - 1)\dot{\varphi} - \varphi - nt \le 0. \tag{3.241}$$

This establishes (ii). Part (iii) follows from Corollary 3.2.3 [or using (i) and (ii) and the fact that $\omega^n / \Omega = e^{\dot{\varphi}+\varphi}$]. $\qquad\square$

We now prove lower bounds for φ and $\dot{\varphi}$ away from a subvariety. To do this we need to use *Tsuji's trick* of applying Kodaira's Lemma [part (iii) of Theorem 3.1.12].

Since K_M is big and nef, there exists an effective divisor E on M with $K_M - \delta[E] > 0$ for some sufficiently small $\delta > 0$. Since $\hat{\omega}_\infty$ lies in the cohomology class $c_1(K_M)$ it follows that for any Hermitian metric h of $[E]$ the cohomology class of $\hat{\omega}_\infty - \delta R_h$ is Kähler. Then by the $\partial\bar{\partial}$-Lemma we may pick a Hermitian metric h on $[E]$ such that

$$\hat{\omega}_\infty - \delta R_h \geq c\omega_0, \tag{3.242}$$

for some constant $c > 0$. Moreover, if we pick any $\varepsilon \in (0, \delta]$ we have

$$\hat{\omega}_\infty - \varepsilon R_h \geq c_\varepsilon \omega_0, \tag{3.243}$$

for $c_\varepsilon = c\varepsilon/\delta > 0$. Indeed, since $\hat{\omega}_\infty$ is semi-positive,

$$\hat{\omega}_\infty - \varepsilon R_h = \frac{\varepsilon}{\delta}(\hat{\omega}_\infty - \delta R_h) + \left(1 - \frac{\varepsilon}{\delta}\right)\hat{\omega}_\infty \geq \frac{\varepsilon}{\delta}(\hat{\omega}_\infty - \delta R_h) \geq \frac{c\varepsilon}{\delta}\omega_0. \tag{3.244}$$

Now fix a holomorphic section σ of $[E]$ which vanishes to order 1 along the divisor E. It follows that

$$\hat{\omega}_\infty + \varepsilon \frac{\sqrt{-1}}{2\pi} \partial\bar{\partial} \log |\sigma|_h^2 \geq c_\varepsilon \omega_0, \qquad \text{on } M \setminus E, \tag{3.245}$$

since $\partial\bar{\partial}\log|\sigma|_h^2 = \partial\bar{\partial}\log h$ away from E. Note that here (and henceforth) we are writing E for the support of the divisor E.

We can then prove:

Lemma 3.5.3. *With the notation above, for every $\varepsilon \in (0, \delta]$ there exists a constant $C_\varepsilon > 0$ such that on $(M \setminus E) \times [0, \infty)$,*

(i) $\varphi \geq \varepsilon \log |\sigma|_h^2 - C_\varepsilon$.
(ii) $\dot\varphi \geq \varepsilon \log |\sigma|_h^2 - C_\varepsilon$.
(iii) $\omega^n \geq \dfrac{1}{C_\varepsilon} |\sigma|_h^{2\varepsilon} \Omega$.

Proof. It suffices to prove the estimate

$$\varphi + \dot\varphi \geq \varepsilon \log |\sigma|_h^2 - C_\varepsilon, \quad \text{on } M \setminus E, \tag{3.246}$$

where we write C_ε for a constant that depends only on ε and the fixed data. Indeed this inequality immediately implies (iii). The estimates (i) and (ii) follow from (3.246) together with the upper bounds of $\dot\varphi$ and φ given by Lemma 3.5.2.

To establish (3.246), we will bound from below the quantity Q defined by

$$Q = \dot\varphi + \varphi - \varepsilon \log |\sigma|_h^2 = \log \frac{\omega^n}{|\sigma|_h^{2\varepsilon} \Omega}, \quad \text{on } M \setminus E. \tag{3.247}$$

Observe that for any fixed time t, $Q(x, t) \to \infty$ as x approaches E. Hence for each time t, Q attains a minimum (in space) in the interior of the set $M \setminus E$. Now from (3.239) we have

$$\left(\frac{\partial}{\partial t} - \Delta\right)(\dot\varphi + \varphi) = \operatorname{tr}_\omega \hat{\omega}_\infty - n. \tag{3.248}$$

Using this we compute on $M \setminus E$,

$$\left(\frac{\partial}{\partial t} - \Delta\right) Q = \text{tr}_\omega \hat{\omega}_\infty - n + \varepsilon \text{tr}_\omega \left(\frac{\sqrt{-1}}{2\pi} \partial \bar{\partial} \log |\sigma|_h^2\right) \qquad (3.249)$$

$$= \text{tr}_\omega \left(\hat{\omega}_\infty + \varepsilon \frac{\sqrt{-1}}{2\pi} \partial \bar{\partial} \log |\sigma|_h^2\right) - n \qquad (3.250)$$

$$\geq c_\varepsilon \text{tr}_\omega \omega_0 - n, \qquad (3.251)$$

where for the last line we used (3.245).

Then if Q achieves a minimum at (x_0, t_0) with x_0 in $M \setminus E$ and $t_0 > 0$ then at (x_0, t_0) we have

$$\text{tr}_\omega \omega_0 \leq \frac{n}{c_\varepsilon}. \qquad (3.252)$$

By the geometric–arithmetic means inequality, at (x_0, t_0),

$$\left(\frac{\omega_0^n}{\omega^n}\right)^{1/n} \leq \frac{1}{n} \text{tr}_\omega \omega_0 \leq \frac{1}{c_\varepsilon}, \qquad (3.253)$$

which gives a uniform lower bound for the volume form $\omega^n(x_0, t_0)$. Hence

$$Q(x_0, t_0) = \log \frac{\omega^n}{|\sigma|_h^2 \Omega}(x_0, t_0) \geq -C_\varepsilon, \qquad (3.254)$$

and since Q is bounded below at time $t = 0$ we obtain the desired lower bound for Q. □

Next we prove estimates for $g(t)$ away from a divisor. First, *we from now on fix an ε in $(0, \delta]$ sufficiently small so that $\omega_0 + \varepsilon \frac{\sqrt{-1}}{2\pi} \partial \bar{\partial} \log h$ is Kähler.* We will need the following lemma.

Lemma 3.5.4. *For the $\varepsilon > 0$ fixed as above, the metrics $\hat{\omega}_{t,\varepsilon}$ defined by*

$$\hat{\omega}_{t,\varepsilon} := \hat{\omega}_t + \varepsilon \frac{\sqrt{-1}}{2\pi} \partial \bar{\partial} \log h = \hat{\omega}_\infty + \varepsilon \frac{\sqrt{-1}}{2\pi} \partial \bar{\partial} \log h + e^{-t}(\omega_0 - \hat{\omega}_\infty). \qquad (3.255)$$

give a smooth family of Kähler metrics for $t \in [0, \infty)$. Moreover there exists a constant $C > 0$ such that for all t,

$$\frac{1}{C} \omega_0 \leq \hat{\omega}_{t,\varepsilon} \leq C \omega_0. \qquad (3.256)$$

Proof. From (3.245) we see that $\hat{\omega}_\infty + \varepsilon \frac{\sqrt{-1}}{2\pi} \partial\bar{\partial} \log h$ is Kähler. Hence we may choose $T_0 > 0$ sufficiently large so that, for $C > 0$ large enough,

$$\frac{1}{C}\omega_0 \leq \hat{\omega}_\infty + \varepsilon \frac{\sqrt{-1}}{2\pi} \partial\bar{\partial} \log h + e^{-t}(\omega_0 - \hat{\omega}_\infty) \leq C\omega_0, \tag{3.257}$$

for all $t > T_0$. It remains to check that $\hat{\omega}_{t,\varepsilon}$ is Kähler for $t \in [0, T_0]$. But for $t \in [0, T_0]$,

$$\hat{\omega}_{t,\varepsilon} = (1 - e^{-t})\left(\hat{\omega}_\infty + \varepsilon \frac{\sqrt{-1}}{2\pi} \partial\bar{\partial} \log h\right) + e^{-t}\left(\omega_0 + \varepsilon \frac{\sqrt{-1}}{2\pi} \partial\bar{\partial} \log h\right)$$

$$> e^{-T_0}\left(\omega_0 + \varepsilon \frac{\sqrt{-1}}{2\pi} \partial\bar{\partial} \log h\right) > 0, \tag{3.258}$$

by definition of ε. $\qquad\square$

We can now prove bounds for the evolving metric:

Lemma 3.5.5. *There exist uniform constants C and α such that on $(M \setminus E) \times [0, \infty)$,*

$$\mathrm{tr}_{\omega_0} \omega \leq \frac{C}{|\sigma|_h^{2\alpha}}. \tag{3.259}$$

Hence there exist uniform constants $C' > 0$ and α' such that on $(M \setminus E) \times [0, \infty)$,

$$\frac{|\sigma|_h^{2\alpha'}}{C'}\omega_0 \leq \omega(t) \leq \frac{C'}{|\sigma|_h^{2\alpha'}}\omega_0. \tag{3.260}$$

Proof. Define a quantity Q on $M \setminus E$ by

$$Q = \log \mathrm{tr}_{\omega_0}\omega - A\left(\varphi - \varepsilon \log |\sigma|_h^2\right), \tag{3.261}$$

for A a sufficiently large constant to be determined later. For any fixed time t, $Q(x, t) \to -\infty$ as x approaches E. Then compute using Proposition 3.2.5,

$$\left(\frac{\partial}{\partial t} - \Delta\right)Q \leq C_0 \mathrm{tr}_\omega \omega_0 - A\dot{\varphi} + A\Delta\left(\varphi - \varepsilon \log |\sigma|_h^2\right). \tag{3.262}$$

Now at any point of $M \setminus E$,

$$\Delta\left(\varphi - \varepsilon \log |\sigma|_h^2\right) = \mathrm{tr}_\omega\left(\omega - \hat{\omega}_t - \varepsilon \frac{\sqrt{-1}}{2\pi} \partial\bar{\partial} \log |\sigma|_h^2\right) = n - \mathrm{tr}_\omega \hat{\omega}_{t,\varepsilon}. \tag{3.263}$$

Applying Lemma 3.5.4, we may choose A sufficiently large so that $A\hat{\omega}_{t,\varepsilon} \geq (C_0 + 1)\omega_0$ and hence

$$\left(\frac{\partial}{\partial t} - \Delta\right) Q \leq -\mathrm{tr}\,_\omega \omega_0 - A\left(\log\frac{\omega^n}{\Omega} - \varphi\right) + An$$

$$\leq -\mathrm{tr}\,_\omega \omega_0 - A\log\frac{\omega^n}{\omega_0^n} + C, \tag{3.264}$$

where we have used the upper bound on φ from Lemma 3.5.2.

Working in a compact time interval $[0, t']$ say, suppose that Q achieves a maximum at (x_0, t_0) with x_0 in M and $t_0 > 0$. Then at (x_0, t_0) we have

$$\mathrm{tr}\,_\omega \omega_0 + A\log\frac{\omega^n}{\omega_0^n} \leq C. \tag{3.265}$$

By the same argument as in the proof of Lemma 3.3.4 we see that $(\mathrm{tr}\,_{\omega_0}\omega)$ $(x_0, t_0) \leq C$.

Then for any $(x, t) \in (M \setminus E) \times [0, t']$ we have

$$Q(x, t) = (\log \mathrm{tr}\,_{\omega_0}\omega)(x, t) - A\left(\varphi - \varepsilon \log |\sigma|_h^2\right)(x, t)$$

$$\leq Q(x_0, t_0)$$

$$\leq \log C - A\left(\varphi - \varepsilon \log |\sigma|_h^2\right)(x_0, t_0) \leq C', \tag{3.266}$$

where for the last line we used part (i) of Lemma 3.5.3. Since t' is arbitrary, we have on $(M \setminus E) \times [0, \infty)$,

$$\log \mathrm{tr}\,_{\omega_0}\omega \leq C + A\left(\varphi - \varepsilon \log |\sigma|_h^2\right). \tag{3.267}$$

Since φ is bounded from above we obtain (3.259) after exponentiating.

For (3.260), combine (3.259) with part (iii) of Lemma 3.5.3.   □

We now wish to obtain higher order estimates on compact subsets of $M \setminus E$:

Lemma 3.5.6. *For $m = 0, 1, 2, \ldots$, there exist uniform constants C_m and α_m such that on $(M \setminus E) \times [0, \infty)$,*

$$S \leq \frac{C_0}{|\sigma|_h^{2\alpha_0}}, \qquad |\nabla_\mathbb{R}^m \mathrm{Rm}(g)| \leq \frac{C_m}{|\sigma|_h^{2\alpha_m}}, \tag{3.268}$$

where we are using the notation of Sects. 3.2.5 and 3.2.6.

Proof. We prove only the bound on S and leave the bounds on curvature and its derivatives as an exercise to the reader. We will follow quite closely an argument given in [SW10]. From Proposition 3.2.8 and (3.260),

$$\left(\frac{\partial}{\partial t} - \Delta\right) S = -|\overline{\nabla}\Psi|^2 - |\nabla\Psi|^2 + |\Psi|^2 - 2\mathrm{Re}\left(g^{\overline{j}i}g^{\overline{q}p}g_{k\overline{\ell}}\nabla^{\overline{b}}\hat{R}_{i\overline{b}p}{}^k\overline{\Psi_{jq}^{\ell}}\right)$$
$$(3.269)$$

$$\leq -|\overline{\nabla}\Psi|^2 - |\nabla\Psi|^2 + S + C|\sigma|_h^{-K}\sqrt{S},$$
$$(3.270)$$

for a uniform constant K. We have

$$|\partial S| \leq \sqrt{S}(|\overline{\nabla}\Psi| + |\nabla\Psi|).$$
$$(3.271)$$

Moreover,

$$|\partial|\sigma|_h^{4K}| \leq C|\sigma|_h^{3K} \quad \text{and} \quad |\Delta|\sigma|_h^{4K}| \leq C|\sigma|_h^{3K},$$
$$(3.272)$$

where we are increasing K if necessary. Then

$$\left(\frac{\partial}{\partial t} - \Delta\right)(|\sigma|_h^{4K}S) = |\sigma|_h^{4K}\left(\frac{\partial}{\partial t} - \Delta\right)S - 2\mathrm{Re}(g^{\overline{j}i}\partial_i|\sigma|_h^{4K}\partial_{\overline{j}}S) - (\Delta|\sigma|_h^{4K})S$$

$$\leq -|\sigma|_h^{4K}(|\overline{\nabla}\Psi|^2 + |\nabla\Psi|^2) + C|\sigma|_h^{3K}\sqrt{S}(|\overline{\nabla}\Psi| + |\nabla\Psi|)$$

$$+ C|\sigma|_h^{2K}S + C$$

$$\leq C(1 + |\sigma|_h^{2K}S).$$
$$(3.273)$$

But from Proposition 3.2.4 and (3.260)

$$\left(\frac{\partial}{\partial t} - \Delta\right)\mathrm{tr}_{\omega_0}\omega = -\mathrm{tr}_{\omega_0}\omega - g^{\overline{\ell}k}R(g_0)_{k\overline{\ell}}{}^{\overline{j}i}g_{i\overline{j}} - g_0^{\overline{j}i}g^{\overline{q}p}g^{\overline{\ell}k}\nabla_i^0 g_{p\overline{\ell}}\nabla_{\overline{j}}^0 g_{k\overline{q}}$$

$$\leq C|\sigma|_h^{-K} - \frac{1}{C}|\sigma|_h^K S - \frac{1}{2}g_0^{\overline{j}i}g^{\overline{q}p}g^{\overline{\ell}k}\nabla_i^0 g_{p\overline{\ell}}\nabla_{\overline{j}}^0 g_{k\overline{q}}, \quad (3.274)$$

where ∇^0 denotes the covariant derivative with respect to g_0. We may assume that K is large enough so that $|(\Delta|\sigma|_h^K)\mathrm{tr}_{\omega_0}\omega| \leq C$. Then

$$\left(\frac{\partial}{\partial t} - \Delta\right)(|\sigma|_h^K\mathrm{tr}_{\omega_0}\omega) \leq -\frac{1}{C}|\sigma|_h^{2K}S + C - 2\mathrm{Re}(g^{\overline{j}i}\partial_i|\sigma|_h^K\partial_{\overline{j}}\mathrm{tr}_{\omega_0}\omega)$$

$$- \frac{1}{2}|\sigma|_h^K g_0^{\overline{j}i}g^{\overline{q}p}g^{\overline{\ell}k}\nabla_i^0 g_{p\overline{\ell}}\nabla_{\overline{j}}^0 g_{k\overline{q}}$$

$$\leq -\frac{1}{C}|\sigma|_h^{2K}S + C,$$
$$(3.275)$$

where for the last line we have used:

$$|2\text{Re}(g^{\bar{j}i}\partial_i|\sigma|_h^K\partial_{\bar{j}}\text{tr}_{\omega_0}\omega)| \le C + \frac{1}{C}|\partial|\sigma|_h^K|^2|\partial\text{tr}_{\omega_0}\omega|^2 \qquad (3.276)$$

$$\le C + \frac{1}{2}|\sigma|_h^K g_0^{\bar{j}i} g^{\bar{q}p} g^{\bar{\ell}k}\nabla_i^0 g_{p\bar{\ell}}\nabla_{\bar{j}}^0 g_{k\bar{q}}, \qquad (3.277)$$

which follows from (3.72), increasing K if necessary.

Now define $Q = |\sigma|_h^{4K}S + A|\sigma|_h^K\text{tr}_{\omega_0}\omega$ for a constant A. Combining (3.273) and (3.275) we see that for A sufficiently large,

$$\left(\frac{\partial}{\partial t} - \Delta\right)Q \le -|\sigma|_h^{2K}S + C, \qquad (3.278)$$

and then Q is bounded from above by the maximum principle. The bound on S then follows. $\square$

As a consequence:

Lemma 3.5.7. $\varphi = \varphi(t)$ and $\omega = \omega(t)$ are uniformly bounded in $C_{\text{loc}}^\infty(M \setminus E)$.

Proof. Applying Theorem 3.2.15 gives the $C_{\text{loc}}^\infty(M \setminus E)$ bounds for ω. Since by Lemmas 3.5.2 and 3.5.3, φ is uniformly bounded (in C^0) on compact subsets of $M \setminus E$, the $C_{\text{loc}}^\infty(M \setminus E)$ bounds on φ follow from those on ω. $\square$

3.5.3 Convergence of the Flow and Uniqueness of the Limit

We now complete the proof of Theorem 3.5.1. From part (ii) of Lemma 3.5.2 we have $\dot{\varphi} \le Cte^{-t}$ for $t \ge t'$. Hence for $t \ge t'$,

$$\frac{\partial}{\partial t}\left(\varphi + Ce^{-t}(t+1)\right) \le 0. \qquad (3.279)$$

On the other hand, from Lemma 3.5.3, the quantity $\varphi + Ce^{-t}(t+1)$ is uniformly bounded from below on compact subsets of $M \setminus E$. Hence $\varphi(t)$ converges pointwise on $M \setminus E$ to a function φ_∞. Since we have $C_{\text{loc}}^\infty(M \setminus E)$ estimates for $\varphi(t)$ this implies, by a similar argument to that given in the proof of Theorem 3.4.1, that φ converges to φ_∞ in $C_{\text{loc}}^\infty(M \setminus E)$. In particular φ_∞ is smooth on $M \setminus E$. Define $\omega_\infty = \hat{\omega}_\infty + \frac{\sqrt{-1}}{2\pi}\partial\bar{\partial}\varphi_\infty$. Then ω_∞ is a smooth Kähler metric on $M \setminus E$.

Moreover, since $\varphi(t)$ converges to φ_∞ we must have, for each $x \in M \setminus E$, $\dot{\varphi}(x, t_i) \to 0$ for a sequence of times $t_i \to \infty$. But since $\dot{\varphi}(t)$ converges in $C_{\text{loc}}^\infty(M \setminus E)$ as $t \to \infty$ we have by uniqueness of limits that $\dot{\varphi}(t)$ converges to zero in $C_{\text{loc}}^\infty(M \setminus E)$ as $t \to \infty$. Taking the limit of (3.234) as $t \to \infty$ we obtain

$$\log\frac{\omega_\infty^n}{\Omega} - \varphi_\infty = 0 \qquad (3.280)$$

on $M \setminus E$ and applying $\partial\bar{\partial}$ to this equation gives $\mathrm{Ric}(\omega_\infty) = -\omega_\infty$ on $M \setminus E$. This completes the proof of Theorem 3.5.1.

We have now proved the existence of a singular Kähler–Einstein metric on M. We now prove a uniqueness result. Let Ω, $\hat{\omega}_\infty$, σ and h be as above.

Theorem 3.5.8. *There exists a unique smooth Kähler metric ω_{KE} on $M \setminus E$ satisfying*

(i) $\mathrm{Ric}(\omega_{\mathrm{KE}}) = -\omega_{\mathrm{KE}}$ on $M \setminus E$.
(ii) There exists a constant C and for every $\varepsilon > 0$ a constant $C_\varepsilon > 0$ with

$$\frac{1}{C_\varepsilon} |\sigma|_h^{2\varepsilon} \Omega \leq \omega_{\mathrm{KE}}^n \leq C\Omega, \qquad \text{on } M \setminus E. \tag{3.281}$$

Note that although it may appear that condition (ii) depends on the choices of Ω, σ and h, in fact it is easy to see it does not.

Proof of Theorem 3.5.8. The existence part follows immediately from Theorem 3.5.1, Lemmas 3.5.2 and 3.5.3, so it remains to prove uniqueness. Suppose ω_{KE} and $\tilde{\omega}_{\mathrm{KE}}$ are two solutions and define functions ψ and $\tilde{\psi}$ on $M \setminus E$ by

$$\psi = \log \frac{\omega_{\mathrm{KE}}^n}{\Omega} \quad \text{and} \quad \tilde{\psi} = \log \frac{\tilde{\omega}_{\mathrm{KE}}^n}{\Omega}, \tag{3.282}$$

with Ω as in (3.233). Then we have

$$\omega_{\mathrm{KE}} = -\mathrm{Ric}(\omega_{\mathrm{KE}}) = \hat{\omega}_\infty + \frac{\sqrt{-1}}{2\pi}\partial\bar{\partial}\psi, \quad \tilde{\omega}_{\mathrm{KE}} = -\mathrm{Ric}(\tilde{\omega}_{\mathrm{KE}}) = \hat{\omega}_\infty + \frac{\sqrt{-1}}{2\pi}\partial\bar{\partial}\tilde{\psi}. \tag{3.283}$$

Hence it suffices to show that $\psi = \tilde{\psi}$. By symmetry it is enough to show $\psi \geq \tilde{\psi}$.

For any $\varepsilon > 0$ and $\delta > 0$ sufficiently small, define

$$H = \psi - (1-\delta)\tilde{\psi} - \delta\varepsilon \log |\sigma|_h^2. \tag{3.284}$$

From the condition (3.281), $\tilde{\psi}$ is bounded from above and $\psi \geq \varepsilon' \log |\sigma|_h^2 - C_{\varepsilon'}$ for any $\varepsilon' > 0$. Taking $\varepsilon' = \varepsilon\delta/2$ we see that

$$H \geq -\frac{\varepsilon\delta}{2}\log |\sigma|_h^2 - C_{\varepsilon'} - C, \tag{3.285}$$

and hence H is bounded from below by a constant depending on ε and δ and tends to infinity on E. Hence H achieves a minimum at a point $x_0 \in M \setminus E$.

On the other hand, we have

$$\log \frac{\omega_{\mathrm{KE}}^n}{\tilde{\omega}_{\mathrm{KE}}^n} = \psi - \tilde{\psi}, \tag{3.286}$$

which using (3.283) we can rewrite as

$$\log \frac{\left(\hat{\omega}_\infty + (1-\delta)\frac{\sqrt{-1}}{2\pi}\partial\bar{\partial}\tilde{\psi} - \delta\varepsilon R_h + \frac{\sqrt{-1}}{2\pi}\partial\bar{\partial}H\right)^n}{\tilde{\omega}_{KE}^n} = \psi - \tilde{\psi}. \qquad (3.287)$$

Since $\delta\hat{\omega}_\infty - \delta\varepsilon R_h$ is Kähler for ε sufficiently small, we obtain

$$\psi - \tilde{\psi} \geq \log \frac{(1-\delta)^n\left(\tilde{\omega}_{KE} + \frac{\sqrt{-1}}{2\pi}\partial\bar{\partial}\left(\frac{H}{1-\delta}\right)\right)^n}{\tilde{\omega}_{KE}^n}. \qquad (3.288)$$

Hence at the point x_0 at which H achieves a minimum we have

$$\psi - \tilde{\psi} \geq n\log(1-\delta), \qquad (3.289)$$

and so, using the inequality $\tilde{\psi} \geq \varepsilon\log|\sigma|_h^2 - C_\varepsilon$,

$$H(x_0) \geq \delta\tilde{\psi}(x_0) + n\log(1-\delta) - \delta\varepsilon\log|\sigma|_h^2(x_0) \geq -\delta C_\varepsilon + n\log(1-\delta). \qquad (3.290)$$

For any $\varepsilon > 0$ we may choose $\delta = \delta(\varepsilon)$ sufficiently small so that $\delta C_\varepsilon < \varepsilon/2$ and $n\log(1-\delta) > -\varepsilon/2$, giving $H(x_0) \geq -\varepsilon$ and hence $H \geq -\varepsilon$ on $M \setminus E$. It follows that on $M \setminus E$,

$$\psi \geq (1-\delta)\tilde{\psi} + \delta\varepsilon\log|\sigma|_h^2 - \varepsilon. \qquad (3.291)$$

Letting $\varepsilon \to 0$ (so that $\delta \to 0$ too) gives $\psi \geq \tilde{\psi}$ as required.   □

3.5.4 Further Estimates Using Pluripotential Theory

In this section we will show how results from pluripotential theory can be used to improve on the estimates given in the proof of Theorem 3.5.1.

The following *a priori* estimate, extending Yau's zeroth order estimate, was proved by Eyssidieux–Guedj–Zeriahi [EGZ11]. A slightly weaker version of this result, which would also suffice for our purposes, was proved independently by Zhang [Zha06].

Theorem 3.5.9. *Let M be a compact Kähler manifold and ω a closed smooth semipositive $(1,1)$-form with $\int_M \omega^n > 0$. Let f be a smooth nonnegative function. Fix $p > 1$. Then if φ is a smooth function with $\omega + \frac{\sqrt{-1}}{2\pi}\partial\bar{\partial}\varphi \geq 0$ solving the complex Monge–Ampère equation*

$$\left(\omega + \frac{\sqrt{-1}}{2\pi}\partial\bar{\partial}\varphi\right)^n = f\omega^n, \qquad (3.292)$$

then there exists a constant C depending only on M, ω and $\|f\|_{L^p(M,\omega)}$ such that

$$\mathrm{osc}_M \varphi \leq C. \tag{3.293}$$

The differences between this result and Theorem 3.4.6 are that here ω is only required to be semi-positive and the estimate on φ depends only on the L^p bound of the right hand side of the equation. We remark that we have not stated the result in the sharpest possible way. The conditions that φ and f are smooth can be relaxed to φ being bounded with $\omega + \frac{\sqrt{-1}}{2\pi}\partial\bar{\partial}\varphi \geq 0$ and f being in L^p. We have ignored this to avoid technicalities such as defining the Monge–Ampère operator in this more general setting. We omit the proof of this theorem since it goes beyond the scope of these notes. The theorem is a generalization of a seminal work of Kołodziej [Kol98]. For a further generalization, see [BEGZ10].

We will apply Theorem 3.5.9 to show that the solution $\varphi = \varphi(t)$ of the parabolic complex Monge–Ampère equation (3.234) is uniformly bounded, a result first established by Tian–Zhang [Tzha06]. Moreover, we can in addition obtain a bound on $\dot{\varphi}$ [Zha09].

Proposition 3.5.10. *There exists a uniform C such that under the assumptions of Theorem 3.5.1, φ solving (3.234) satisfies for $t \in [0, \infty)$,*

$$\|\varphi\|_{C^0} \leq C \quad \text{and} \quad \|\dot{\varphi}\|_{C^0} \leq C. \tag{3.294}$$

Hence there exists a uniform constant $C' > 0$ such that for $t \in [0, \infty)$,

$$\frac{1}{C'}\Omega \leq \omega^n \leq C'\Omega. \tag{3.295}$$

Proof. First observe that

$$\left(\hat{\omega}_t + \frac{\sqrt{-1}}{2\pi}\partial\bar{\partial}\varphi\right)^n = f\hat{\omega}_t^n, \quad \text{for} \quad f = e^{\dot{\varphi}+\varphi}\frac{\Omega}{\hat{\omega}_t^n} \geq 0. \tag{3.296}$$

From the definition of $\hat{\omega}_t$ and Lemma 3.5.2 we see that f is uniformly bounded from above, and hence bounded in L^p for any p. Applying Theorem 3.5.9 we see that $\mathrm{osc}_M\varphi \leq C$ for some uniform constant.

For the bound on φ, it only remains to check that there exists a constant C' such that for each time t there exists $x \in M$ with $|\varphi(x)| \leq C'$. From Lemma 3.5.2 we have an upper bound for $\varphi(x)$ for all $x \in M$. For the lower bound, observe that

$$\int_M e^{\dot{\varphi}+\varphi}\Omega = \int_M \left(\hat{\omega}_t + \frac{\sqrt{-1}}{2\pi}\partial\bar{\partial}\varphi\right)^n = \int_M \hat{\omega}_t^n \geq c, \tag{3.297}$$

for some uniform constant $c > 0$. It follows that at each time t there exists $x \in M$ with $e^{\dot{\varphi}(x)+\varphi(x)} \geq c/\int_M \Omega$. Since $\dot{\varphi}$ is uniformly bounded from above by Lemma 3.5.2 this gives $\varphi(x) \geq -C'$ for that x, as required.

For the bound on $\dot{\varphi}$ we use an argument due to Zhang [Zha09]. From (3.234) and Theorem 3.2.2,

$$\frac{\partial}{\partial t}(\dot{\varphi} + \varphi) = \frac{\partial}{\partial t}\left(\log \frac{\omega^n}{\Omega}\right) = -R - n \leq C_0 e^{-t} \qquad (3.298)$$

for a uniform constant C_0. We may suppose that $\|\varphi\|_{C^0} \leq C_0$ for the same constant $C_0 > 0$. We claim that $\dot{\varphi} > -4C_0$. Suppose not. Then there exists a point (x_0, t_0) with $\dot{\varphi}(x_0, t_0) \leq -4C_0$. Using (3.298) we have for any $t > t_0$,

$$(\dot{\varphi} + \varphi)(x_0, t) - (\dot{\varphi} + \varphi)(x_0, t_0) \leq C_0 \int_{t_0}^t e^{-s} ds = C_0(e^{-t_0} - e^{-t}). \qquad (3.299)$$

Hence for $t > t_0$,

$$\dot{\varphi}(x_0, t) \leq (\dot{\varphi} + \varphi)(x_0, t_0) + C_0 e^{-t_0} - \varphi(x_0, t) \leq -C_0, \qquad (3.300)$$

using the fact that $\dot{\varphi}(x_0, t_0) \leq -4C_0$. This is a contradiction since $\varphi(x_0, t)$ is uniformly bounded as $t \to \infty$. $\qquad \square$

An immediate consequence is:

Corollary 3.5.11. *The singular Kähler–Einstein metric ω_{KE} constructed in Theorem 3.5.1 satisfies*

$$\frac{1}{C}\Omega \leq \omega_{\mathrm{KE}}^n \leq C\Omega \quad \text{on } M \setminus E, \qquad (3.301)$$

for some $C > 0$.

As another application of Proposition 3.5.10, we use the estimate on φ together with the parabolic Schwarz lemma to obtain a lower bound on the metric ω.

Lemma 3.5.12. *Under the assumptions of Theorem 3.5.1, there exists a uniform constant C such that*

$$\omega \geq \frac{1}{C}\hat{\omega}_\infty, \quad \text{on } M \times [0, \infty). \qquad (3.302)$$

Proof. Recall that $\hat{\omega}_\infty = \frac{1}{m}\Phi^* \omega_{\mathrm{FS}}$ where $\Phi : M \to \mathbb{P}^N$ is a holomorphic map and ω_{FS} is the Fubini–Study metric on $\mathbb{P}^N$. We can then directly apply Theorem 3.2.6 to obtain

$$\left(\frac{\partial}{\partial t} - \Delta\right) \log \mathrm{tr}_\omega \hat{\omega}_\infty \leq C' \mathrm{tr}_\omega \hat{\omega}_\infty + 1, \qquad (3.303)$$

for C' an upper bound for the bisectional curvature of ω_{FS}. Define $Q = \log \text{tr}_\omega \hat{\omega}_\infty - A\varphi$ for A to be determined later. Compute, using Proposition 3.5.10,

$$\left(\frac{\partial}{\partial t} - \Delta\right) Q \leq C' \text{tr}_\omega \hat{\omega}_\infty - A\dot\varphi + An - A\text{tr}_\omega \hat{\omega}_t + 1$$

$$\leq -\text{tr}_\omega \hat{\omega}_\infty + C, \tag{3.304}$$

where we have chosen A to be sufficiently large so that $A\hat{\omega}_t \geq (C' + 1)\hat{\omega}_\infty$. It follows from the maximum principle that Q and hence $\text{tr}_\omega \hat{\omega}_\infty$ is uniformly bounded from above and this completes the proof of the lemma. □

Observe that Lemma 3.5.12 together with the volume upper bound from Lemma 3.5.2 show that the metric $\omega(t)$ is uniformly bounded above and below on compact subsets of $M \setminus S$, for S the set of points where $\hat{\omega}_\infty$ is degenerate. Thus we can obtain an alternative proof of Theorem 3.5.1 which avoids the use of Lemmas 3.5.3 and 3.5.5.

Finally we mention that Zhang [Zha09] also proved a uniform bound for the scalar curvature of the evolving metric in this setting.

3.6 Kähler–Ricci Flow on a Product Elliptic Surface

In this section we investigate *collapsing* along the Kähler–Ricci flow. We study this behavior in the simple case of a product of two Riemann surfaces.

3.6.1 Elliptic Surfaces and the Kähler–Ricci Flow

Let M now have complex dimension two. An *elliptic curve* E is a compact Riemann surface with $c_1(E) = 0$ (by the Gauss–Bonnet formula this is equivalent to having genus equal to 1). We say that M is an *elliptic surface* if there exists a surjective holomorphic map $\pi : M \to S$ onto a Riemann surface S such that the fiber $\pi^{-1}(s)$ is an elliptic curve for all but finitely many $s \in S$. In particular, the product of an elliptic curve and any Riemann surface is an elliptic surface, which we will call a *product elliptic surface*.

In [ST07], the Kähler–Ricci flow was studied on a general minimal elliptic surface (see Sect. 3.8 for a definition of *minimal*). In this case there are finitely many singular fibers of the map π. It was shown that the Kähler–Ricci flow converges in $C^{1+\beta}$ for any $\beta \in (0, 1)$ at the level of potentials away from the singular fibers, and also converges on M in the sense of currents, to a *generalized Kähler–Einstein metric* on the base S. A higher dimensional analogue was given in [ST12].

Here we study the behavior of the Kähler–Ricci flow in the more elementary case of a product elliptic surface $M = E \times S$, where E is an elliptic curve and S is a Riemann surface with $c_1(S) < 0$ (genus greater than 1). Because of the simpler structure of the manifold, we can obtain stronger estimates than in [ST07].

By the uniformization theorem for Riemann surfaces (or the results of Sect. 3.4), S and E admit Kähler metrics of constant curvature which are unique up to scaling. Hence we can define Kähler metrics ω_S on S and ω_E on E by

$$\text{Ric}(\omega_S) = -\omega_S, \quad \text{Ric}(\omega_E) = 0, \quad \int_E \omega_E = 1. \tag{3.305}$$

Denote by π_S and π_E the projection maps $\pi_S : M \to S$ and $\pi_E : M \to E$.

As in the case of the previous section we consider the normalized Kähler–Ricci flow

$$\frac{\partial}{\partial t}\omega = -\text{Ric}(\omega) - \omega, \quad \omega|_{t=0} = \omega_0, \tag{3.306}$$

The first Chern class of M is given by $c_1(M) = -[\pi^*\omega_S]$, which can be seen from the equation

$$\text{Ric}(\pi_S^*\omega_S + \pi_E^*\omega_E) = -\pi_S^*\omega_S. \tag{3.307}$$

Since $\pi^*\omega_S$ is a nonnegative (1,1) form on M, it follows from Theorem 3.3.1 that a solution to (3.306) exists for all time for any initial Kähler metric ω_0.

As a simple example, first consider the case when the initial metric ω_0 splits as a product. Suppose $\omega_0 = \pi_E^*\omega_E^0 + \pi_S^*\omega_S^0$, where ω_E^0 and ω_S^0 are smooth metrics on E and S respectively. Then the Kähler–Ricci flow splits into the Kähler–Ricci flows on E and S, with $\omega(t) = \pi_E^*\omega_{E,t} + \pi_S^*\omega_{S,t}$ where $\omega_{E,t}$ and $\omega_{S,t}$ solve the Kähler–Ricci flow on E and S respectively. Since $c_1(E) = 0$ and $c_1(S) < 0$ we can apply the results of Sect. 3.4 to see that $\omega_{E,t}$ converges in C^∞ to 0 (because of the normalization) as $t \to \infty$ and $\omega_{S,t}$ converges in C^∞ to ω_S. Hence the solution to the original normalized Kähler–Ricci flow converges in C^∞ to $\pi_S^*\omega_S$.

We now turn back to the general case of a non-product metric. For convenience, here and henceforth we will drop the π_S^* and π_E^* and write ω_S and ω_E for the (1, 1)-forms pulled back to M. We prove:

Theorem 3.6.1. *Let $\omega(t)$ be the solution of the normalized Kähler–Ricci flow (3.306) on $M = E \times S$ with initial Kähler metric ω_0. Then*

(i) *For any $\beta \in (0, 1)$, $\omega(t)$ converges to ω_S in $C^\beta(M, g_0)$.*
(ii) *The curvature tensors of $\omega(t)$ and their derivatives are uniformly bounded along the flow.*
(iii) *For any fixed fiber $E = \pi_S^{-1}(s)$, we have*

$$\|e^t\omega(t)|_E - \omega_{\text{flat}}\|_{C^0(E)} \to 0 \quad \text{as} \quad t \to \infty, \tag{3.308}$$

where ω_{flat} is the Kähler-Ricci-flat metric on E with $\int_E \omega_{\text{flat}} = \int_E \omega_0$.

Remark 3.6.2. We conjecture that in (i), the convergence in $C^\beta(M)$ can be replaced by $C^\infty(M)$ convergence. Such a result is contained in the work of Gross–Tosatti–Zhang [GTZ11] for the case of a family of Ricci-flat metrics. It seems likely that their methods could be extended to cover this case too. It would also be interesting to find a proof of C^∞ convergence using only the maximum principle.

Since the normalized Kähler–Ricci flow exists for all time we can compute, as in (3.230) and (3.231), the evolution of the Kähler class to be

$$[\omega(t)] = e^{-t}[\omega_0] + (1 - e^{-t})[\omega_S]. \tag{3.309}$$

Before proving Theorem 3.6.1 we will, as in the sections above, reduce (3.306) to a parabolic complex Monge–Ampère equation. Define reference metrics $\hat{\omega}_t \in [\omega(t)]$ by

$$\hat{\omega}_t = e^{-t}\omega_0 + (1 - e^{-t})\omega_S, \qquad \text{for } t \in [0, \infty). \tag{3.310}$$

Define a smooth volume form Ω on M by

$$\frac{\sqrt{-1}}{2\pi}\partial\bar{\partial}\log\Omega = \omega_S \in -c_1(M), \quad \int_M \Omega = 2\int_M \omega_0 \wedge \omega_S. \tag{3.311}$$

In fact, from (3.307) one can see that Ω is a constant multiple of $\omega_S \wedge \omega_E$. We consider the parabolic complex Monge–Ampère equation

$$\frac{\partial}{\partial t}\varphi = \log\frac{e^t(\hat{\omega}_t + \frac{\sqrt{-1}}{2\pi}\partial\bar{\partial}\varphi)^2}{\Omega} - \varphi, \qquad \hat{\omega}_t + \frac{\sqrt{-1}}{2\pi}\partial\bar{\partial}\varphi > 0, \qquad \varphi|_{t=0} = 0. \tag{3.312}$$

As in earlier sections, a solution $\varphi = \varphi(t)$ of (3.312) exists for all time and $\omega = \hat{\omega}_t + \frac{\sqrt{-1}}{2\pi}\partial\bar{\partial}\varphi$ solves the normalized Kähler–Ricci flow. Note that we insert the factor of e^t in the equation to ensure that φ is uniformly bounded (see Lemma 3.6.3 below) but of course it does not change the evolution of the metric along the flow.

3.6.2 Estimates

In this section we establish uniform estimates for the solution $\varphi = \varphi(t)$ of (3.312), which we know exists for all time.

Lemma 3.6.3. *There exists $C > 0$ such that on $M \times [0, \infty)$,*

  (i) $|\varphi| \leq C$.
  (ii) $|\dot{\varphi}| \leq C$.
  (iii) $\dfrac{1}{C}\hat{\omega}_t^2 \leq \omega^2 \leq C\hat{\omega}_t^2$.

Proof. For (i), first note that since $e^t \hat{\omega}_t^2 = e^{-t}\omega_0^2 + 2(1 - e^{-t})\omega_0 \wedge \omega_S$ we have

$$\frac{1}{C}\Omega \le e^t \hat{\omega}_t^2 \le C\Omega. \tag{3.313}$$

Hence if φ achieves a maximum at (x_0, t_0) with $t_0 > 0$ then at that point,

$$0 \le \frac{\partial}{\partial t}\varphi \le \log \frac{e^t \hat{\omega}_t^2}{\Omega} - \varphi \le \log C - \varphi, \tag{3.314}$$

giving $\varphi \le \log C$. The lower bound of φ follows similarly.

For (ii) observe that $\frac{\partial}{\partial t}\hat{\omega}_t = \omega_S - \hat{\omega}_t$ and hence

$$\left(\frac{\partial}{\partial t} - \Delta\right)\dot{\varphi} = \mathrm{tr}_\omega(\omega_S - \hat{\omega}_t) + 1 - \dot{\varphi}. \tag{3.315}$$

By definition of $\hat{\omega}_t$ there exists a uniform constant $C_0 > 0$ such that $C_0\hat{\omega}_t \ge \omega_S$. For the upper bound of $\dot{\varphi}$, we apply the maximum principle to $Q_1 = \dot{\varphi} - (C_0 - 1)\varphi$. Compute

$$\left(\frac{\partial}{\partial t} - \Delta\right)Q_1 = \mathrm{tr}_\omega(\omega_S - \hat{\omega}_t) + 1 - C_0\dot{\varphi} + (C_0 - 1)\mathrm{tr}_\omega(\omega - \hat{\omega}_t)$$

$$\le 1 - C_0\dot{\varphi} + 2(C_0 - 1), \tag{3.316}$$

and we see that $\dot{\varphi}$ is uniformly bounded from above at a point where Q_1 achieves a maximum. Since φ is bounded by (i) we obtain the required upper bound of $\dot{\varphi}$.

For the lower bound of $\dot{\varphi}$, let $Q_2 = \dot{\varphi} + 2\varphi$ and compute

$$\left(\frac{\partial}{\partial t} - \Delta\right)Q_2 = \mathrm{tr}_\omega(\omega_S - \hat{\omega}_t) + 1 + \dot{\varphi} - 2\mathrm{tr}_\omega(\omega - \hat{\omega}_t)$$

$$\ge \mathrm{tr}_\omega\hat{\omega}_t + \dot{\varphi} - 3. \tag{3.317}$$

Using (3.312), (3.313) and the arithmetic–geometric means inequality, we have at a point (x_0, t_0) where Q_2 achieves a minimum,

$$e^{-(\dot{\varphi}+\varphi)/2} = \left(\frac{\Omega}{e^t\omega^2}\right)^{1/2} \le C\left(\frac{\hat{\omega}_t^2}{\omega^2}\right)^{1/2} \le \frac{C}{2}\mathrm{tr}_\omega\hat{\omega}_t \le C' - \dot{\varphi}. \tag{3.318}$$

Hence $\dot{\varphi}$ is uniformly bounded from below at (x_0, t_0), giving (ii). Part (iii) follows from (i) and (ii). □

Next we estimate ω in terms of $\hat{\omega}_t$. It is convenient to define another family of reference metrics $\tilde{\omega}_t$ whose curvature we can control more precisely. Define $\tilde{\omega}_0 = \omega_E + \omega_S$ and

$$\tilde{\omega}_t = e^{-t}\tilde{\omega}_0 + (1 - e^{-t})\omega_S = \omega_S + e^{-t}\omega_E. \tag{3.319}$$

Observe that $\tilde{\omega}_t$ and $\hat{\omega}_t$ are uniformly equivalent.

Lemma 3.6.4. *There exists $C > 0$ such that on $M \times [0, \infty)$,*

$$\frac{1}{C}\tilde{\omega}_t \le \omega \le C\tilde{\omega}_t \tag{3.320}$$

Proof. From part (iii) of Lemma 3.6.3 it suffices to obtain an upper bound of the quantity $\operatorname{tr}_{\tilde{\omega}_t}\omega$ from above. Compute using the argument of Proposition 3.2.4,

$$\left(\frac{\partial}{\partial t} - \Delta\right)\operatorname{tr}_{\tilde{\omega}_t}\omega \le -\operatorname{tr}_{\tilde{\omega}_t}\omega - g^{\bar{j}i}\,\tilde{R}_{i\bar{j}}{}^{\bar{\ell}k}g_{k\bar{\ell}}$$

$$- \tilde{g}^{\bar{j}i}g^{\bar{q}p}g^{\bar{\ell}k}\tilde{\nabla}_i g_{p\bar{\ell}}\tilde{\nabla}_{\bar{j}}g_{k\bar{q}} + \left(\frac{\partial}{\partial t}\tilde{g}^{\bar{j}i}\right)g_{i\bar{j}}, \tag{3.321}$$

where we are using $\tilde{R}_{i\bar{j}}{}^{\bar{\ell}k}$ and $\tilde{\nabla} = \nabla_{\tilde{g}_t}$ to denote the curvature and covariant derivative with respect to $\tilde{g}_t$. Since $\frac{\partial}{\partial t}\tilde{\omega}_t = -\tilde{\omega}_t + \omega_S \ge -\tilde{\omega}_t$, we have

$$\left(\frac{\partial}{\partial t}\tilde{g}_t^{\bar{j}i}\right)g_{i\bar{j}} \le \operatorname{tr}_{\tilde{\omega}_t}\omega. \tag{3.322}$$

Hence, from the argument of Proposition 3.2.5,

$$\left(\frac{\partial}{\partial t} - \Delta\right)\log\operatorname{tr}_{\tilde{\omega}_t}\omega \le -\frac{1}{\operatorname{tr}_{\tilde{\omega}_t}\omega}g^{\bar{j}i}\,\tilde{R}_{i\bar{j}}{}^{\bar{\ell}k}g_{k\bar{\ell}}. \tag{3.323}$$

Next we claim that

$$-g^{\bar{j}i}\,\tilde{R}_{i\bar{j}}{}^{\bar{\ell}k}g_{k\bar{\ell}} = (\operatorname{tr}_\omega\omega_S)\frac{2\,\omega_E \wedge \omega}{\tilde{\omega}_0^2}$$

$$\le (\operatorname{tr}_\omega\omega_S)(\operatorname{tr}_{\tilde{\omega}_0}\omega) \le (\operatorname{tr}_\omega\omega_S)(\operatorname{tr}_{\tilde{\omega}_t}\omega). \tag{3.324}$$

To see (3.324), compute in a local holomorphic product coordinate system (z^1, z^2) with z^1 a normal coordinate for $\omega_S|_S$ in the base S direction and z^2 a normal coordinate for $\omega_E|_E$ in the fiber E direction. In these coordinates $\tilde{g}_t$ is diagonal and $(\tilde{g}_t)_{1\bar{1}} = (g_S)_{1\bar{1}}$. Since the curvature of ω_E vanishes, we have from (3.307)

$$\tilde{R}_{1\bar{1}1\bar{1}} = -(g_S)_{1\bar{1}}(g_S)_{1\bar{1}}, \tag{3.325}$$

and $\tilde{R}_{i\bar{j}k\bar{\ell}} = 0$ if i, j, k and ℓ are not all equal to 1. Hence the only non-zero component of the curvature of $\tilde{\omega}_t$ appearing in (3.324) is $\tilde{R}_{1\bar{1}}{}^{1\bar{1}} = -1$. This gives the first equality of (3.324), and the next two inequalities follow from the definition of $\tilde{\omega}_0$ and $\tilde{\omega}_t$.

Combining (3.323), (3.324) we have

$$\left(\frac{\partial}{\partial t} - \Delta\right) \log \operatorname{tr}_{\tilde{\omega}_t} \omega \leq \operatorname{tr}_\omega \omega_S. \tag{3.326}$$

Now define

$$Q_3 = \log \operatorname{tr}_{\tilde{\omega}_t} \omega - A\varphi, \tag{3.327}$$

for $A = C_0 + 1$ where C_0 is the positive constant with $C_0 \hat{\omega}_t \geq \omega_S$ and compute

$$\left(\frac{\partial}{\partial t} - \Delta\right) Q_3 \leq \operatorname{tr}_\omega \omega_S - A\dot{\varphi} + A\operatorname{tr}_\omega(\omega - \hat{\omega}_t)$$

$$\leq C - \operatorname{tr}_\omega \hat{\omega}_t$$

$$\leq C - \frac{1}{C'}\operatorname{tr}_{\tilde{\omega}_t}\omega, \tag{3.328}$$

for some $C' > 0$. For the last line we have used the estimate (iii) of Lemma 3.6.3 and the fact that $\tilde{\omega}_t$ and $\hat{\omega}_t$ are uniformly equivalent. Since φ is uniformly bounded from part (i) of Lemma 3.6.3 we see that Q_3 is bounded from above by the maximum principle, completing the proof of the lemma.   □

Next we prove an estimate on the derivative of ω using an argument similar to that of Theorem 3.2.9.

Lemma 3.6.5. *There exists a uniform constant C such that on $M \times [0, \infty)$,*

$$S := |\nabla_{\tilde{g}_0} g|^2 \leq C \quad \text{and} \quad |\nabla_{\tilde{g}_0} g|^2_{\tilde{g}_0} \leq C, \tag{3.329}$$

where $|\cdot|$, $|\cdot|_{\tilde{g}_0}$ denote the norms with respect to the metrics $g = g(t)$ and $\tilde{g}_0$ respectively. Moreover, we have

$$\left(\frac{\partial}{\partial t} - \Delta\right) S \leq -\frac{1}{2}|\operatorname{Rm}(g)|^2 + C' \tag{3.330}$$

for a uniform constant C'.

Proof. First we show that

$$\left(\frac{\partial}{\partial t} - \Delta\right) \operatorname{tr}_{\tilde{\omega}_t} \omega \leq C - \frac{1}{C'}|\nabla_{\tilde{g}_0} g|^2, \tag{3.331}$$

for uniform constants C, C'. From (3.321), (3.322), (3.324) and part (iii) of Lemma 3.6.3,

$$\left(\frac{\partial}{\partial t} - \Delta\right) \text{tr}_{\tilde{\omega}_t}\omega \leq C - \tilde{g}_t^{\bar{j}i} g^{\bar{q}p} g^{\bar{\ell}k} \tilde{\nabla}_i g_{p\bar{\ell}} \overline{\tilde{\nabla}_j g_{p\bar{q}}} \leq C - \frac{1}{C'}S. \qquad (3.332)$$

For the last inequality we are using the fact that $\nabla_{\tilde{g}_t} = \nabla_{\tilde{g}_0}$ which can be seen by choosing a coordinate system at a point in which $\partial_i \tilde{g}_t = 0$ for all i and any $t \geq 0$. This establishes (3.331).

Using the notation of Proposition 3.2.8, write $\Psi_{ij}^k = \Gamma_{ij}^k - \Gamma(\tilde{g}_0)_{ij}^k$ so that $S = |\Psi|^2$. Then

$$\left(\frac{\partial}{\partial t} - \Delta\right) S = -|\overline{\nabla}\Psi|^2 - |\nabla\Psi|^2 + |\Psi|^2 - 2\text{Re}\left(g^{\bar{j}i} g^{\bar{q}p} g_{k\bar{\ell}} \nabla^{\bar{b}} R(\tilde{g}_0)_{i\bar{b}p}{}^k \overline{\Psi_{jq}^\ell}\right).$$

We have

$$\nabla^{\bar{b}} R(\tilde{g}_0)_{i\bar{b}p}{}^k = -g^{\bar{b}a}\Psi_{ia}^m R(\tilde{g}_0)_{m\bar{b}p}{}^k - g^{\bar{b}a}\Psi_{pa}^m R(\tilde{g}_0)_{i\bar{b}m}{}^k + g^{\bar{b}a}\Psi_{ma}^k R(\tilde{g}_0)_{i\bar{b}p}{}^m. \tag{3.333}$$

Indeed, as in the proof of Lemma 3.6.4, this can be seen by choosing a local holomorphic product coordinate system (z^1, z^2) centered at a point x with z^1 normal for ω_S and z^2 normal for ω_E. Using the argument of (3.324) and the result of Lemma 3.6.4 we have

$$|\text{Rm}(\tilde{g}_0)|_g^2 := g^{\bar{j}i} g^{\bar{\ell}k} g^{\bar{q}p} g_{a\bar{b}} R(\tilde{g}_0)_{i\bar{\ell}p}{}^a \overline{R(\tilde{g}_0)_{j\bar{k}q}{}^b}$$

$$= (\text{tr}_\omega\omega_S)^3 \frac{2\,\omega_E \wedge \omega}{\tilde{\omega}_0^2}$$

$$\leq (\text{tr}_\omega\tilde{\omega}_t)^3 \text{tr}_{\tilde{\omega}_t}\omega \leq C. \tag{3.334}$$

Combining (3.333) and (3.334),

$$\left|2\text{Re}\left(g^{\bar{j}i} g^{\bar{q}p} g_{k\bar{\ell}} \nabla^{\bar{b}} R(\tilde{g}_0)_{i\bar{b}p}{}^k \overline{\Psi_{jq}^\ell}\right)\right| \leq CS. \tag{3.335}$$

Since $|\overline{\nabla}\Psi|^2 = |\text{Rm}(\tilde{g}_0) - \text{Rm}(g)|_g^2$, we compute

$$\left(\frac{\partial}{\partial t} - \Delta\right) S \leq -|\overline{\nabla}\Psi|^2 - |\nabla\Psi|^2 + CS$$

$$\leq -\frac{1}{2}|\text{Rm}(g)|^2 + CS + C' \tag{3.336}$$

Then the upper bound on S follows from (3.332) and (3.336) by applying the maximum principle to $S + A \operatorname{tr}_{\tilde{\omega}_t} \omega$ for sufficiently large A. The inequality $|\nabla_{\tilde{g}_0} g|^2_{g_0} \leq C$ follows from the fact that the metric $g(t)$ is bounded from above by g_0 (Lemma 3.6.4). The inequality (3.330) follows from (3.336).   □

We then easily obtain estimates for curvature and all covariant derivatives of curvature, establishing part (ii) of Theorem 3.6.1.

Lemma 3.6.6. *There exist uniform constants C_m for $m = 0, 1, 2, \ldots$ such that on $M \times [0, \infty)$,*

$$|\nabla^m_{\mathbb{R}} \operatorname{Rm}(g)|^2 \leq C_m. \tag{3.337}$$

Proof. This follows from Lemma 3.6.5 and the arguments of Theorems 3.2.13 and 3.2.14.   □

3.6.3 Fiber Collapsing and Convergence

In this subsection, we complete the proof of Theorem 3.6.1.

First we define a closed $(1, 1)$ form ω_{flat} on M with the properties that $[\omega_{\text{flat}}] = [\omega_0]$ and for each $s \in S$, ω_{flat} restricted to the fiber $\pi_S^{-1}(s)$ is a Kähler–Ricci flat metric. To do this, fix $s \in S$ and define a function ρ_s on $\pi_S^{-1}(s)$ by

$$\omega_0|_{\pi_S^{-1}(s)} + \frac{\sqrt{-1}}{2\pi} \partial\bar{\partial}\rho_s > 0, \quad \operatorname{Ric}\left(\omega_0|_{\pi_S^{-1}(s)} + \frac{\sqrt{-1}}{2\pi} \partial\bar{\partial}\rho_s\right) = 0,$$

$$\int_{\pi_S^{-1}(s)} \rho_s \, \omega_0 = 0. \tag{3.338}$$

Since ρ_s satisfies a partial differential equation with parameters depending smoothly on $s \in S$, it follows that ρ_s varies smoothly with s and hence defines a smooth function on M, which we will call ρ. Now set $\omega_{\text{flat}} := \omega_0 + \frac{\sqrt{-1}}{2\pi} \partial\bar{\partial}\rho$. This is a closed $(1, 1)$ form with the desired properties. Note that for each s in S, $\omega_{\text{flat}}|_{\pi_S^{-1}(s)}$ is a metric, but ω_{flat} may not be positive definite as a $(1, 1)$ form on M.

We make use of ω_{flat} to prove the following estimate on φ.

Lemma 3.6.7. *There exists $C > 0$ such that on $M \times [0, \infty)$,*

$$|\varphi| \leq C(1 + t)e^{-t}. \tag{3.339}$$

Proof. Since ω_{flat} is a constant multiple of ω_E when restricted to each fiber, we see from the definition of Ω that

$$\Omega = 2\omega_S \wedge \omega_{\text{flat}}. \tag{3.340}$$

Let $Q = \varphi - e^{-t}\rho$. Then

$$\frac{\partial}{\partial t}Q = \log \frac{e^t(e^{-t}\omega_{\text{flat}} + (1 - e^{-t})\omega_S + \frac{\sqrt{-1}}{2\pi}\partial\bar{\partial}Q)^2}{2\omega_S \wedge \omega_{\text{flat}}} - Q. \qquad (3.341)$$

For a positive constant A, consider the quantity $Q_1 = e^t Q - At$. At a point (x_0, t_0) with $t_0 > 0$ where Q_1 achieves a maximum, we have

$$0 \le \frac{\partial}{\partial t}Q_1 \le e^t \log \frac{e^t(e^{-t}\omega_{\text{flat}} + (1 - e^{-t})\omega_S)^2}{2\omega_S \wedge \omega_{\text{flat}}} - A$$

$$\le e^t \log(1 + Ce^{-t}) - A \le C' - A, \qquad (3.342)$$

for uniform constants C, C'. Choosing $A > C'$ gives a contradiction. Hence Q_1 is bounded from above. It follows that $\varphi \le C(1+t)e^{-t}$ for a uniform constant C. The lower bound for φ is similar. $\qquad \square$

Lemma 3.6.8. *Fix $\beta \in (0, 1)$. We have*

(i) $\varphi(t) \to 0$ in $C^{2+\beta}(M)$ as $t \to \infty$.
(ii) $\omega(t) \to \omega_S$ in $C^\beta(M)$ as $t \to \infty$.
(iii) $\dfrac{\partial}{\partial t}\varphi \to 0$ in $C^0(M)$ as $t \to \infty$.

Proof. From Lemma 3.6.5 the tensor $\nabla_{\tilde{g}_0}g$ is bounded with respect to the fixed metric $\tilde{g}_0$. Moreover, $g \le C\tilde{g}_0$ for some uniform C. It follows that $\Delta_{\tilde{g}_0}\varphi$ is bounded in $C^1(M, \tilde{g}_0)$. Since φ is bounded in C^0, we can apply the standard Schauder estimates for Poisson's equation [GT01], to see that φ is bounded in $C^{2+\alpha}$ for any $\alpha \in (0, 1)$. Choosing $\alpha > \beta$, part (i) follows from this together with Lemma 3.6.7. Part (ii) follows from part (i) and the fact that $\hat{\omega}_t$ converges in C^∞ to ω_S as $t \to \infty$.

For part (iii), suppose for a contradiction that there exist $\varepsilon > 0$ and a sequence $\{(x_i, t_i)\}_{i \in \mathbb{N}} \subset M \times [0, \infty)$ with $t_i \to \infty$ and

$$|\dot{\varphi}|(x_i, t_i) > \varepsilon. \qquad (3.343)$$

From Lemmas 3.6.3 and 3.6.6, the quantity

$$\frac{\partial}{\partial t}\dot{\varphi} = -R(\omega) - 1 - \dot{\varphi} \qquad (3.344)$$

is uniformly bounded in $C^0(M \times [0, \infty))$. Hence there exists a uniform constant $\delta > 0$ such that for each i,

$$|\dot{\varphi}|(x_i, t) \ge \frac{\varepsilon}{2} \quad \text{for all } t \in [t_i, t_i + \delta]. \qquad (3.345)$$

Hence

$$\frac{\varepsilon\delta}{2} \leq \int_{t_i}^{t_i+\delta} |\dot{\varphi}|(t,x_i)dt = \left|\int_{t_i}^{t_i+\delta} \dot{\varphi}(x_i,t)dt\right|$$

$$= |\varphi(x_i,t_i+\delta) - \varphi(x_i,t_i)|$$

$$\leq \sup_{x\in M} |\varphi(x,t_i+\delta) - \varphi(x,t_i)|, \qquad (3.346)$$

a contradiction since $\varphi(t)$ converges uniformly to 0 in $C^0(M)$ as $t \to \infty$.   □

Finally, we prove part (iii) of Theorem 3.6.1.

Lemma 3.6.9. *Fix $s \in S$ and write $E = \pi_S^{-1}(s)$ for the fiber over s. Write $\omega_{\text{flat}} = \omega_{\text{flat}}|_E$. Then on E,*

$$e^t\omega(t)|_E \to \omega_{\text{flat}} \quad \text{as } t \to \infty, \qquad (3.347)$$

where the convergence is uniform on $C^0(E)$. Moreover, the convergence is uniform in $s \in S$.

Proof. We use here an argument similar to one found in [Tos10b]. Applying Lemma 3.6.5 we have

$$|\nabla_{g_E}(g|_E)|^2_{g|_E} \leq |\nabla_{\tilde{g}_0}g|^2 \leq C. \qquad (3.348)$$

From Lemma 3.6.4, we see that $g|_E$ is uniformly equivalent to $e^{-t}g_E$. It follows that

$$|\nabla_{g_E}(e^t g|_E)|^2_{g_E} = e^{-t}|\nabla_{g_E}(g|_E)|^2_{e^{-t}g_E} \leq Ce^{-t}|\nabla_{g_E}(g|_E)|^2_{g|_E} \leq C'e^{-t}. \qquad (3.349)$$

Since g_{flat} is a constant multiple of g_E we see that

$$|\nabla_{g_E}(e^t g|_E - g_{\text{flat}})|^2_{g_E} \leq C'e^{-t}. \qquad (3.350)$$

Moreover, $[e^t\omega|_E] = [\omega_{\text{flat}}]$. It is now straightforward to complete the proof of the lemma. Indeed, any two Kähler metrics on the Riemann surface E are conformally equivalent and hence we can write $e^t\omega|_E = e^\sigma\omega_{\text{flat}}$ for a smooth function $\sigma = \sigma(x,t)$ on $E \times [0,\infty)$. We have

$$|d(e^\sigma - 1)|^2_{g_E} \to 0 \quad \text{as } t \to \infty, \quad \text{and} \quad \int_E (e^\sigma - 1)\omega_E = 0. \qquad (3.351)$$

From the second condition, for each time t there exists $y(t) \in E$ with $\sigma(y(t),t) = 0$ and hence by the Mean Value Theorem for manifolds,

$$|e^{\sigma(x,t)} - 1| = |(e^{\sigma(x,t)} - 1) - (e^{\sigma(y(t),t)} - 1)| \to 0 \quad \text{as } t \to \infty, \tag{3.352}$$

uniformly in $x \in E$. This says precisely that $e^t \omega(t)|_E \to \omega_{\text{flat}}$ uniformly as $t \to \infty$. Moreover, none of our constants depend on the choice of $s \in S$. This completes the proof of the lemma. $\qquad\qquad\square$

Combining Lemma 3.6.6 with Lemmas 3.6.8 and 3.6.9 completes the proof of Theorem 3.6.1.

3.7 Finite Time Singularities

In this section, we describe some behaviors of the Kähler–Ricci flow in the case of a finite time singularity. The complete behavior of the flow is far from understood, and is the subject of current research. In Sect. 3.7.1, we prove some basic estimates, most of which hold under fairly weak hypotheses. Next, in Sect. 3.7.2, we describe a result of [Zha10] on the behavior of the scalar curvature and discuss some speculations. In Sects. 3.7.3 and 3.7.4 we describe, without proof, some recent results [SSW11, SW10] and illustrate with an example.

3.7.1 Basic Estimates

We now consider the Kähler–Ricci flow

$$\frac{\partial}{\partial t}\omega = -\text{Ric}(\omega), \qquad \omega|_{t=0} = \omega_0, \tag{3.353}$$

in the case when $T < \infty$. The cohomology class $[\omega_0] - Tc_1(M)$ is a limit of Kähler classes but is itself no longer Kähler. The behavior of the Kähler–Ricci flow as t tends towards the singular time T will depend crucially on properties of this cohomology class.

We first observe that since $T < \infty$ we immediately have from Corollary 3.2.3 the estimate

$$\omega^n \le C\Omega, \tag{3.354}$$

for a uniform constant C.

As in Sect. 3.3 we reduce (3.135) to a parabolic complex Monge–Ampère equation. Choose a closed $(1,1)$ form $\hat{\omega}_T$ in the cohomology class $[\omega_0] - Tc_1(M)$. Given this we can define a family of reference forms $\hat{\omega}_t$ by

$$\hat{\omega}_t = \frac{1}{T}((T-t)\omega_0 + t\hat{\omega}_T) \in [\omega_0] - tc_1(M). \tag{3.355}$$

Observe that $\hat{\omega}_t$ is *not* necessarily a metric, since $\hat{\omega}_T$ may have negative eigenvalues. Write $\chi = \frac{1}{T}(\hat{\omega}_T - \omega_0) = \frac{\partial}{\partial t}\hat{\omega}_t \in -c_1(M)$ and define Ω to be the volume form with

$$\frac{\sqrt{-1}}{2\pi}\partial\bar{\partial}\log\Omega = \chi \in -c_1(M), \qquad \int_M \Omega = \int_M \omega_0^n. \qquad (3.356)$$

We then consider the parabolic complex Monge–Ampère equation

$$\frac{\partial}{\partial t}\varphi = \log\frac{(\hat{\omega}_t + \frac{\sqrt{-1}}{2\pi}\partial\bar{\partial}\varphi)^n}{\Omega}, \qquad \hat{\omega}_t + \frac{\sqrt{-1}}{2\pi}\partial\bar{\partial}\varphi > 0, \qquad \varphi|_{t=0} = 0.$$
$$(3.357)$$

From (3.354) we immediately have:

Lemma 3.7.1. *For a uniform constant C we have on $M \times [0, T)$,*

$$\dot{\varphi} \leq C. \qquad (3.358)$$

If we assume that $\hat{\omega}_T \geq 0$ then the next result shows that the potential φ is bounded [Tzha06] (see also [SW10]). Note that since $[\omega_0] - Tc_1(M)$ is on the boundary of the Kähler cone, one would expect in many cases that this class contains a nonnegative representative $\hat{\omega}_T$.

Proposition 3.7.2. *Assume that $\hat{\omega}_T$ is nonnegative. Then for a uniform constant C we have on $M \times [0, T)$,*

$$|\varphi| \leq C. \qquad (3.359)$$

Proof. The upper bound of φ follows from Lemma 3.7.1. Alternatively, use the same argument as in the upper bound of φ in Lemma 3.3.2. For the lower bound, observe that

$$\hat{\omega}_t^n = \frac{1}{T^n}((T-t)\omega_0 + t\hat{\omega}_T)^n \geq \frac{1}{T^n}(T-t)^n\omega_0^n \geq c_0(T-t)^n\Omega, \qquad (3.360)$$

for some uniform constant $c_0 > 0$. Here we are using the fact that $\hat{\omega}_T$ is nonnegative. Define

$$\psi = \varphi + n(T-t)(\log(T-t) - 1) - (\log c_0 - 1)t, \qquad (3.361)$$

and compute

$$\frac{\partial}{\partial t}\psi = \log\frac{(\hat{\omega}_t + \frac{\sqrt{-1}}{2\pi}\partial\bar{\partial}\varphi)^n}{\Omega} - n\log(T-t) - (\log c_0 - 1). \qquad (3.362)$$

At a point where ψ achieves a minimum in space we have $\frac{\sqrt{-1}}{2\pi}\partial\bar{\partial}\psi = \frac{\sqrt{-1}}{2\pi}\partial\bar{\partial}\varphi \geq 0$ and hence from (3.360),

$$\frac{\partial}{\partial t}\psi \geq \log(c_0(T-t)^n) - n\log(T-t) - (\log c_0 - 1) = 1. \tag{3.363}$$

It follows from the minimum principle that ψ cannot achieve a minimum after time $t = 0$, and so ψ is uniformly bounded from below. Hence φ is bounded from below. □

If $\hat{\omega}_T$ is the pull-back of a Kähler metric from another manifold via a holomorphic map (so in particular $\hat{\omega}_T \geq 0$), we have by the parabolic Schwarz lemma (Theorem 3.2.6) a lower bound for $\omega(t)$:

Lemma 3.7.3. *Suppose there exists a holomorphic map $f : M \to N$ to a compact Kähler manifold N and let ω_N be a Kähler metric on N. We assume that*

$$[\omega_0] - Tc_1(M) = [f^*\omega_N]. \tag{3.364}$$

Then on $M \times [0, T)$,

$$\omega \geq \frac{1}{C}f^*\omega_N, \tag{3.365}$$

for a uniform constant C.

Proof. The method is similar to that of Lemma 3.5.12. We take $\hat{\omega}_T = f^*\omega_N \geq 0$. Define $u = \mathrm{tr}_\omega f^*\omega_N$. We apply the maximum principle to the quantity

$$Q = \log u - A\varphi - An(T-t)(\log(T-t) - 1), \tag{3.366}$$

for A to be determined later, and where we assume without loss of generality that $u > 0$. Compute using (3.77)

$$\left(\frac{\partial}{\partial t} - \Delta\right) Q \leq C_0 u - A\dot{\varphi} + An\log(T-t) + A\mathrm{tr}_\omega(\omega - \hat{\omega}_t)$$

$$= \mathrm{tr}_\omega(C_0 f^*\omega_N - (A-1)\hat{\omega}_t) - A\log\frac{\omega^n}{\Omega(T-t)^n} - \mathrm{tr}_\omega\hat{\omega}_t + An. \tag{3.367}$$

Now choose A sufficiently large so that $(A-1)\hat{\omega}_t - C_0 f^*\omega_N \geq f^*\omega_N$ for all $t \in [0, T]$. By the geometric–arithmetic means inequality, there exists a constant $c > 0$ such that

$$\mathrm{tr}_\omega\hat{\omega}_t \geq \frac{(T-t)}{T}\mathrm{tr}_\omega\omega_0 \geq c\left(\frac{(T-t)^n\Omega}{\omega^n}\right)^{1/n}. \tag{3.368}$$

Then, arguing as in the proof of Lemma 3.3.4,

$$\left(\frac{\partial}{\partial t} - \Delta\right) Q \leq -u + A \log \frac{(T-t)^n \Omega}{\omega^n} - c \left(\frac{(T-t)^n \Omega}{\omega^n}\right)^{1/n} + An \leq -u + C,$$

for a uniform constant C, since the map $\mu \mapsto A \log \mu - c \mu^{1/n}$ is uniformly bounded from above for $\mu > 0$. Hence at a maximum point of Q we see that u is bounded from above by C. Since φ and $(T-t) \log(T-t)$ are uniformly bounded this shows that Q is uniformly bounded from above. Hence u is uniformly bounded from above. □

A natural question is: when is the limiting class $[\omega_0] - T c_1(M)$ represented by the pull-back of a Kähler metric from another manifold via a holomorphic map? It turns out that this always occurs if the initial data is appropriately "algebraic".

Proposition 3.7.4. *Assume there exists a line bundle L on M such that $k[\omega_0] = c_1(L)$ for some positive integer k. Then there exists a holomorphic map $f : M \to \mathbb{P}^N$ to some projective space $\mathbb{P}^N$ and*

$$[\omega_0] - T c_1(M) = [f^* \omega], \tag{3.369}$$

for some Kähler metric ω on $\mathbb{P}^N$.

Proof. We give a sketch of the proof. Note that by the assumption on L, the manifold M is a smooth projective variety. From the Rationality Theorem of Kawamata and Shokurov [KMM87, KolMori98], T is rational. The class $[\omega_0] - T c_1(M)$ is nef since it is the limit of Kähler classes. From the Base Point Free Theorem [part (ii) of Theorem 3.1.12], $[\omega_0] - T c_1(M)$ is semi-ample, and the result follows. □

If we make a further assumption on the map f then we can obtain C^∞ estimates for the evolving metric away from a subvariety.

Theorem 3.7.5. *Suppose there exists a holomorphic map $f : M \to N$ to a compact Kähler manifold N which is a biholomorphism outside a subvariety $E \subset M$. Let ω_N be a Kähler metric on N. We assume that*

$$[\omega_0] - T c_1(M) = [f^* \omega_N]. \tag{3.370}$$

Then on any compact subset K of $M \setminus E$ there exists a constant $c_K > 0$ such that

$$\omega \geq c_K \omega_0, \quad \text{on} \quad K \times [0, T). \tag{3.371}$$

Moreover we have uniform C^∞_{loc} estimates for $\omega(t)$ on $M \setminus E$.

Proof. The inequality (3.371) is immediate from Lemma 3.7.3 and the fact that $f^* \omega_N$ is a Kähler metric on $M \setminus E$. From the volume form bound (3.354), we immediately obtain uniform upper and lower bounds for ω on compact subsets of $M \setminus E$. The higher order estimates follow from the same arguments as in Lemmas 3.5.6 and 3.5.7.   □

We will see in Sect. 3.7.3 that the situation of Theorem 3.7.5 arises in the case of blowing down an exceptional divisor.

3.7.2 Behavior of the Scalar Curvature

In this section we give prove the following result of Zhang [Zha10] on the behavior of the scalar curvature. Given the estimates we have developed so far, we can give quite a short proof. Recall that we have a lower bound of the scalar curvature from Theorem 3.2.2.

Theorem 3.7.6. *Let $\omega = \omega(t)$ be a solution of the Kähler–Ricci flow (3.353) on the maximal time interval $[0, T)$. If $T < \infty$ then*

$$\limsup_{t \to T} \left(\sup_M R(g(t)) \right) = \infty. \tag{3.372}$$

In the case of the general Ricci flow with a singularity at time $T < \infty$ it is known that $\sup_M |\mathrm{Ric}(g(t))| \to \infty$ as $t \to T$ [Se05].

Proof of Theorem 3.7.6. We will assume that (3.372) does not hold and obtain a contradiction. Since we know from Theorem 3.2.2 that the scalar curvature has a uniform lower bound, we may assume that $\|R(t)\|_{C^0(M)}$ is uniformly bounded for $t \in [0, T)$. Let φ solve the parabolic complex Monge–Ampère equation (3.357). First note that

$$\left| \frac{\partial}{\partial t} \log \left(\frac{\omega^n}{\Omega} \right) \right| = |R| \leq C. \tag{3.373}$$

Integrating in time we see that $|\dot{\varphi}| = |\log \frac{\omega^n}{\Omega}|$ is uniformly bounded. Integrating in time again, we obtain a uniform bound for φ. Define $H = t\dot{\varphi} - \varphi - nt$, which is a bounded quantity. Then using (3.150) we obtain [cf. (3.240)],

$$\left(\frac{\partial}{\partial t} - \Delta \right) H = t \, \mathrm{tr}_\omega \chi - n + \mathrm{tr}_\omega (\omega - \hat{\omega}_t) = \mathrm{tr}_\omega (t\chi - \hat{\omega}_t) = -\mathrm{tr}_\omega \omega_0. \tag{3.374}$$

Apply Proposition 3.2.5 to see that

$$\left(\frac{\partial}{\partial t} - \Delta \right) \mathrm{tr}_{\omega_0} \omega \leq C_0 \mathrm{tr}_\omega \omega_0, \tag{3.375}$$

for a uniform constant C_0 depending only on ω_0. Define $Q = \log \text{tr}_{\omega_0} \omega + AH$ for $A = C_0 + 1$. Combining (3.374) and (3.375), compute

$$\left(\frac{\partial}{\partial t} - \Delta\right) Q \leq -\text{tr}_\omega \omega_0 < 0, \tag{3.376}$$

and hence by the maximum principle Q is bounded from above by its value at time $t = 0$. It follows that $\text{tr}_{\omega_0} \omega$ is uniformly bounded from above. Since we have a lower bound for $\dot{\varphi} = \log \frac{\omega^h}{\Omega}$, we see that for a uniform constant C,

$$\frac{1}{C}\omega_0 \leq \omega \leq C\omega_0, \quad \text{on} \quad M \times [0, T). \tag{3.377}$$

Applying Corollary 3.2.16, we obtain uniform estimates for $\omega(t)$ and all of its derivatives. Hence $\omega(t)$ converges to a smooth Kähler metric $\omega(T)$ which is contained in $[\omega_0] - Tc_1(M)$. Thus $[\omega_0] - Tc_1(M)$ is a Kähler class, contradicting the definition of T.   □

We remark that Theorem 3.7.6 can be proved just as easily using the parabolic Schwarz lemma instead of Proposition 3.2.5. Indeed one can replace Q with $Q = \log \text{tr}_\omega \omega_0 + AH$ and apply the Schwarz lemma with the holomorphic map f being the identity map and $\omega_N = \omega_0$. This was the method in [Zha10]. Also, one can find in [Zha10] a different way of obtaining a contradiction, one which avoids the higher order estimates.

We finish this section by mentioning a couple of "folklore conjectures":

Conjecture 3.7.7. Let $\omega = \omega(t)$ be a solution of the Kähler–Ricci flow (3.135) on the maximal time interval $[0, T)$. If $T < \infty$ then

$$R \leq \frac{C}{T - t}, \tag{3.378}$$

for some uniform constant C.

This conjecture has been established in dimension 1 by Hamilton and Chow [Chow91, Ham88] and by Perelman in higher dimensions if $[\omega_0] = c_1(M) > 0$ (Perelman, unpublished work on the Kähler–Ricci flow; see also [SeT08]). Perelman's result makes use of the functionals he introduced in [Per02]. In [Zha10], it was shown in a quite general setting, that $R \leq C/(T - t)^2$.

A stronger version of Conjecture 3.7.7 is:

Conjecture 3.7.8. Let $\omega = \omega(t)$ be a solution of the Kähler–Ricci flow (3.135) on the maximal time interval $[0, T)$. If $T < \infty$ then

$$|\text{Rm}| \leq \frac{C}{T - t}, \tag{3.379}$$

for some uniform constant C.

Another way of saying this is that all finite time singularities along the Kähler–Ricci flow are of *Type I*. This is related to a conjecture of Hamilton and Tian that the (appropriately normalized) Kähler–Ricci flow on a manifold with positive first Chern class converges to a Kähler–Ricci soliton, with a possibly different complex structure in the limit.

3.7.3 Contracting Exceptional Curves

In this section we briefly describe, without proof, the example of *blowing-down exceptional curves* on a Kähler surface in finite time. We begin by defining what is meant by a *blowing-down* and *blowing-up* (see for example [GH78]).

First, we define the *blow-up* of the origin in $\mathbb{C}^2$. Let z^1, z^2 be coordinates on $\mathbb{C}^2$, and let U be a open neighborhood of the origin. Define

$$\tilde{U} = \{(z, \ell) \in U \times \mathbb{P}^1 \mid z \in \ell\}, \tag{3.380}$$

where we are considering ℓ as a line in $\mathbb{C}^2$ through the origin. One can check that $\tilde{U}$ is a 2-dimensional complex submanifold of $U \times \mathbb{P}^1$. There is a holomorphic map $\pi : \tilde{U} \to U$ given by $(z, \ell) \mapsto z$ which maps $\tilde{U} \setminus \pi^{-1}(0)$ biholomorphically onto $U \setminus \{0\}$. The set $\pi^{-1}(0)$ is a one-dimensional submanifold of $\tilde{U}$, isomorphic to $\mathbb{P}^1$.

Given a point p in a Kähler surface N we can use local coordinates to construct the *blow up* $\pi : M \to N$ of p, by replacing a neighborhood U of p with the blow up $\tilde{U}$ as above. Thus M is a Kähler surface and π a holomorphic map extending the local map given above. Up to isomorphism, this construction is independent of choice of coordinates. The curve $E = \pi^{-1}(p)$ is called the *exceptional curve*. Since $\pi(E) = p$, the map π contracts or *blows down* the curve E. Moreover, π is an isomorphism from $M \setminus E$ to $N \setminus \{p\}$. From the above we see that E is a smooth curve which is isomorphic to $\mathbb{P}^1$. Moreover, the reader can check that it satisfies $E \cdot E = -1$.

Conversely, given a curve E on a surface M with these properties we can define a map blowing down E. More precisely, we define an irreducible curve E in M to be a (-1)-*curve* if it is smooth, isomorphic to $\mathbb{P}^1$ and has $E \cdot E = -1$. If M admits a (-1)-curve E then there exists a holomorphic map $\pi : M \to N$ to a smooth Kähler surface N and a point $p \in Y$ such that π is precisely the blow down of E to p, as constructed above. Note that if E is a (-1) curve then by the Adjunction Formula, $K_E \cdot E = -1$.

The main result of [SW10] says that, under appropriate hypotheses on the initial Kähler class, the Kähler–Ricci flow will blow down (-1)-curves on M and then continue on the new manifold. To make this more precise, we need a definition.

Definition 3.7.9. We say that the solution $g(t)$ of the Kähler–Ricci flow (3.353) on a compact Kähler surface M performs a **canonical surgical contraction** if the following holds. There exist distinct (-1) curves $E_1, \ldots, E_k$ of M, a compact

Kähler surface N and a blow-down map $\pi : M \to N$ with $\pi(E_i) = y_i \in N$
and $\pi|_{M \setminus \bigcup_{i=1}^{k} E_i}$ a biholomorphism onto $N \setminus \{y_1, \ldots, y_k\}$ such that:

(i) As $t \to T^-$, the metrics $g(t)$ converge to a smooth Kähler metric g_T on $M \setminus \bigcup_{i=1}^{k} E_i$ smoothly on compact subsets of $M \setminus \bigcup_{i=1}^{k} E_i$.

(ii) $(M, g(t))$ converges to a unique compact metric space $(\hat{N}, d_T)$ in the Gromov–Hausdorff sense as $t \to T^-$. In particular, $(\hat{N}, d_T)$ is homeomorphic to the Kähler surface N.

(iii) There exists a unique maximal smooth solution $g(t)$ of the Kähler–Ricci flow on N for $t \in (T, T_N)$, with $T < T_N \le \infty$, such that $g(t)$ converges to $(\pi^{-1})^* g_N$ as $t \to T^+$ smoothly on compact subsets of $N \setminus \{y_1, \ldots, y_k\}$.

(iv) $(N, g(t))$ converges to (N, d_T) in the Gromov–Hausdorff sense as $t \to T^+$.

The following theorem is proved in [SW10]. It essentially says that whenever the evolution of the Kähler classes along the Kähler–Ricci flow indicate that a blow down should occur at the singular time $T < \infty$, then the Kähler–Ricci flow carries out a canonical surgical contraction at time T.

Theorem 3.7.10. *Let $g(t)$ be a smooth solution of the Kähler–Ricci flow (3.353) on a Kähler surface M for t in $[0, T)$ and assume $T < \infty$. Suppose there exists a blow-down map $\pi : M \to N$ contracting disjoint (-1) curves $E_1, \ldots, E_k$ on M with $\pi(E_i) = y_i \in N$, for a smooth compact Kähler surface (N, ω_N) such that the limiting Kähler class satisfies*

$$[\omega_0] - Tc_1(M) = [\pi^* \omega_N]. \tag{3.381}$$

Then the Kähler–Ricci flow $g(t)$ performs a canonical surgical contraction with respect to the data $E_1, \ldots, E_k$, N and π.

Note that from Theorem 3.7.5, we have $C_{\mathrm{loc}}^{\infty}$ estimates for $g(t)$ on $M \setminus \bigcup_{i=1}^{k} E_i$, and thus part (i) in the definition of canonical surgical contraction follows immediately. For the other parts, estimates are needed for $g(t)$ near the subvariety E. To continue the flow on the new manifold, some techniques are adapted from [ST09]. We refer the reader to [SW10] for the details. In fact, the same result is shown to hold in [SW10] for blowing up points in higher dimensions, and in [SW11] the results are extended to the case of an exceptional divisor E with normal bundle $\mathcal{O}(-k)$, which blows down to an orbifold point. See also [LaNT09] for a different approach to the study of blow-downs.

In Sect. 3.8.3, we will show how Theorem 3.7.10 can be applied quite generally for the Kähler–Ricci flow on a Kähler surface.

3.7.4 Collapsing in Finite Time

In this section, we briefly describe, again without proof, another example of a finite time singularity.

Let M be a projective bundle over a smooth projective variety B. That is, $M = \mathbb{P}(E)$, where $\pi : E \to B$ is a holomorphic vector bundle which we can take to have rank r. Write π also for the map $\pi : M \to B$. Of course, the simplest example of this would be a product $B \times \mathbb{P}^{r-1}$. We consider the Kähler–Ricci flow (3.353) on M. The flow will always develop a singularity in finite time. This is because

$$\int_F (c_1(M))^{r-1} > 0, \tag{3.382}$$

for any fiber F, whereas if $T = \infty$ then $\frac{1}{t}[\omega_0] - c_1(M) > 0$ for all $t > 0$. The point is that the fibers $F \cong \mathbb{P}^{r-1}$ must shrink to zero in finite time along the Kähler–Ricci flow.

In [SSW11], it is shown that:

Theorem 3.7.11. *Assume that*

$$[\omega_0] - Tc_1(M) = [\pi^*\omega_B], \tag{3.383}$$

for some Kähler metric ω_B on B. Then there exists a sequence of times $t_i \to T$ and a distance function d_B on B, which is uniformly equivalent to the distance function induced by ω_B, such that $(M, \omega(t_i))$ converges to (B, d_B) in the Gromov–Hausdorff sense.

Note that from Lemma 3.7.3 we immediately have $\omega(t) \geq \frac{1}{C}\pi^*\omega_B$ for some uniform $C > 0$. The key estimates proved in [SSW11] are:

(i) $\omega(t) \leq C\omega_0$.
(ii) $\mathrm{diam}_{\omega(t)} F \leq C(T - t)^{1/3}$, for every fiber F.

Thus we see that the metrics are uniformly bounded from above along the flow and the fibers collapse. Given (i) and (ii) it is fairly straightforward to establish Theorem 3.7.11. We refer the reader to [SSW11] for the details.

The following conjectures are made in [SSW11]:

Conjecture 3.7.12. With the assumptions above:

(a) There exists unique distance function d_B on B such that $(M, \omega(t))$ converges in the Gromov–Hausdorff sense to (B, d_B), without taking subsequences.
(b) The estimate (ii) above can be strengthened to $\mathrm{diam}_{\omega(t)} F \leq C(T - t)^{1/2}$, for every fiber F.
(c) Theorem 3.7.11 [and parts (a) and (b) of this conjecture] should hold more generally for a bundle $\pi : M \to B$ over a Kähler base B with fibers $\pi^{-1}(b)$ being Fano manifolds admitting metrics of nonnegative bisectional curvature.

We end this section by describing an example which illustrates both the case of contracting an exceptional curve and the case of collapsing the fibers of a projective bundle. Let M be the blow up of $\mathbb{P}^2$ at one point $p \in \mathbb{P}^2$. Let $f : M \to \mathbb{P}^2$ be the map blowing down the exceptional curve E to the point p. To see the bundle structure on M, note that the blow-up of $\mathbb{C}^2$ at the origin can be identified with $M \setminus f^{-1}(H)$ for H a hyperplane in $\mathbb{P}^2$. We have a map π from the blow up of $\mathbb{C}^2$, which is $\{(z, \ell) \in \mathbb{C}^2 \times \mathbb{P}^1 \mid z \in \ell\}$, to $\mathbb{P}^1$ given by projection onto the second factor. This extends to a holomorphic bundle map $\pi : M \to \mathbb{P}^1$ which has $\mathbb{P}^1$ fibers. We refer the reader to [Cal82, SW09] for more details.

Writing ω_1 and ω_2 for the Fubini–Study metrics on $\mathbb{P}^1$ and $\mathbb{P}^2$ respectively, we see that every Kähler class α on M can be written as a linear combination $\alpha = \beta[\pi^*\omega_1] + \gamma[f^*\omega_2]$ for $\beta, \gamma > 0$. The boundary of the Kähler cone is spanned by the two rays $\mathbb{R}^{\geq 0}[\pi^*\omega_1]$ and $\mathbb{R}^{\geq 0}[f^*\omega_2]$. The first Chern class of M is given by

$$c_1(M) = [\pi^*\omega_1] + 2[f^*\omega_2] > 0. \qquad (3.384)$$

Hence if the initial Kähler metric ω_0 is in the cohomology class $\alpha_0 = \beta_0[\pi^*\omega_1] + \gamma_0[f^*\omega_2]$ then the solution $\omega(t)$ of the Kähler–Ricci flow (3.353) has cohomology class

$$[\omega(t)] = \beta(t)[\pi^*\omega_1] + \gamma(t)[f^*\omega_2], \quad \text{with} \quad \beta(t) = \beta_0 - t, \ \gamma(t) = \gamma_0 - 2t. \qquad (3.385)$$

There are three different behaviors of the Kähler–Ricci flow according to whether the cohomology class $[\omega(t)]$ hits the boundary of the Kähler cone at a point on $\mathbb{R}^{>0}[\pi^*\omega_1]$, at a point on $\mathbb{R}^{>0}[f^*\omega_2]$ or at zero. Namely:

(i) If $\gamma_0 > 2\beta_0$ then a singularity occurs at time $T = \beta_0$ and

$$[\omega_0] - Tc_1(M) = \gamma(T)[f^*\omega_2], \quad \text{with} \quad \gamma(T) = \gamma_0 - 2\beta_0 > 0. \quad (3.386)$$

Thus we are in the case of Theorem 3.7.10 and the Kähler–Ricci flow performs a canonical surgical contraction at time T.

(ii) If $\gamma_0 < 2\beta_0$ then a singularity occurs at time $T = \gamma_0/2$ and

$$[\omega_0] - Tc_1(M) = \beta(T)[\pi^*\omega_1], \quad \text{with} \quad \beta(T) = \beta_0 - \gamma_0/2 > 0. \quad (3.387)$$

Thus we are in the case of Theorem 3.7.11 and the Kähler–Ricci flow will collapse the $\mathbb{P}^1$ fibers and converge in the Gromov–Hausdroff sense, after passing to a subsequence, to a metric on the base $\mathbb{P}^1$.

(iii) If $\gamma_0 = 2\beta_0$ then the cohomology class changes by a rescaling. It was shown by Perelman (unpublished work on the Kähler–Ricci flow) [SeT08] that $(M, \omega(t))$ converges in the Gromov–Hausdorff sense to a point.

The behavior of the Kähler–Ricci flow on this manifold M, and higher dimensional analogues, was analyzed in detail by Feldman–Ilmanen–Knopf [FIK03]. They constructed self-similar solutions of the Kähler–Ricci flow through such singularities (see also [Cao94]) and carried out a careful study of their properties. Moreover, they posed a number of conjectures, some of which were established in [SW09].

Indeed if we make the assumption that the initial metric ω_0 is invariant under a maximal compact subgroup of the automorphism group of M, then stronger results than those given in Theorems 3.7.10 and 3.7.11 were obtained in [SW09]. In particular, in the situation of case (ii), it was shown in [SW09] that $(M, \omega(t))$ converges in the Gromov–Hausdorff sense (without taking subsequences) to a multiple of the Fubini–Study metric on $\mathbb{P}^1$ (see also [Fo11]).

One can see from the above some general principles for what we expect with the Kähler–Ricci flow. Namely, the behavior of the flow ought to be able to be read from the behavior of the cohomology classes $[\omega(t)]$ as t tends to the singular time T. If the limiting class $[\omega_0] - T c_1(M) = [\pi^* \omega_N]$ for some $\pi : M \to N$ with ω_N Kähler on N, then we expect geometric convergence of $(M, \omega(t))$ to (N, ω_N) in some appropriate sense. This philosophy was discussed by Feldman–Ilmanen–Knopf [FIK03].

3.8 The Kähler–Ricci Flow and the Analytic MMP

In this section, we begin by discussing, rather informally, some of the basic ideas behind the minimal model program (MMP) with scaling. Next we discuss the program of Song–Tian relating this to the Kähler–Ricci flow. Finally, we describe the case of Kähler surfaces.

3.8.1 Introduction to the Minimal Model Program with Scaling

In this section, we give a brief introduction of Mori's minimal model program (MMP) in birational geometry. For more extensive references on this subject, see [CKL11, Deb01, KMM87, KolMori98], for example. We also refer the reader to [Siu08] for a different analytic approach to some of these questions.

We begin with a definition. Let X and Y be projective varieties. A *rational map* from X to Y is given by an algebraic map $f : X \setminus V \to Y$, where V is a subvariety of X. We identify two such maps if they agree on $X - W$ for some subvariety W. Thus a rational map is really an equivalence class of pairs (f_U, U) where U is the complement of a variety in X (i.e. a Zariski open subset of X) and $f_U : U \to Y$ is a holomorphic map.

We say that a rational map f from X to Y is *birational* if there exists a rational map from Y to X such that $f \circ g$ is the identity as a rational map. If a birational map from X to Y exists then we say that X and Y are *birationally equivalent* (or *birational* or *in the same birational class*).

Although birational varieties agree only on a dense open subset, they share many properties (see e.g. [GH78, Ha77]). The minimal model program is concerned with finding a "good" representative of a variety within its birational class. A "good" variety X is one satisfying either:

(i) K_X is nef; or

(ii) There exists a holomorphic map $\pi : X \to Y$ to a lower dimensional variety Y such that the generic fiber $X_y = \pi^{-1}(y)$ is a manifold with $K_{X_y} < 0$.

In the first case, we say that X is a *minimal model* and in the second case we say that X is a *Mori fiber space* (or *Fano fiber space*). Roughly speaking, since K_X nef can be thought of as a "nonpositivity" condition on $c_1(X) = [\text{Ric}(\omega)]$, (i) implies that X is "nonpositively curved" in some weak sense. Condition (ii) says rather that X has a "large part" which is "positively curved". The two cases (i) and (ii) are mutually exclusive.

The basic idea of the MMP is to find a finite sequence of birational maps $f_1, \ldots, f_k$ and varieties $X_1, \ldots, X_k$,

$$X = X_0 \xdashrightarrow{f_1} X_1 \xdashrightarrow{f_2} X_2 \xdashrightarrow{f_2} \cdots \xdashrightarrow{f_k} X_k \qquad (3.388)$$

so that X_k is our "good" variety: either of type (i) or type (ii). Recall that K_X nef means that $K_X \cdot C \geq 0$ for all curves C. Thus we want to find maps f_i which "remove" curves C with $K_X \cdot C < 0$, in order to make the canonical bundle "closer" to being nef.

If the complex dimension is 1 or 2, then we can carry this out in the category of smooth varieties. In the case of complex dimension 1, no birational maps are needed and case (i) corresponds to $c_1(X) < 0$ or $c_1(X) = 0$ while case (ii) corresponds to $X = \mathbb{P}^1$. Note that in case (i), X admits a metric of negative or zero curvature, while in case (ii) X has a metric of positive curvature.

In complex dimension two, by the Enriques–Kodaira classification (see [BHPV]), we can obtain our "good" variety X via a finite sequence of blow downs (see Sect. 3.8.3).

Unfortunately, in dimensions three and higher, it is not possible to find such a sequence of birational maps if we wish to stay within the category of smooth varieties. Thus to carry out the minimal model program, it is necessary to consider varieties with singularities. This leads to all kind of complications, which go well beyond the scope of these notes. For the purposes of this discussion, we will restrict ourselves to smooth varieties except where it is absolutely impossible to avoid mentioning singularities.

We need some further definitions. Let X be a smooth projective variety. As we have discussed in Sect. 3.1.7, there is a natural pairing between divisors and curves. A *1-cycle* C on X is a formal finite sum $C = \sum_i a_i C_i$, for $a_i \in \mathbb{Z}$ and C_i irreducible curves. We say that 1-cycles C and C' are *numerically equivalent* if $D \cdot C = D \cdot C'$ for all divisors D, and in this case we write $C \sim C'$. We denote by $N_1(X)_\mathbb{Z}$ the space of 1-cycles modulo numerical equivalence. Write

$$N_1(X)_\mathbb{Q} = N_1(X)_\mathbb{Z} \otimes_\mathbb{Z} \mathbb{Q} \quad \text{and} \quad N_1(X)_\mathbb{R} = N_1(X)_\mathbb{Z} \otimes_\mathbb{Z} \mathbb{R}. \qquad (3.389)$$

Similarly, we say that divisors D and D' are *numerically equivalent* if $D \cdot C = D' \cdot C$ for all curves C. Write $N^1(X)_\mathbb{Z}$ for the set of divisors modulo numerical equivalence. Define $N^1(X)_\mathbb{Q}$, $N^1(X)_\mathbb{R}$ similarly. One can check that $N_1(X)_\mathbb{R}$ and $N^1(X)_\mathbb{R}$ are vector spaces of the same (finite) dimension. In the obvious way, we can talk about 1-cycles with coefficients in $\mathbb{Q}$ or $\mathbb{R}$ (and correspondingly, $\mathbb{Q}$- or $\mathbb{R}$-divisors) and we can talk about numerical equivalence of such objects.

Within the vector space $N_1(X)_\mathbb{R}$ is a cone NE(X) which we will now describe. We say that an element of $N_1(X)_\mathbb{R}$ is *effective* if it is numerically equivalent to a 1-cycle of the form $C = \sum_i a_i C_i$ with $a_i \in \mathbb{R}^{\geq 0}$ and C_i irreducible curves. Write NE(X) for the cone of effective elements of $N_1(X)_\mathbb{R}$, and write $\overline{\text{NE}}(X)$ for its closure in the vector space $N_1(X)_\mathbb{R}$. The importance of $\overline{\text{NE}}(X)$ can be seen immediately from the following theorem, known as *Kleiman's criterion*:

Theorem 3.8.1. *A divisor D is ample if and only if $D \cdot w > 0$ for all nonzero $w \in \overline{\text{NE}}(X)$.*

We can now begin to describe the *MMP with scaling* of [BCHM10]. This is an algorithm for finding a specific sequence of birational maps $f_1, \ldots f_k$. First, choose an ample divisor H on X. Then define

$$T = \sup\{t > 0 \mid H + tK_X > 0\}. \tag{3.390}$$

If $T = \infty$, then we have nothing to show since K_X is already nef and we are in case (i). Indeed, if C is any curve in X then

$$K_X \cdot C = \frac{1}{t}(H + tK_X) \cdot C - \frac{1}{t} H \cdot C \geq -\frac{1}{t} H \cdot C \to 0 \quad \text{as} \quad t \to \infty. \tag{3.391}$$

We can assume then that $T < \infty$. We can apply the Rationality Theorem of Kawamata and Shokurov [KMM87, KolMori98] to see that T is rational, and hence $H + TK_X$ defines a $\mathbb{Q}$-line bundle.

Next we apply the Base Point Free Theorem [part (ii) of Theorem 3.1.12] to $L = H + TK_X$ to see that for sufficiently large $m \in \mathbb{Z}^{\geq 0}$, L^m is globally generated and $H^0(X, L^m)$ defines a holomorphic map $\pi : X \to \mathbb{P}^N$ such that $L^m = \pi^*\mathcal{O}(1)$. We write Y for the image of π. This variety Y is uniquely determined for m sufficiently large. The next step is to establish properties of this map π.

Define a subcone NE(π) of $\overline{\text{NE}}(X)$ by

$$\text{NE}(\pi) = \{w \in \overline{\text{NE}}(X) \mid L \cdot w = 0\}, \tag{3.392}$$

which is nonempty by Theorem 3.8.1. We now make the following:

Simplifying assumption: NE(π) is an *extremal ray* of $\overline{\text{NE}}(X)$.

A *ray* R of $\overline{\text{NE}}(X)$ is a subcone of the form $R = \{\lambda w \mid \lambda \in [0, \infty)\}$ for some $w \in \overline{\text{NE}}(X)$. We say that a subcone C in $\overline{\text{NE}}(X)$ is *extremal* if $a, b \in \overline{\text{NE}}(X)$, $a + b \in C$ implies that $a, b \in C$. In general, NE(π) is an extremal subcone but not necessarily a ray. However, it is expected that it will be an extremal ray for generic choice of initial ample divisor H (see the discussion in [ST09]).

Remark 3.8.2. In the case that $\mathrm{NE}(\pi)$ is not an extremal ray, one can still continue the MMP with scaling by applying Mori's Cone Theorem [KolMori98] to find such an extremal ray contained in $\mathrm{NE}(\pi)$.

The extremal ray $R = \mathrm{NE}(\pi)$ has the additional property of being K_X-*negative*. We say that a ray is K_X-*negative* if $K_X \cdot w < 0$ for all nonzero w in the ray. Clearly this is true in this case since $0 = L \cdot w = H \cdot w + T K_X \cdot w$ and therefore $K_X \cdot w = -T^{-1} H \cdot w < 0$ if w is a nonzero element of R. Thus from the point of view of the minimal model program, R contains "bad" curves (those with negative intersection with K_X) which we want to remove.

Moreover, the map π contracts all curves whose class lies in the extremal ray $R = \mathrm{NE}(\pi)$. The union of these curves is called the *locus* of R. In fact, the locus of $R = \mathrm{NE}(\pi)$ is exactly the set of points where the map $\pi : X \to Y$ is not an isomorphism. Moreover, R is a subvariety of X [Deb01, KolMori98]. There are three cases:

Case 1. The locus of R is equal to X. In this case π is a fiber contraction and X is a Mori fiber space.
Case 2. The locus of R is an irreducible divisor D. In this case π is called a *divisorial contraction*.
Case 3. The locus of R has codimension at least 2. In this case, π is called a *small contraction*.

The process of the MMP with scaling is then as follows: if we are in case 1, we stop, since X is already of type (ii). In case 2 we have a map $\pi : X \to Y \subset \mathbb{P}^N$ to a subvariety Y. Let H_Y on Y be restriction of $\mathcal{O}(1)|_Y$. We can then repeat the process of the minimal model program with scaling with (Y, H_Y) instead of (X, H).

The serious difficulties occur in case 3. Here the image Y of π will have very bad singularities and it will not be possible to continue this process on Y. Instead we have to work on a new space given by a procedure known as a *flip*. Let $\pi : X \to Y$ be a small contraction as in case 3. The *flip* of $\pi : X \to Y$ is a variety X^+ together with a holomorphic birational map $\pi^+ : X^+ \to Y$ satisfying the following conditions:

(a) The *exceptional locus* of π^+ (that is, the set of points in X^+ on which π^+ is not an isomorphism) has codimension strictly larger than 1.
(b) If C is a curve contracted by π^+ then $K_{X^+} \cdot C > 0$.

Thus we have a diagram

$$
\begin{array}{ccc}
X & \xrightarrow{\;(\pi^+)^{-1}\circ\pi\;} & X^+ \\
 & {\pi}\searrow \quad \swarrow{\pi^+} & \\
 & Y &
\end{array}
\qquad (3.393)
$$

The composition $(\pi^+)^{-1} \circ \pi$ is a birational map from X to X^+, and is also sometimes called a *flip*. In going from X to Y we have contracted curves C with $K_X \cdot C < 0$. The point of (b) in the definition above is that in going from Y to X^+

we do not wish to "gain" any curves C of negative intersection with the canonical bundle. The process of the flip replaces curves C on X with $K_X \cdot C < 0$ with curves C' on X^+ with $K_{X^+} \cdot C' > 0$. This fits into the strategy of trying to make the canonical bundle "more nef".

Given a small contraction $\pi : X \to Y$, the question of whether there actually exists a flip $\pi^+ : X^+ \to Y$ is a difficult one. It has been established for the MMP with scaling [BCHM10, HM10]. Returning now to the MMP with scaling: if we are in case 3 we replace X by its flip X^+ and we denote by L^+ the strict transform of $\mathcal{O}(1)|_Y$ via π^+ (see for example [Ha77]). We can now repeat the process with (X^+, H^+) instead of (X, H).

We have described now the basic process of the MMP with scaling. Start with (X, H) and find $\pi : X \to Y$ contracting the extremal ray R on which $H + TK_X$ is zero. In case 1, we stop. In case 2 we carry out a divisorial contraction and restart the process. In case 3, we replace X by its flip X^+ and again restart the process. A question is now: does this process terminate in finitely many steps? It was proved in [BCHM10, HM10] that the answer to this is yes, at least in the case of varieties of general type. If we have not already obtained a Mori fiber space, then the final variety X_k contains no curves C with $K_X \cdot C < 0$, and we are done.

We conclude this section with an example of a flip (see [Deb01, SY10]). Let $X_{m,n} = \mathbb{P}(\mathcal{O}_{\mathbb{P}^n} \oplus \mathcal{O}_{\mathbb{P}^n}(-1)^{\oplus (m+1)})$ be the $\mathbb{P}^{m+1}$ bundle over $\mathbb{P}^n$. Let $Y_{m,n}$ be the projective cone over $\mathbb{P}^m \times \mathbb{P}^n$ in $\mathbb{P}^{(m+1)(n+1)}$ by the Segre embedding

$$[Z_0, \ldots, Z_m] \times [W_0, \ldots, W_n] \to [Z_0 W_0, \ldots, Z_i W_j, \ldots, Z_m W_n] \in \mathbb{P}^{(m+1)(n+1)-1}.$$

Note that $Y_{m,n} = Y_{n,m}$. Then there exists a holomorphic map $\Phi_{m,n} : X_{m,n} \to Y_{m,n}$ for $m \geq 1$ contracting the zero section of $X_{m,n}$ of codimension $m + 1$ to the cone singularity of $Y_{m,n}$. The following diagram gives a flip from $X_{m,n}$ to $X_{n,m}$ for $1 \leq m < n$,

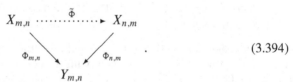

$$(3.394)$$

3.8.2 The Kähler–Ricci Flow and the MMP with Scaling

Let X be a smooth projective variety with an ample divisor H. We now relate the MMP with scaling to the (unnormalized) Kähler–Ricci flow

$$\frac{\partial}{\partial t}\omega = -\mathrm{Ric}(\omega), \qquad \omega|_{t=0} = \omega_0, \tag{3.395}$$

We assume that the initial metric ω_0 lies in the cohomology class $c_1([H])$ associated to the divisor H. As we have seen from Sect. 3.3.1, a smooth solution $\omega(t)$ of the Kähler–Ricci flow exists precisely on the time interval $[0, T)$, with T defined by (3.390). In general, we expect that as $t \to T$, the Kähler–Ricci flow carries out a "surgery", which is equivalent to the algebraic procedure of contracting an extremal ray, as discussed above.

The following is a (rather sketchy) conjectural picture for the behavior of the Kähler–Ricci flow, as proposed by Song and Tian in [ST07, ST09, Tian08].

Step 1. We start with a metric ω_0 in the class of a divisor H on a variety X. We then consider the solution $\omega(t)$ of the Kähler–Ricci flow (3.395) on X starting at ω_0. The flow exists on $[0, T)$ with $T = \sup\{t > 0 \mid H + tK_X > 0\}$.

Step 2. If $T = \infty$, then K_X is nef and the Kähler–Ricci flow exists for all time. The flow $\omega(t)$ should converge, after an appropriate normalization, to a canonical "generalized Kähler–Einstein metric" on X as $t \to \infty$.

Step 3. If $T < \infty$, the Kähler–Ricci flow deforms X to (Y, g_Y) with a possibly singular metric g_Y as $t \to T$.

(a) If $\dim X = \dim Y$ and Y differs from X by a subvariety of codimension 1, then we return to Step 1, replacing (X, g_0) by (Y, g_Y).

(b) If $\dim X = \dim Y$ and Y differs from X by a subvariety of codimension greater than 1, we are in the case of a small contraction. Y will be singular. By considering an appropriate notion of weak Kähler–Ricci flow on Y, starting at g_Y, the flow should immediately resolve the singularities of Y and replace Y by its flip X^+ (see [SY10]). Then we return to Step 1 with X^+.

(c) If $0 < \dim Y < \dim X$, then we return to Step 1 with (Y, g_Y).

(d) If $\dim Y = 0$, X should have $c_1(X) > 0$. Moreover, after appropriate normalization, the solution $(X, \omega(t))$ of the Kähler–Ricci flow should deform to (X', ω') where X' is possibly a different manifold and ω' is either a Kähler–Einstein metric or a Kähler–Ricci soliton [i.e. $\mathrm{Ric}(\omega') = \omega' + \mathcal{L}_V(\omega')$ for a holomorphic vector field V]. See the discussion after Conjecture 3.7.8.

Namely, the Kähler–Ricci flow should construct the sequence of manifolds $X_1, \ldots, X_k$ of the MMP with scaling, with X_k either nef (as in Step 2) or a Mori fiber space [as in Step 3, part (c) or (d)]. If we have a Mori fiber space, then we can continue the flow on the lower dimensional manifold Y, which would correspond to a lower dimensional MMP with scaling. At the very last step, we expect the Kähler–Ricci flow to converge, after an appropriate normalization, to a canonical metric.

In [ST09], Song–Tian constructed weak solutions for the Kähler–Ricci flow through the finite time singularities if the flips exist a priori (see Chap. 4 in the present volume). Such a weak solution is smooth outside the singularities of X and the exceptional locus of the contractions and flips, and it is a nonnegative closed $(1, 1)$-current with locally bounded potentials. Furthermore, the weak solution of the Kähler–Ricci flow is unique.

In Step 2, when $T = \infty$, one can say more about the limiting behavior of the Kähler–Ricci flow. The abundance conjecture in birational geometry predicts that K_X is semi-ample whenever it is nef. Assuming this holds, the pluricanonical system $H^0(X, K_X^m)$ for sufficiently large m induces a holomorphic map $\phi : X \to X_{\text{can}}$. X_{can} is called the canonical model of X and it is uniquely determined by the canonical ring of X. If we assume that X is nonsingular and K_X is semi-ample, then normalized solution $g(t)/t$ always converges weakly in the sense of distributions. Moreover:

- If $\text{kod}(X) = \dim X$, then X_{can} is birationally equivalent to X and the limit of $g(t)/t$ is the unique singular Kähler–Einstein metric on X_{can} [Tzha06, Tsu88]. If X is a singular minimal model, we expect the Kähler–Ricci flow to converge to the singular Kähler–Einstein metric of Guedj–Eyssidieux–Zeriahi [EGZ11].
- If $0 < \text{kod}(X) < \dim X$, then X admits a Calabi–Yau fibration over X_{can}, and the limit of $g(t)/t$ is the unique generalized Kähler–Einstein metric (possibly singular) g_{can} on X_{can} defined by $\text{Ric}(g_{\text{can}}) = -g_{\text{can}} + g_{\text{WP}}$ away from a subvariety of X_{can}, where g_{WP} is the Weil–Petersson metric induced from the Calabi–Yau fibration of X over X_{can} [ST07, ST12].
- If $\text{kod}(X) = 0$, then X itself is a Calabi–Yau manifold and so the limit of $g(t)$ is the unique Ricci flat Kähler metric in its initial Kähler class [Cao85, Yau78].

A deeper question to ask is whether such a weak solution is indeed a geometric solution of the Kähler–Ricci flow in the Gromov–Hausdorff topology. One would like to show that the Kähler–Ricci flow performs geometric surgeries in Gromov–Hausdorff topology at each singular time and replaces the previous projective variety by a "better" model. Such a model is again a projective variety and the geometric surgeries coincide with the algebraic surgeries such as contractions and flips. If this picture holds, the Kähler–Ricci flow gives a continuous path from X to its canonical model X_{can} coupled with a canonical metric in the moduli space of Gromov–Hausdorff. We can further ask: how does the curvature behave near the (finite) singular time? Is the singularity is always of Type I (see Conjecture 3.7.8)? Will the flow give a complete or compact shrinking soliton after rescaling (cf. [Cao94, FIK03])?

3.8.3 The Kähler–Ricci Flow on Kähler Surfaces

In this section, we describe the behavior of the Kähler–Ricci flow on Kähler surfaces, and how it relates to the MMP. For the purpose of this section, X will be a Kähler surface.

We begin by discussing the minimal model program for surfaces. It turns out to be relatively straightforward. We obtain a sequence of smooth manifolds $X_1, \ldots, X_k$ and holomorphic maps $f_1, \ldots, f_k$,

$$X = X_0 \xrightarrow{f_1} X_1 \xrightarrow{f_2} X_2 \xrightarrow{f_3} \cdots \cdots \xrightarrow{f_k} X_k \qquad (3.396)$$

with X_k "minimal" in the sense described below. Moreover, each of the maps f_i is a blow down of a curve to a point.

We say that a Kähler surface X is a *minimal surface* if it contains no (-1)-curve. By the Adjunction Formula, a surface with K_X nef is minimal. On the other hand, a minimal surface may not have K_X nef (an example is $\mathbb{P}^2$) and hence this definition of *minimal surface* differs from the notion of "minimal model" discussed above.

The minimal model program for surfaces is simply as follows: given a surface X, contract all the (-1)-curves to arrive at a minimal surface. The Kodaira–Enriques classification can then be used to deduce that one either obtains a minimal surface with K_X nef, or a minimal Mori fiber space. A minimal Mori fiber space is either $\mathbb{P}^2$ or a *ruled surface*, i.e. a $\mathbb{P}^1$ bundle over a Riemann surface (in the literature, sometimes a broader definition for ruled surface is used). Dropping the minimality condition, Mori fiber spaces in dimension two are precisely those surfaces birational to a ruled surface. Note that since $\mathbb{P}^2$ is birational to $\mathbb{P}^1 \times \mathbb{P}^1$, every surface birational to $\mathbb{P}^2$ is birational to a ruled surface.

We wish to see whether the Kähler–Ricci flow on a Kähler surface will carry out this "minimal model program". The Kähler–Ricci flow should carry out the algebraic procedure of contracting (-1)-curves. Recall that in Sect. 3.7.3 we defined the notion of *canonical surgical contraction* for the Kähler–Ricci flow.

Starting at any Kähler surface X, we will use Theorem 3.7.10 to show that the Kähler–Ricci flow will always carry out a finite sequence of canonical surgical contractions until it either arrives at a minimal surface or the flow collapses the manifold.

Theorem 3.8.3. *Let (X, ω_0) be a Kähler surface with a smooth Kähler metric ω_0. Then there exists a unique maximal Kähler–Ricci flow $\omega(t)$ on $X_0, X_1, \ldots, X_k$ with canonical surgical contractions starting at (X, ω_0). Moreover, each canonical surgical contraction corresponds to a blow-down $\pi : X_i \to X_{i+1}$ of a finite number of disjoint (-1) curves on X_i. In addition we have:*

(a) Either $T_k < \infty$ and the flow $\omega(t)$ collapses X_k, in the sense that

$$\mathrm{Vol}_{\omega(t)} X_k \to 0, \quad \text{as } t \to T_k^-.$$

Then X_k is birational to a ruled surface.
(b) Or $T_k = \infty$ and X_k has no (-1) curves.

Proof. Let T_1 be the first singular time. If $T_1 = \infty$ then K_X is nef and hence X has no (-1)-curves, giving case (b).

Assume then that $T_1 < \infty$. The limiting class at time T_1 is given by $\alpha = [\omega_0] - T_1 c_1(X)$. Suppose that

$$\alpha^2 = \lim_{t \to T_1} ([\omega_0] - t c_1(X))^2 = \lim_{t \to T_1} \mathrm{Vol}_{g(t)} X > 0, \qquad (3.397)$$

so that we are not in case (a). Thus the class α is nef and big. On the other hand, α cannot be a Kähler class by Theorem 3.3.1.

We further notice that $\alpha + \varepsilon[\omega_0]$ is Kähler for all $\varepsilon > 0$ by Theorem 3.3.1. Then

$$\alpha \cdot (\alpha + \varepsilon[\omega_0]) = \alpha^2 + \varepsilon\alpha \cdot [\omega_0] > 0 \tag{3.398}$$

if we choose $\varepsilon > 0$ sufficiently small.

We now apply the Nakai–Moishezon criterion for Kähler surfaces (Theorem 3.1.15) to see that there must exist an irreducible curve C on X such that $\alpha \cdot C = 0$. Let $\mathcal{E}$ be the space of all irreducible curves E on X with $\alpha \cdot E = 0$. Then $\mathcal{E}$ is non-empty and every E in $\mathcal{E}$ has $E^2 < 0$ by the Hodge Index Theorem (Theorem 3.1.14). Moreover, if $E \in \mathcal{E}$,

$$E \cdot K_X = \frac{1}{T_1} E \cdot (\alpha - [\omega_0]) = -\frac{1}{T_1} E \cdot [\omega_0] < 0$$

since $[\omega_0]$ is Kähler. It then follows from the Adjunction formula (Theorem 3.1.13) that E must be a (-1) curve.

We claim that if E_1 and E_2 are distinct elements of $\mathcal{E}$ then they must be disjoint. Indeed, since E_1, E_2 are irreducible and distinct we have $E_1 \cdot E_2 \geq 0$. Moreover, $(E_1 + E_2) \cdot \alpha = 0$ and applying the Hodge Index Theorem again, we see that $0 > (E_1 + E_2)^2 = -2 + 2E_1 \cdot E_2$, so that the only possibility is $E_1 \cdot E_2 = 0$, proving the claim. It follows that $\mathcal{E}$ consists of finitely many disjoint (-1) curves $E_1, \ldots, E_k$.

Let $\pi : X \to Y$ be the blow-down map contracting $E_1, \ldots, E_k$ on X. Then Y is again a smooth Kähler surface. Since $H^{1,1}(X, \mathbb{R})$ is generated by $H^{1,1}(Y, \mathbb{R})$ and the $c_1([E_i])$ for $i = 1, \ldots, k$ (see for example Theorem I.9.1 in [BHPV]), there exists $\beta \in H^{1,1}(X, \mathbb{R})$ and $a_i \in \mathbb{R}$ such that

$$\alpha = \pi^*\beta + \sum_{i=1}^{k} a_i c_1([E_i]). \tag{3.399}$$

Since $\pi^*\beta \cdot E_i = 0$ for each $i = 1, \ldots, k$, we have $\alpha \cdot E_i = a_i = 0$ for all i and hence $\alpha = \pi^*\beta$.

We claim that β is a Kähler class on Y. First, for any curve C on Y, we have $\beta \cdot C = \alpha \cdot \pi^*C > 0$. Moreover, $\beta^2 = \alpha^2 > 0$.

By the Nakai–Moishezon criterion, it remains to show that $\beta \cdot \gamma > 0$ for γ some fixed Kähler class on Y. Now $\beta \cdot \gamma = \alpha \cdot \pi^*\gamma = \lim_{t \to T^-} [\omega(t)] \cdot \pi^*\gamma \geq 0$. Then put $\tilde{\gamma} = \gamma + c\beta$ for $c > 0$. If c is sufficiently small then $\tilde{\gamma}$ is Kähler and since $\beta^2 > 0$ we have $\beta \cdot \tilde{\gamma} > 0$, as required.

We now apply Theorem 3.7.10 to see that the Kähler–Ricci flow performs a canonical surgical contraction. We repeat the above procedure until either the volume tends to 0 or the flow exists for all time. This proves that either $T_k < \infty$ and $\mathrm{Vol}_{g(t)} X_k \to 0$ as $t \to T_k^-$ or $T_k = \infty$ and X_k has no (-1) curves.

Finally, in the case (a), the theorem follows from Proposition 3.8.4 below. □

We make use of Enriques–Kodaira classification for complex surfaces (see [BHPV]) to prove:

Proposition 3.8.4. *Let (X, ω_0) be a Kähler surface with a smooth Kähler metric ω_0. Let T be the first singular time of the Kähler–Ricci flow (3.395). If $T < \infty$ and $\mathrm{Vol}_{g(t)} X \to 0$, as $t \to T$. Then X is birational to a ruled surface. Moreover:*

(a) Either there exists $C > 0$ such that

$$C^{-1} \le \frac{\mathrm{Vol}_{g(t)} X}{(T - t)^2} \le C, \qquad (3.400)$$

and X is a Fano surface (in particular, is birational to $\mathbb{P}^2$) and $\omega_0 \in T c_1(X)$.
(b) Or there exists $C > 0$ such that

$$C^{-1} \le \frac{\mathrm{Vol}_{g(t)} X}{T - t} \le C. \qquad (3.401)$$

If X is Fano then ω_0 is not in a multiple of $c_1(X)$.

Proof. We first show that X is birational to a ruled surface. Suppose for a contradiction that $\mathrm{kod}(X) \ge 0$. Then some multiple of K_X has a global holomorphic section and hence is effective. In particular, $([\omega_0] + T K_X) \cdot K_X \ge 0$, since $[\omega_0] + T K_X$ is a limit of Kähler classes. Then

$$0 = ([\omega_0] + T K_X)^2 = T([\omega_0] + T K_X) \cdot K_X + ([\omega_0] + T K_X) \cdot [\omega_0]$$
$$\ge ([\omega_0] + T K_X) \cdot [\omega_0] \ge 0, \qquad (3.402)$$

which implies that $([\omega_0] + T K_X) \cdot [\omega_0] = 0$. Using the Index Theorem and the fact that $[\omega_0]^2 > 0$ and $([\omega_0] + T K_X)^2 = 0$ we have $[\omega_0] + T K_X = 0$. But this implies that X is Fano, contradicting the assumption $\mathrm{kod}(X) \ge 0$. Thus we have shown that X must have $\mathrm{kod}(X) = -\infty$. By the Enriques–Kodaira classification for complex surfaces which are Kähler (see Chap. VI of [BHPV]), X is birational to a ruled surface.

Since $\mathrm{Vol}_{g(t)} X = ([\omega_0] + t K_X)^2$ is a quadratic polynomial in t which is positive for $t \in [0, T)$ and tends to zero as t tends to T, we have

$$\mathrm{Vol}_{g(t)} X = [\omega_0]^2 + 2t[\omega_0] \cdot K_X + t^2 K_X^2 = C_1(T - t) + C_2(T - t)^2, \quad (3.403)$$

for constants $C_1 \ge 0$ and C_2. First assume $C_1 = 0$. Then $C_2 > 0$ and we are in case (a). From (3.403) we obtain

$$K_X^2 = C_2 > 0, \quad [\omega_0]^2 = K_X^2 T^2, \quad [\omega_0] \cdot K_X = -K_X^2 T < 0. \qquad (3.404)$$

In particular, $([\omega_0] + TK_X) \cdot [\omega_0] = 0$ and hence by the Index Theorem, $[\omega_0] + TK_X = 0$. Thus X is Fano and $\omega_0 \in Tc_1(X)$. Note that by the classification of surfaces, X is either $\mathbb{P}^2$, $\mathbb{P}^1 \times \mathbb{P}^1$ or $\mathbb{P}^2$ blown-up at k points for $1 \leq k \leq 8$.

Finally, if $C_1 > 0$ then we are in case (b). If $[\omega_0]$ is a multiple of $c_1(X)$ then the volume $\mathrm{Vol}_{g(t)}X$ tends to zero of order $(T - t)^2$, a contradiction. $\qquad \square$

We now discuss the long time behavior of the Kähler–Ricci flow when we are in case (b) of Theorem 3.8.3. There are three different behaviors of the Kähler–Ricci flow as $t \to \infty$ depending on whether X has Kodaira dimension equal to 0, 1 or 2:

- If $\mathrm{kod}(X) = 0$, then the minimal model of X is a Calabi–Yau surface with $c_1(X) = 0$. The flow $g(t)$ converges smoothly to a Ricci-flat Kähler metric as $t \to \infty$, as shown in Sect. 3.6.
- If $\mathrm{kod}(X) = 1$, then $\frac{1}{t}g(t)$ converges in the sense of currents to the pullback of the unique generalized Kähler–Einstein metric on the canonical model of X as $t \to \infty$ [ST07]. A simple example of this is given in Sect. 3.6 in the case of a product elliptic surface.
- If $\mathrm{kod}(X) = 2$, $\frac{1}{t}g(t)$ converges in the sense of currents (and smoothly outside a subvariety) to the pullback of the unique smooth orbifold Kähler–Einstein metric on the canonical model of X as $t \to \infty$ [Kob85, Tzha06, Tsu96]. In the case that $c_1(X) < 0$, we showed in Sect. 3.6 that $\frac{1}{t}g(t)$ converges smoothly to a smooth Kähler–Einstein metric.

In fact, in the case when $T_k = \infty$, the scalar curvature of $\frac{1}{t}g(t)$ is uniformly bounded as $t \to \infty$ [ST11, Zha09]. Furthermore, if we assume that X_k is a minimal surface of general type, and it admits only irreducible (-2)-curves, then $(X_k, \frac{1}{t}g(t))$ converges in Gromov–Hausdorff sense to its canonical model with the unique smooth orbifold Kähler–Einstein metric [SW11].

Acknowledgements Jian Song is supported in by an NSF CAREER grant DMS-08-47524 and a Sloan Research Fellowship. Ben Weinkove is supported by the NSF grants DMS-08-48193 and DMS-11-05373.

References

[Aub78] T. Aubin, Equation de type Monge-Ampère sur les variétés kählériennes compactes. Bull. Sci. Math. **102**, 63–95 (1978)

[Aub82] T. Aubin, in *Nonlinear Analysis on Manifolds. Monge-Ampère Equations*. Grundlehren der Mathematischen Wissenschaften, vol. 252 (Springer, New York, 1982)

[Bando84] S. Bando, On the classification of three-dimensional compact Kaehler manifolds of nonnegative bisectional curvature. J. Differ. Geom. **19**(2), 283–297 (1984)

[Bando87] S. Bando, The K-energy map, almost Einstein Kähler metrics and an inequality of the Miyaoka-Yau type. Tohoku Math. J. **39**, 231–235 (1987)

[BHPV] W.P. Barth, K. Hulek, C. Peters, A. Van de Ven, in *Compact Complex Surfaces*, 2nd edn. Ergebnisse der Mathematik und ihrer Grenzgebiete (Springer, Berlin, 2004)

[BCHM10] C. Birkar, P. Cascini, C. Hacon, J. McKernan, Existence of minimal models for varieties of log general type. J. Am. Math. Soc. **23**(2), 405–468 (2010)

[BW08] C. Böhm, B. Wilking, Manifolds with positive curvature operators are space forms. Ann. Math. (2) **167**(3), 1079–1097 (2008)

[BEGZ10] S. Boucksom, P. Eyssidieux, V. Guedj, A. Zeriahi, Monge-Ampère equations in big cohomology classes. Acta Math. **205**, 199–262 (2010)

[BS08] S. Brendle, R. Schoen, Classification of manifolds with weakly 1/4-pinched curvatures. Acta Math. **200**(1), 1–13 (2008)

[Buch99] N. Buchdahl, On compact Kähler surfaces. Ann. Inst. Fourier **49**(1), 287–302 (1999)

[Cal57] E. Calabi, On Kähler manifolds with vanishing canonical class, in *Algebraic Geometry and Topology.* A Symposium in Honor of S. Lefschetz (Princeton University Press, Princeton, 1957), pp. 78–89

[Cal58] E. Calabi, Improper affine hyperspheres of convex type and a generalization of a theorem by K. Jörgens. Mich. Math. J. **5**, 105–126 (1958)

[Cal82] E. Calabi, Extremal Kähler metrics, in *Seminar on Differential Geometry.* Annals of Mathematics Studies, vol. 102, (Princeton University Press, Princeton, 1982), pp. 259–290

[Cao85] H.D. Cao, Deformation of Kähler metrics to Kähler-Einstein metrics on compact Kähler manifolds. Invent. Math. **81**(2), 359–372 (1985)

[Cao92] H.D. Cao, On Harnack's inequalities for the Kähler–Ricci flow. Invent. Math. **109**(2), 247–263 (1992)

[Cao94] H.-D. Cao, in *Existence of Gradient Kähler–Ricci Solitons.* Elliptic and Parabolic Methods in Geometry (Minneapolis, MN, 1994) (A.K. Peters, Wellesley, 1996), pp. 1–16

[Cao] H.-D. Cao, The Kähler–Ricci flow on Fano manifolds, in *An Introduction to the Kähler–Ricci Flow*, ed. by S. Boucksom, P. Eyssidieux, V. Guedj. Lecture Notes in Mathematics (Springer, Heidelberg, 2013)

[CZ09] H.-D. Cao, M. Zhu, A note on compact Kähler–Ricci flow with positive bisectional curvature. Math. Res. Lett. **16**(6), 935–939 (2009)

[CZ06] H.-D. Cao, X.-P. Zhu, A complete proof of the Poincaré and geometrization conjectures - application of the Hamilton-Perelman theory of the Ricci flow. Asian J. Math. **10**(2), 165–492 (2006)

[ChauT06] A. Chau, L.-F. Tam, On the complex structure of Kähler manifolds with nonnegative curvature. J. Differ. Geom. **73**(3), 491–530 (2006)

[CST09] X. Chen, S. Sun, G. Tian, A note on Kähler–Ricci soliton. Int. Math. Res. Not. IMRN **2009**(17), 3328–3336 (2009)

[CheT06] X.X. Chen, G. Tian, Ricci flow on Kähler-Einstein manifolds. Duke Math. J. **131**(1), 17–73 (2006)

[ChW09] X. Chen, B. Wang, The Kähler–Ricci flow on Fano manifolds (I). J. Eur. Math. Soc. (JEMS) **14**(6), 2001–2038 (2012)

[ChengYau75] S.Y. Cheng, S.T. Yau, Differential equations on Riemannian manifolds and their geometric applications. Comm. Pure Appl. Math. **28**(3), 333–354 (1975)

[Cher87] P. Cherrier, Équations de Monge-Ampère sur les variétés hermitiennes compactes. Bull. Sci. Math. (2) **111**(4), 343–385 (1987)

[Chow91] B. Chow, The Ricci flow on the 2-sphere. J. Differ. Geom. **33**(2), 325–334 (1991)

[ChowKnopf] B. Chow, D. Knopf, in *The Ricci Flow: An Introduction.* Mathematical Surveys and Monographs, vol. 110 (American Mathematical Society, Providence, 2004), xii+325 pp.

[CLN06] B. Chow, P. Lu, L. Ni, in *Hamilton's Ricci Flow.* Graduate Studies in Mathematics, vol. 77 (American Mathematical Society/Science Press, Providence/New York, 2006), xxxvi+608 pp.

[CKL11] A. Corti, A.-S. Kaloghiros, V. Lazić, Introduction to the Minimal Model Program and the existence of flips. Bull. Lond. Math. Soc. **43**(3), 415–418 (2011)

[Deb01] O. Debarre, in *Higher-Dimensional Algebraic Geometry*. Universitext (Springer, New York, 2001)

[Dem96] J.-P. Demailly, in L^2 *Vanishing Theorems for Positive Line Bundles and Adjunction Theory*. Transcendental Methods in Algebraic Geometry (Cetraro, 1994). Lecture Notes in Mathematics, vol. 1646 (Springer, Berlin, 1996), pp. 1–97

[DemPaun04] J.-P. Demailly, M. Paun, Numerical characterization of the Kähler cone of a compact Kähler manifold. Ann. Math. (2) **159**(3), 1247–1274 (2004)

[Det83] D.M. DeTurck, Deforming metrics in the direction of their Ricci tensors. J. Differ. Geom. **18** (1), 157–162 (1983)

[DT92] W.-Y. Ding, G. Tian, Kähler-Einstein metrics and the generalized Futaki invariant. Invent. Math. **110**(2), 315–335 (1992)

[Don02] S.K. Donaldson, Scalar curvature and stability of toric varieties. J. Differ. Geom. **62**, 289–349 (2002)

[Eva82] L.C. Evans, Classical solutions of fully nonlinear, convex, second-order elliptic equations. Comm. Pure Appl. Math. **35**(3), 333–363 (1982)

[EGZ11] P. Eyssidieux, V. Guedj, A. Zeriahi, Viscosity solutions to degenerate complex Monge-Ampère equations. Comm. Pure Appl. Math. **64**, 1059–1094 (2011)

[FIK03] M. Feldman, T. Ilmanen, D. Knopf, Rotationally symmetric shrinking and expanding gradient Kähler–Ricci solitons. J. Differ. Geom. **65**(2), 169–209 (2003)

[Fo11] T.-H.F. Fong, Kähler–Ricci flow on projective bundles over Kähler-Einstein manifolds (2011). Preprint (arXiv:1104.3924 [math.DG])

[GT01] D. Gilbarg, N.S. Trudinger, in *Elliptic Partial Differential Equations of Second Order*, Reprint of the 1998 edn. Classics in Mathematics (Springer, Berlin, 2001), xiv+517 pp.

[Gill11] M. Gill, Convergence of the parabolic complex Monge-Ampère equation on compact Hermitian manifolds. Comm. Anal. Geom. **19**(2), 277–303 (2011)

[GH78] P. Griffiths, J. Harris, in *Principles of Algebraic Geometry*. Pure and Applied Mathematics (Wiley-Interscience, New York, 1978)

[GTZ11] M. Gross, V. Tosatti, Y. Zhang, Collapsing of abelian fibred Calabi-Yau manifolds. Duke Math. J. **162**(3), 517–551 (2013)

[Gu09] H.-L. Gu, A new proof of Mok's generalized Frankel conjecture theorem. Proc. Am. Math. Soc. **137**(3), 1063–1068 (2009)

[HM10] C.D. Hacon, J. McKernan, Existence of minimal models for varieties of log general type, II. J. Am. Math. Soc. **23**(2), 469–490 (2010)

[Ham82] R.S. Hamilton, Three-manifolds with positive Ricci curvature. J. Differ. Geom. **17**(2), 255–306 (1982)

[Ham86] R.S. Hamilton, Four-manifolds with positive curvature operator. J. Differ. Geom. **24**(2), 153–179 (1986)

[Ham88] R. Hamilton, in *The Ricci Flow on Surfaces*. Mathematics and General Relativity (Santa Cruz, CA, 1986). Contemporary Mathematics, vol. 71 (American Mathematical Society, Providence, 1988), pp. 237–262

[Ham95a] R.S. Hamilton, in *The Formation of Singularities in the Ricci Flow*. Surveys in Differential Geometry, vol. II (Cambridge, MA, 1993) (International Press, Cambridge, 1995), pp. 7–136

[Ha77] R. Hartshorne, in *Algebraic Geometry*. Graduate Texts in Mathematics, vol. 52 (Springer, New York, 1977)

[KMM87] Y. Kawamata, K. Matsuda, K. Matsuki, in *Introduction to the Minimal Model Problem*. Algebraic Geometry (Sendai, 1985). Advanced Studies in Pure Mathematics, vol. 10 (North-Holland, Amsterdam, 1987), pp. 283–360

[KL08] B. Kleiner, J. Lott, Notes on Perelman's papers. Geom. Topol. **12**(5), 2587–2855 (2008)

[Kob85] R. Kobayashi, Einstein-Kähler V-metrics on open Satake V-surfaces with isolated quotient singularities. Math. Ann. **272**(3), 385–398 (1985)

[KodMor71] K. Kodaira, J. Morrow, *Complex Manifolds* (Holt, Rinehart and Winston, Inc., New York, 1971), vii+192 pp.

[KolMori98] J. Kollár, S. Mori, in *Birational Geometry of Algebraic Varieties*. Cambridge Tracts in Mathematics, vol. 134 (Cambridge University Press, Cambridge, 1998)

[Kol98] S. Kołodziej, The complex Monge-Ampère equation. Acta Math. **180**(1), 69–117 (1998); Math. Ann. **342**(4), 773–787 (2008); Differ. Geom. **29**, 665–683 (1989)

[Kryl82] N.V. Krylov, Boundedly nonhomogeneous elliptic and parabolic equations. Izvestia Akad. Nauk. SSSR **46**, 487–523 (1982). English translation in Math. USSR Izv. **20**(3), 459–492 (1983)

[LaNT09] G. La Nave, G. Tian, Soliton-type metrics and Kähler–Ricci flow on symplectic quotients (2009). Preprint (arXiv: 0903.2413 [math.DG])

[Lam99] A. Lamari, Le cône Kählérien d'une surface. J. Math. Pure Appl. **78**, 249–263 (1999)

[Laz04] R. Lazarsfeld, in *Positivity in Algebraic Geometry. I. Classical Setting: Line Bundles and Linear Series*. Ergebnisse der Mathematik und ihrer Grenzgebiete, 3. Folge. A Series of Modern Surveys in Mathematics, vol. 48 (Springer, Berlin, 2004), xviii+387 pp.

[Lieb96] G.M. Lieberman, *Second Order Parabolic Differential Equations* (World Scientific, River Edge, 1996)

[Mab86] T. Mabuchi, K-energy maps integrating Futaki invariants. Tohoku Math. J. (2) **38**(4), 575–593 (1986)

[Mok88] N. Mok, The uniformization theorem for compact Kähler manifolds of nonnegative holomorphic bisectional curvature. J. Differ. Geom. **27**(2), 179–214 (1988)

[MT07] J. Morgan, G. Tian, in *Ricci Flow and the Poincaré Conjecture*. Clay Mathematics Monographs, vol. 3 (American Mathematical Society/Clay Mathematics Institute, Providence/Cambridge, 2007)

[MT08] J. Morgan, G. Tian, Completion of the proof of the geometrization conjecture (2008). Preprint (arXiv: 0809.4040 [math.DG])

[MSz09] O. Munteanu, G. Székelyhidi, On convergence of the Kähler–Ricci flow. Commun. Anal. Geom. **19**(5), 887–903 (2011)

[Ni04] L. Ni, A monotonicity formula on complete Kähler manifolds with nonnegative bisectional curvature. J. Am. Math. Soc. **17**(4), 909–946 (2004)

[NiW10] L. Ni, B. Wilking, Manifolds with 1/4-pinched flag curvature. Geom. Funct. Anal. **20**(2), 571–591 (2010)

[Per02] G. Perelman, The entropy formula for the Ricci flow and its geometric applications (2002). Preprint (arXiv: math.DG/0211159)

[Per03q] G. Perelman, Ricci flow with surgery on three-manifolds (2003). Preprint (arXiv:math.DG/0303109)

[Per03b] G. Perelman, Finite extinction time for the solutions to the Ricci flow on certain three-manifolds (2003). Preprint [arXiv:math.DG/0307245]

[PSS07] D.H. Phong, N. Sesum, J. Sturm, Multiplier ideal sheaves and the Kähler–Ricci flow. Comm. Anal. Geom. **15**(3), 613–632 (2007)

[PSSW08b] D.H. Phong, J. Song, J. Sturm, B. Weinkove, The Kähler–Ricci flow with positive bisectional curvature. Invent. Math. **173**(3), 651–665 (2008)

[PSSW09] D.H. Phong, J. Song, J. Sturm, B. Weinkove, The Kähler–Ricci flow and the $\bar{\partial}$ operator on vector fields. J. Differ. Geom. **81**(3), 631–647 (2009)

[PSSW11] D.H. Phong, J. Song, J. Sturm, B. Weinkove, On the convergence of the modified Kähler–Ricci flow and solitons. Comment. Math. Helv. **86**(1), 91–112 (2011)

[PS05] D.H. Phong, J. Sturm, On the Kähler–Ricci flow on complex surfaces. Pure Appl. Math. Q. **1**(2), Part 1, 405–413 (2005)

[PS06] D.H. Phong, J. Sturm, On stability and the convergence of the Kähler–Ricci flow. J. Differ. Geom. **72**(1), 149–168 (2006)

[PS10] D.H. Phong, J. Sturm, Lectures on stability and constant scalar curvature, in *Handbook of Geometric Analysis*, No. 3. Advanced Lectures in Mathematics (ALM), vol. 14 (International Press, Somerville, 2010), pp. 357–436

[Rub09] Y. Rubinstein, On the construction of Nadel multiplier ideal sheaves and the limiting behavior of the Ricci flow. Trans. Am. Math. Soc. **361**(11), 5839–5850 (2009)

[Se05] N. Šešum, Curvature tensor under the Ricci flow. Am. J. Math. **127**(6), 1315–1324 (2005)

[SeT08] N. Sesum, G. Tian, Bounding scalar curvature and diameter along the Kähler–Ricci flow (after Perelman). J. Inst. Math. Jussieu **7**(3), 575–587 (2008)

[ShW11] M. Sherman, B. Weinkove, Interior derivative estimates for the Kähler–Ricci flow. Pac. J. Math. **257**(2), 491–501 (2012)

[Shi89] W.-X. Shi, Deforming the metric on complete Riemannian manifolds. J. Differ. Geom. **30**(1), 223–301 (1989)

[Shok85] V.V. Shokurov, A nonvanishing theorem. Izv. Akad. Nauk SSSR Ser. Mat. **49**(3), 635–651 (1985)

[Siu87] Y.T. Siu, in *Lectures on Hermitian-Einstein Metrics for Stable Bundles and Kähler-Einstein Metrics*. DMV Seminar, vol. 8 (Birkhäuser, Basel, 1987)

[Siu08] Y.-T. Siu, Finite generation of canonical ring by analytic method. Sci. China Ser. A **51**(4), 481–502 (2008)

[SSW11] J. Song, G. Székelyhidi, B. Weinkove, The Kähler–Ricci flow on projective bundles. Int. Math. Res. Not. IMRN **2013**(2), 243–257 (2013)

[ST07] J. Song, G. Tian, The Kähler–Ricci flow on surfaces of positive Kodaira dimension. Invent. Math. **170**(3), 609–653 (2007)

[ST12] J. Song, G. Tian, Canonical measures and Kähler–Ricci flow. J. Am. Math. Soc. **25**, 303–353 (2012)

[ST09] J. Song, G. Tian, The Kähler–Ricci flow through singularities (2009). Preprint [arXiv:0909.4898]

[ST11] J. Song, G. Tian, Bounding scalar curvature for global solutions of the Kähler–Ricci flow (2011). Preprint

[SW09] J. Song, B. Weinkove, The Kähler–Ricci flow on Hirzebruch surfaces. J. Reine Ange. Math. **659**, 141–168 (2011)

[SW10] J. Song, B. Weinkove, Contracting exceptional divisors by the Kähler–Ricci flow. Duke Math. J. **162**(2), 367–415 (2013)

[SW11] J. Song, B. Weinkove, Contracting exceptional divisors by the Kähler–Ricci flow II (2011). Preprint (arXiv:1102.1759 [math.DG])

[SY10] J. Song, Y. Yuan, Metric flips with Calabi ansatz. Geom. Funct. Anal. **22**(1), 240–265 (2012)

[StT10] J. Streets, G. Tian, A parabolic flow of pluriclosed metrics. Int. Math. Res. Not. IMRN **2010**(16), 3101–3133 (2010)

[Sz10] G. Székelyhidi, The Kähler–Ricci flow and K-stability. Am. J. Math. **132**(4), 1077–1090 (2010)

[Tian97] G. Tian, Kähler-Einstein metrics with positive scalar curvature. Invent. Math. **130**, 239–265 (1997)

[Tian] G. Tian, in *Canonical Metrics in Kähler Geometry*. Lectures in Mathematics ETH Zürich (Birkhäuser, Basel, 2000)

[Tian02] G. Tian, in *Geometry and Nonlinear Analysis*. Proceedings of the International Congress of Mathematicians, vol. I (Beijing, 2002) (Higher Ed. Press, Beijing, 2002), pp. 475–493

[Tian08] G. Tian, New results and problems on Kähler–Ricci flow. Géométrie différentielle, physique mathématique, mathématiques et société, II. Astérisque **322**, 71–92 (2008)

[Tzha06] G. Tian, Z. Zhang, On the Kähler–Ricci flow on projective manifolds of general type. Chin. Ann. Math. Ser. B **27**(2), 179–192 (2006)

[TZ07] G. Tian, X. Zhu, Convergence of Kähler–Ricci flow. J. Am. Math. Soc. **20**(3), 675–699 (2007)

[Tos10a] V. Tosatti, Kähler–Ricci flow on stable Fano manifolds. J. Reine Angew. Math. **640**, 67–84 (2010)

[Tos10b] V. Tosatti, Adiabatic limits of Ricci-flat Kähler metrics. J. Differ. Geom. **84**(2), 427–453 (2010)

[TWY08] V. Tosatti, B. Weinkove, S.-T. Yau, Taming symplectic forms and the Calabi-Yau equation. Proc. Lond. Math. Soc. (3) **97**(2), 401–424 (2008)

[Tsu88] H. Tsuji, Existence and degeneration of Kähler-Einstein metrics on minimal algebraic varieties of general type. Math. Ann. **281**(1), 123–133 (1988)

[Tsu96] H. Tsuji, Generalized Bergmann metrics and invariance of plurigenera (1996). Preprint [arXiv:math.CV/9604228]

[Yau78] S.T. Yau, On the Ricci curvature of a compact Kähler manifold and the complex Monge-Ampère equation, I. Comm. Pure Appl. Math. **31**(3), 339–411 (1978)

[Yau78b] S.T. Yau, A general Schwarz lemma for Kähler manifolds. Am. J. Math. **100**(1), 197–203 (1978)

[Yau93] S.-T. Yau, Open problems in geometry. Proc. Symp. Pure Math. **54**, 1–28 (1993)

[ZhaZha11] X. Zhang, X. Zhang, Regularity estimates of solutions to complex Monge-Ampère equations on Hermitian manifolds. J. Funct. Anal. **260**(7), 2004–2026 (2011)

[Zha06] Z. Zhang, On degenerate Monge-Ampère equations over closed Kähler manifolds. Int. Math. Res. Not. Art. ID 63640, 18 pp. (2006)

[Zha09] Z. Zhang, Scalar curvature bound for Kähler–Ricci flows over minimal manifolds of general type. Int. Math. Res. Not. IMRN **2009**(20), 3901–3912 (2009)

[Zha10] Z. Zhang, Scalar curvature behavior for finite-time singularity of Kähler–Ricci flow. Mich. Math. J. **59**(2), 419–433 (2010)

[Zhu07] X. Zhu, Kähler–Ricci flow on a toric manifold with positive first Chern class (2007). Preprint [arXiv:math.DG/0703486]

Chapter 4
Regularizing Properties of the Kähler–Ricci Flow

Sébastien Boucksom and Vincent Guedj

Abstract These notes present a general existence result for degenerate parabolic complex Monge–Ampère equations with continuous initial data, slightly generalizing the work of Song and Tian on this topic. This result is applied to construct a Kähler–Ricci flow on varieties with log terminal singularities, in connection with the Minimal Model Program. The same circle of ideas is also used to prove a regularity result for elliptic complex Monge–Ampère equations, following Székelyhidi–Tosatti.

Introduction

As we saw in Chap. 3, each initial Kähler form ω_0 on a compact Kähler manifold X uniquely determines a solution $(\omega_t)_{t \in [0,T_0)}$ to the (unnormalized) Kähler–Ricci flow

$$\frac{\partial \omega_t}{\partial t} = -\mathrm{Ric}(\omega_t).$$

Along the flow, the cohomology class $[\omega_t] = [\omega_0] + t[K_X]$ must remain in the Kähler cone, and this is in fact the only obstruction to the existence of the flow. In other words, the maximal existence time T_0 is either infinite, in which case K_X is nef and X is thus a *minimal model* by definition, or T_0 is finite and $[\omega_0] + T_0[K_X]$ lies on the boundary of the Kähler cone.

S. Boucksom (✉)
Institut de Mathématiques de Jussieu, CNRS-Université Pierre et Marie Curie, 4 place Jussieu, 75251, Paris, France
e-mail: boucksom@math.jussieu.fr

V. Guedj
Institut de Mathématiques de Toulouse and Institut Universitaire de France, Université Paul Sabatier, 118 route de Narbonne, 31062, Toulouse Cedex 9, France
e-mail: vincent.guedj@math.univ-toulouse.fr

S. Boucksom et al. (eds.), *An Introduction to the Kähler–Ricci Flow*,
Lecture Notes in Mathematics 2086, DOI 10.1007/978-3-319-00819-6_4,
© Springer International Publishing Switzerland 2013

In [ST09], J. Song and G. Tian proposed to use the Minimal Model Program (MMP for short) to continue the flow beyond time T_0. At least when $[\omega_0]$ is a rational cohomology class (and hence X is projective), the MMP allows to find a mildly singular projective variety X' birational to X such that $[\omega_0] + t[K_X]$ induces a Kähler class on X' for $t > T_0$ sufficiently close to T_0. It is therefore natural to try and continue the flow on X', but new difficulties arise due to the singularities of X'. After blowing-up X' to resolve these singularities, the problem boils down to showing the existence of a unique solution to a certain degenerate parabolic complex Monge–Ampère equation, whose initial data is furthermore singular.

The primary purpose of this chapter is to present a detailed account of Song and Tian's solution to this problem. Along the way, a regularizing property of parabolic complex Monge–Ampère equations is exhibited, which can in turn be applied to prove the regularity of weak solutions to certain elliptic Monge–Ampère equations, following [SzTo11].

The chapter is organized as follows. In Sect. 4.1 we gather the main analytic tools to be used in the proof: a Laplacian inequality, the maximum principle, and Evans–Krylov type estimates for parabolic complex Monge–Ampère equations. In Sect. 4.2, we first consider the simpler case of non-degenerate parabolic complex Monge–Ampère equations involving a time-independent Kähler form. We show that such equations smooth out continuous initial data, and give a proof of the main result of [SzTo11]. Sections 4.3–4.5 contain the main result of the chapter, dealing with the general case of degenerate parabolic complex Monge–Ampère equations, basically following [ST09] (and independently of Sect. 4.2). In the final Sect. 4.6, we apply the previous results to study the Kähler–Ricci flow on varieties with log terminal singularities.

Nota Bene. This text is an expanded version of a series of lectures delivered by the two authors during the second ANR-MACK meeting (8–10 June 2011, Toulouse, France). As the audience mostly consisted of non specialists, we have tried to make these lecture notes accessible with only few prerequisites.

4.1 An Analytic Toolbox

4.1.1 A Laplacian Inequality

If θ and ω are $(1, 1)$-forms on a complex manifold X with $\omega > 0$, θ can be diagonalized with respect to ω at each point of X, with real eigenvalues $\lambda_1 \leq \ldots \leq \lambda_n$, and the *trace* of θ with respect to ω is defined as $\mathrm{tr}_\omega(\theta) = \sum_i \lambda_i$. More invariantly, we have

$$\mathrm{tr}_\omega(\theta) = n \frac{\theta \wedge \omega^{n-1}}{\omega^n}.$$

The Laplacian of a function φ with respect to ω is given by

$$\Delta_\omega \varphi = \operatorname{tr}_\omega(dd^c\varphi).$$

For later use, we record an elementary eigenvalue estimate:

Lemma 4.1.1. *If ω and ω' are two positive $(1,1)$-forms on a complex manifold X, then*

$$\left(\frac{\omega'^n}{\omega^n}\right)^{\frac{1}{n}} \le \frac{1}{n}\operatorname{tr}_\omega(\omega') \le \left(\frac{\omega'^n}{\omega^n}\right)(\operatorname{tr}_{\omega'}(\omega))^{n-1}.$$

Proof. In terms of the eigenvalues $0 < \lambda_1 \le \ldots \le \lambda_n$ of ω' with respect to ω (at a given point of X), the assertion writes

$$\left(\prod_i \lambda_i\right)^{1/n} \le \frac{1}{n}\sum_i \lambda_i \le \left(\prod_i \lambda_i\right)\left(\sum_i \lambda_i^{-1}\right)^{n-1}.$$

The left-hand inequality is nothing but the arithmetico-geometric inequality. By homogeneity, we may assume that $\prod_i \lambda_i = 1$ in proving the right-hand inequality. We then have

$$\left(\sum_i \lambda_i^{-1}\right)^{n-1} \ge \lambda_1^{-1}\ldots\lambda_{n-1}^{-1} = \lambda_n \ge \frac{1}{n}\sum_i \lambda_i. \qquad \square$$

The next result is a Laplacian inequality, which basically goes back to [Aub78, Yau78] and is the basic tool for establishing second order a priori estimates for elliptic and parabolic complex Monge–Ampère equations. In its present form, the result is found in [Siu87, pp. 97–99]; we include a proof for the reader's convenience.

Proposition 4.1.2. *Let ω, ω' be two Kähler forms on a complex manifold X. If the holomorphic bisectional curvature of ω is bounded below by a constant $B \in \mathbb{R}$ on X, then*

$$\Delta_{\omega'} \log \operatorname{tr}_\omega(\omega') \ge -\frac{\operatorname{tr}_\omega \operatorname{Ric}(\omega')}{\operatorname{tr}_\omega(\omega')} + B\operatorname{tr}_{\omega'}(\omega).$$

Proof. Since this is a pointwise inequality, we can choose normal holomorphic coordinates (z_j) at a given point $p \in X$ so that $\omega = i\sum_{k,l}\omega_{kl}dz_k \wedge d\bar{z}_l$ and $\omega' = i\sum_{k,l}\omega'_{kl}dz_k \wedge d\bar{z}_l$ satisfy

$$\omega_{kl} = \delta_{kl} - \sum_{i,j} R_{ijkl}z_i\bar{z}_j + O(|z|^3)$$

and

$$\omega'_{kl} = \lambda_k\delta_{kl} + O(|z|)$$

near p. Here R_{ijkl} denotes the curvature tensor of ω, δ_{kl} stands for the Kronecker symbol, and $\lambda_1 \leq \ldots \leq \lambda_n$ are the eigenvalues of ω' with respect to ω at p.

Observe that the inverse matrix $(\omega^{kl}) = (\omega_{kl})^{-1}$ satisfies

$$\omega^{kl} = \delta_{kl} + \sum_{i,j} R_{ijkl} z_i \bar{z}_j + O(|z|^3). \tag{4.1}$$

Recall also that the curvature tensor of ω' is given in the local coordinates (z_i) by

$$R'_{ijkl} = -\partial_i \bar{\partial}_j \omega'_{kl} + \sum_{p,q} \omega'_{pq} \partial_i \omega'_{kq} \bar{\partial}_j \omega'_{pl},$$

hence

$$R'_{ijkl} = -\partial_i \bar{\partial}_j \omega'_{kl} + \sum_p \lambda_p^{-1} \partial_i \omega_{kp} \bar{\partial}_j \omega'_{pl} \tag{4.2}$$

at p. Set $u := \mathrm{tr}_\omega(\omega')$, and note that

$$\Delta_{\omega'} \log u = u^{-1} \Delta_{\omega'} u - u^{-2} \mathrm{tr}_{\omega'}(du \wedge d^c u).$$

At the point p we have

$$\Delta_{\omega'} u = \sum_{ik} \lambda_i^{-1} \partial_i \bar{\partial}_i (\omega^{kk} \omega'_{kk})$$

and

$$\mathrm{tr}_{\omega'}(du \wedge d^c u) = \sum_{i,k,l} \lambda_i^{-1} \partial_i \omega'_{kk} \partial_i \omega'_{ll},$$

with

$$\partial_i \bar{\partial}_i (\omega^{kk} \omega'_{kk}) = \lambda_k R_{iikk} + \partial_i \bar{\partial}_i \omega'_{kk}$$

thanks to (4.1). It follows that

$$\Delta_{\omega'} \log u = u^{-1} \left(\sum_{ik} \lambda_i^{-1} \lambda_k R_{iikk} + \sum_{i,k} \lambda_i^{-1} \partial_i \bar{\partial}_i \omega'_{kk} \right)$$

$$- u^{-2} \left(\sum_{i,k,l} \lambda_i^{-1} \partial_i \omega'_{kk} \partial_i \omega'_{ll} \right) \tag{4.3}$$

holds at p. On the one hand, the assumption on the holomorphic bisectional curvature of ω reads $R_{iikk} \geq B$ for all i, k, hence

$$\sum_{ik} \lambda_i^{-1} \lambda_k R_{iikk} \geq B \left(\sum_i \lambda_i^{-1} \right) \left(\sum_k \lambda_k \right) = B \operatorname{tr}_{\omega'}(\omega) u. \qquad (4.4)$$

On the other hand, (4.2) yields

$$\sum_{i,k} \lambda_i^{-1} \partial_i \overline{\partial}_i \omega'_{kk} = - \sum_{i,k} \lambda_i^{-1} R'_{iikk} + \sum_{i,k,p} \lambda_i^{-1} \lambda_p^{-1} |\partial_i \omega'_{kp}|^2.$$

Note that $\sum_{i,k} \lambda_i^{-1} R'_{iikk} = \operatorname{tr}_\omega \operatorname{Ric}(\omega')$, while

$$\sum_{i,k,p} \lambda_i^{-1} \lambda_p^{-1} |\partial_i \omega'_{kp}|^2 \geq \sum_{i,k} \lambda_i^{-1} \lambda_k^{-1} |\partial_i \omega'_{kk}|^2 \geq u^{-1} \sum_{i,k,l} \lambda_i^{-1} \partial_i \omega'_{kk} \partial_i \omega'_{ll}$$

by the Cauchy–Schwarz inequality. Combining this with (4.3) and (4.4) yields the desired inequality.   □

4.1.2 The Maximum Principle

The following simple maximum principle (or at least its proof) will be systematically used in what follows to establish a priori estimates.

Proposition 4.1.3. *Let U be a complex manifold and $0 < T \leq +\infty$. Let $(\omega_t)_{t \in [0,T)}$ be a smooth path of Kähler metrics on U, and denote by $\Delta_t = \operatorname{tr}_{\omega_t} dd^c$ the Laplacian with respect to ω_t. Assume that*

$$H \in C^0(U \times [0, T)) \cap C^\infty(U \times (0, T))$$

satisfies either

$$\left(\frac{\partial}{\partial t} - \Delta_t \right) H \leq 0,$$

or

$$\frac{\partial H}{\partial t} \leq \log \left[\frac{(\omega_t + dd^c H)^n}{\omega_t^n} \right]$$

on $U \times (0, T)$. When U is non compact, assume further that $H \to -\infty$ at infinity on $U \times [0, T']$, for each $T' < T$. Then we have

$$\sup_{U \times [0,T)} H = \sup_{U \times \{0\}} H.$$

If we replace $\leq$ with $\geq$ in the above differential inequalities and assume that $H \to +\infty$ in the non compact case, then the conclusion is that

$$\inf_{U \times [0,T)} H = \inf_{U \times \{0\}} H.$$

Proof. Upon replacing H with $H - \varepsilon t$ (resp. $H + \varepsilon t$ for the reverse inequality) with $\varepsilon > 0$ and then letting $\varepsilon \to 0$, we may assume in each case that the differential inequality is strict. It is enough to show that $\sup_{U \times [0,T']} H = \sup_{U \times \{0\}} H$ for each $T' < T$. The properness assumption guarantees that H achieves its maximum (resp. minimum) on $U \times [0, T']$, at some point $(x_0, t_0) \in U \times [0, T']$, and the strict differential inequality forces $t_0 = 0$. Indeed, we would otherwise have $dd^c H \leq 0$ at (x_0, t_0), $\frac{\partial H}{\partial t} = 0$ if $t_0 < T'$, and at least $\frac{\partial H}{\partial t} \geq 0$ if $t_0 = T'$, which would at any rate contradict the strict differential inequality.   □

4.1.3   Evans–Krylov Type Estimates for Parabolic Complex Monge–Ampère Equations

Since it will play a crucial in what follows, we want to give at least a brief idea of the proof of the next result, which says in essence that it is enough to control the time derivative and the Laplacian to get smooth solutions to parabolic complex Monge–Ampère equations.

Theorem 4.1.4. *Let $U \Subset \mathbb{C}^n$ be an open subset and $T \in (0, +\infty)$. Suppose that $u, f \in C^\infty (\bar{U} \times [0, T])$ satisfy*

$$\frac{\partial u}{\partial t} = \log \det \left(\frac{\partial^2 u}{\partial z_j \, \partial \bar{z}_k} \right) + f, \tag{4.5}$$

and assume also given a constant $C > 0$ such that

$$\sup_{U \times (0,T)} \left(\left| \frac{\partial u}{\partial t} \right| + |\Delta u| \right) \leq C.$$

For each compact $K \Subset U$, each $\varepsilon > 0$ and each $p \in \mathbb{N}$, the C^p norm of u on $K \times [\varepsilon, T]$ can then be bounded in terms of the constant C and of the C^q norm of f on $\bar{U} \times [0, T]$ for some $q \geq p$.

The first ingredient in the proof are the Schauder estimates for linear parabolic equations. If f is a function on the cylinder $Q = U \times (0, T)$, recall from Chap. 2 that for $0 < \alpha < 1$ the *parabolic α-Hölder norm* of f on Q is defined as

$$\|f\|_{C_P^\alpha(Q)} := \|f\|_{C^0(Q)} + [f]_{\alpha;Q},$$

where $[f]_{\alpha;Q}$ denotes the α-Hölder seminorm with respect to the parabolic distance

$$d_P\left((z,t),(z',t')\right) = \max\left\{|z-z'|, |t-t'|^{1/2}\right\}.$$

For each $k \in \mathbb{N}$, the *parabolic $C^{k,\alpha}$-norm* is then defined as

$$\|f\|_{C_P^{k,\alpha}(Q)} := \sum_{|\beta|+2j \leq k} \|D_x^\beta D_t^j f\|_{C_P^\alpha(Q)}.$$

If $(\theta_t)_{t\in(0,T)}$ is a path of differential forms on U, we can similarly consider $[\theta_t]_{\alpha,Q}$ and $\|\theta_t\|_{C_P^{k,\alpha}(Q)}$, with respect to the flat metric ω_U on U. In our context, the parabolic Schauder estimates can then be stated as:

Lemma 4.1.5. *Let $(\omega_t)_{t\in(0,T)}$ be a smooth path of Kähler metrics on U, and assume that $u, f \in C^\infty(Q)$ satisfy*

$$\left(\frac{\partial}{\partial t} - \Delta_t\right)u = f,$$

with Δ_t the Laplacian with respect to ω_t, and setting as above $Q = U \times (0,T)$. Suppose also given $C > 0$ and $0 < \alpha < 1$ such that on Q we have

$$C^{-1}\omega_U \leq \omega_t \leq C\omega_U \text{ and } [\omega_t]_{\alpha,Q} \leq C.$$

For each $Q' = U' \times (\varepsilon, T)$ with $U' \Subset U$ and $\varepsilon \in (0,T)$, we can then find a constant $A > 0$ only depending on U', ε and C such that

$$\|u\|_{C_P^{2,\alpha}(Q')} \leq A\left(\|u\|_{C^0(Q)} + \|f\|_{C_P^\alpha(Q)}\right).$$

This result follows for instance from [Lieb96, Theorem 4.9] (see also Chap. 2 in the present volume). Note that these estimates are interior only with respect to the parabolic boundary, i.e. the upper limit of the time interval is the same on both sides of the estimates.

The second ingredient in the proof of Theorem 4.1.4 is the following version of the Evans–Krylov estimates for parabolic complex Monge–Ampère equations. We refer to [Gill11, Theorem 4.9] for the proof, which relies on a Harnack estimate for linear parabolic equations.

Lemma 4.1.6. *Suppose that $u, f \in C^\infty(Q)$ satisfy*

$$\frac{\partial u}{\partial t} = \log \det\left(\frac{\partial^2 u}{\partial z_j \partial \bar{z}_k}\right) + f,$$

and assume also given a constant $C > 0$ such that

$$C^{-1} \leq \left(\frac{\partial^2 u}{\partial z_j \, \partial \bar{z}_k} \right) \leq C \text{ and } \left| \frac{\partial f}{\partial t} \right| + |dd^c f| \leq C.$$

For each $Q' = U' \times (\varepsilon, T)$ with $U' \Subset U$ an open subset and $\varepsilon \in (0, T)$, we can then find $A > 0$ and $0 < \alpha < 1$ only depending on U', ε and C such that

$$[dd^c u]_{\alpha, Q'} \leq A.$$

Proof of Theorem 4.1.4. The proof consists in a standard boot-strapping argument. Consider the path $\omega_t := dd^c u_t$ of Kähler forms on U. By assumption, we have $\omega_t \leq C_1 \omega_U$ with $C_1 > 0$ under control. Since

$$\omega_t^n = \exp \left(\frac{\partial u}{\partial t} - f \right) \omega_U^n$$

where $\frac{\partial u}{\partial t} - f$ is bounded below by a constant under control thanks to the assumptions, simple eigenvalue considerations show that $\omega_t \geq c \omega_U$ with $c > 0$ under control. We can thus apply the Evans–Krylov estimates of Lemma 4.1.6 and assume, after perhaps slightly shrinking Q, that $[\omega_t]_{\alpha, Q}$ is under control for some $0 < \alpha < 1$.

Now let D be any first order differential operator with constant coefficients. Differentiating (4.5), we get

$$\left(\frac{\partial}{\partial t} - \Delta_t \right) Du = Df. \tag{4.6}$$

Since $\left| \frac{\partial u}{\partial t} \right| + |\Delta u|$ is under control, the elliptic Schauder estimates (for the flat Laplacian Δ) show in particular, after perhaps shrinking U (but not the time interval), that the C^0 norm of Du is under control. By the parabolic Schauder estimates of Lemma 4.1.5, the parabolic $C^{2,\alpha}$ norm of Du is thus under control as well. Applying D to (4.6) we find

$$\left(\frac{\partial}{\partial t} - \Delta_t \right) D^2 u = D^2 f + \sum_{j,k} \left(D\omega_t^{jk} \right) \frac{\partial^2 Du}{\partial z_j \, \partial \bar{z}_k},$$

where the parabolic C^α norm of the right-hand side is under control. By the parabolic Schauder estimates, the parabolic $C^{2,\alpha}$ norm of $D^2 u$ is in turn under control, and iterating this procedure concludes the proof of Theorem 4.1.4. $\qquad \square$

4.2   Smoothing Properties of the Kähler–Ricci Flow

By analogy with the regularizing properties of the heat equation, it is natural to expect that the Kähler–Ricci flow can be started from a singular initial data (say a positive current, rather than a Kähler form), instantaneously smoothing out the latter.

The goal of this section is to illustrate positively this expectation by explaining the proof of the following result of Szekelyhidi–Tosatti [SzTo11]:

Theorem 4.2.1. *Let (X, ω) be a n-dimensional compact Kähler manifold. Let $F : \mathbb{R} \times X \to \mathbb{R}$ be a smooth function and assume $\psi_0 \in PSH(X, \omega)$ is continuous[1] and satisfies*

$$(\omega + dd^c \psi_0)^n = e^{-F(\psi_0, x)} \omega^n.$$

Then $\psi_0 \in C^\infty(X)$ is smooth.

As the reader will realize later on, the proof is a good warm up, as the arguments are similar to the ones we are going to use when proving Theorem 4.3.3.

Let us recall that such equations contain as a particular case the Kähler–Einstein equation. Namely when the cohomology class $\{\omega\}$ is proportional to the first Chern class of X,[2] $\lambda\{\omega\} = c_1(X)$ for some $\lambda \in \mathbb{R}$, then the above equation is equivalent to

$$\mathrm{Ric}(\omega + dd^c \psi_0) = \lambda(\omega + dd^c \psi_0),$$

when taking

$$F(\varphi, x) = \lambda \varphi + h(x)$$

with $h \in C^\infty(X)$ such that $\mathrm{Ric}(\omega) = \lambda\omega + dd^c h$. Szekelyhidi and Tosatti's result is thus particularly striking since the solutions to such equations, if any, are in general not unique.[3]

The interest in such regularity results stems for example from the recent works [BBGZ13, EGZ11] which provide new tools to construct weak solutions to such complex Monge–Ampère equations.

The idea of the proof is both simple and elegant, and goes as follows: assume we can run a complex Monge–Ampère flow

$$\frac{\partial \varphi}{\partial t} = \log\left[\frac{(\omega + dd^c \varphi)^n}{\omega^n}\right] + F(\varphi, x)$$

with an initial data $\varphi_0 \in PSH(X, \omega) \cap C^0(X)$ in such a way that

$$\varphi \in C^0(X \times [0, T]) \cap C^\infty(X \times (0, T]).$$

---

[1] The authors state their result assuming that ψ_0 is merely bounded, but they use in an essential way the continuity of ψ_0, which is nevertheless known in this context by Kołodziej [Kol98].

[2] This of course assumes that $c_1(X)$ has a definite sign.

[3] In the Kähler–Einstein Fano case, a celebrated result of Bando and Mabuchi [BM87] asserts that any two solutions are connected by the flow of a holomorphic vector field.

Then ψ_0 will be a fixed point of such a flow hence if ψ_t denotes the flow originating from ψ_0, $\psi_0 \equiv \psi_t$ has to be smooth!

To simplify our task, we will actually give full details only in case

$$F(s, x) = -G(s) + h(x) \text{ with } s \mapsto G(s) \text{ being } convex$$

and merely briefly indicate what extra work has to be done to further establish the most general result. Note that this particular case nevertheless covers the Kähler–Einstein setting.

In the sequel we consider the above flow starting from a smooth initial potential φ_0 and establish various a priori estimates that eventually will allow us to start from a poorly regular initial data. We fix once and for all a finite time $T > 0$ (independent of φ_0) such that all flows to be considered are well defined on $X \times [0, T]$: it is standard that the maximal interval of time on which such a flow is well defined can be computed in cohomology, hence depends on the cohomology class of the initial data rather than on the (regularity properties of the) chosen representative.

4.2.1 A Priori Estimate on φ_t

We consider in this section on $X \times [0, T]$ the complex Monge–Ampère flow (*CMAF*)

$$\frac{\partial \varphi}{\partial t} = \log \left[\frac{(\omega + dd^c \varphi)^n}{\omega^n} \right] + F(\varphi, x)$$

with initial data $\varphi_0 \in \mathrm{PSH}(X, \omega) \cap C^\infty(X)$. Our aim is to bound $\|\varphi\|_{L^\infty(X \times [0,T])}$ in terms of $\|\varphi_0\|_{L^\infty(X)}$ and T.

4.2.1.1 Heuristic Control

Set $M(t) = \sup_X \varphi_t$. It suffices to bound $M(t)$ from above, the bound from below for $m(t) := \inf_X \varphi_t$ will follow by symmetry.

Assume that we can find $t \in [0, T] \mapsto x(t) \in X$ a differentiable map such that $M(t) = \varphi_t(x(t))$. Then M is differentiable and satisfies

$$M'(t) = \frac{\partial \varphi_t}{\partial t}(x(t)) \leq F(\varphi_t(x(t)), x(t)) \leq \overline{F}(M(t)),$$

where

$$\overline{F}(s) := \sup_{x \in X} F(s, x)$$

is a Lipschitz map.

It follows therefore from the Cauchy–Lipschitz theory of ODE's that $M(t)$ is bounded from above on $[0, T]$ in terms of T, $M(0) = \sup_X \varphi_0$ and $\overline{F}$ (hence F).

4.2.1.2 A Precise Bound

We now would like to establish a more precise control under a simplifying assumption:

Lemma 4.2.2. *Assume that $\varphi, \psi \in C^\infty (X \times [0, T])$ define ω-psh functions φ_t, ψ_t for all t and satisfy*

$$\frac{\partial \varphi}{\partial t} \leq \log \left[\frac{(\omega + dd^c \varphi)^n}{\omega^n} \right] + F(\varphi, x)$$

and

$$\frac{\partial \psi}{\partial t} \geq \log \left[\frac{(\omega + dd^c \psi)^n}{\omega^n} \right] + F(\psi, x)$$

on $X \times [0, T]$, where

$$F(s, x) = \lambda s - G(s, x) \text{ with } s \mapsto G(s, \cdot) \text{ non-decreasing.}$$

Then we have

$$\sup_{X \times [0,T]} (\varphi - \psi) \leq e^{\lambda T} \max\{\sup_X (\varphi_0 - \psi_0), 0\}.$$

Proof. Set $u(x, t) := e^{-\lambda t} (\varphi_t - \psi_t)(x) - \varepsilon t \in C^0 (X \times [0, T])$, where $\varepsilon > 0$ is fixed (arbitrary small). Let $(x_0, t_0) \in X \times [0, T]$ be a point at which u is maximal.

If $t_0 = 0$, then $u(x, t) \leq (\varphi_0 - \psi_0)(x_0) \leq \sup_X (\varphi_0 - \psi_0)$ and we obtain the desired upper bound by letting $\varepsilon > 0$ decrease to zero.

Assume now that $t_0 > 0$. Then $\dot{u} \geq 0$ at this point, hence

$$0 \leq -\varepsilon - \lambda e^{-\lambda t} (\varphi_t - \psi_t) + e^{-\lambda t} (\dot{\varphi}_t - \dot{\psi}_t).$$

On the other hand $dd_x^c u \leq 0$, hence $dd_x^c \varphi_t \leq dd_x^c \psi_t$ and

$$\dot{\varphi}_t - \dot{\psi}_t \leq F(\varphi_t, x) - F(\psi_t, x) + \log \left[\frac{(\omega + dd^c \varphi_t)^n}{(\omega + dd^c \psi_t)^n} \right]$$

$$\leq F(\varphi_t, x) - F(\psi_t, x).$$

Recall now that $F(s, x) = \lambda s - G(s, x)$. Previous inequalities therefore yield

$$G(\varphi_t, x) < G(\psi_t, x) \text{ at point } (x, t) = (x_0, t_0).$$

Since $s \mapsto G(s, \cdot)$ is assumed to be non-decreasing, we infer $\varphi_{t_0}(x_0) \leq \psi_{t_0}(x_0)$, so that for all $(x, t) \in X \times [0, T]$,

$$u(x, t) \leq u(x_0, t_0) \leq 0.$$

Letting ε decrease to zero yields the second possibility for the upper bound. $\square$

By reversing the roles of φ_t, ψ_t, we obtain the following useful:

Corollary 4.2.3. *Assume φ, ψ are solutions of (CMAF) with F as above. Then*

$$\|\varphi - \psi\|_{L^\infty(X \times [0,T])} \leq e^{\lambda T} \|\varphi_0 - \psi_0\|_{L^\infty(X)}.$$

As a consequence, if $\varphi_{0,j}$ is a sequence of smooth ω-psh functions decreasing to $\varphi_0 \in \mathrm{PSH}(X, \omega) \cap C^0(X)$, and φ_j are the corresponding solutions to (CMAF) on $X \times [0, T]$, then the sequence φ_j converges uniformly on $X \times [0, T]$ to some $\varphi \in C^0(X \times [0, T])$ as $j \to +\infty$.

4.2.2 A Priori Estimate on $\frac{\partial \varphi}{\partial t}$

We assume here again that on $X \times [0, T]$

$$\frac{\partial \varphi}{\partial t} = \log \left[\frac{(\omega + dd^c \varphi)^n}{\omega^n} \right] + F(\varphi, x)$$

with initial data $\varphi_0 \in \mathrm{PSH}(X, \omega) \cap C^\infty(X)$.

Lemma 4.2.4. *There exists $C > 0$ which only depends on $\|\varphi_0\|_{L^\infty(X)}$ such that for all $t \in [0, T]$,*

$$\|\dot{\varphi}_t\|_{L^\infty(X)} \leq e^{CT} \|\dot{\varphi}_0\|_{L^\infty(X)}.$$

Let us stress that such a bound requires both that the initial potential φ_0 is uniformly bounded and that the initial density

$$f_0 = \frac{(\omega + dd^c \varphi_0)^n}{\omega^n} = \log \dot{\varphi}_0 - F(\varphi_0, x)$$

is uniformly bounded away from zero and infinity. We shall consider in the sequel more general situations with no a priori control on the initial density f_0.

Proof. Observe that

$$\frac{\partial \dot{\varphi}}{\partial t} = \Delta_t \dot{\varphi} + \frac{\partial F}{\partial s}(\varphi, x)\dot{\varphi},$$

where Δ_t denotes the Laplace operator associated to $\omega_t = \omega + dd^c\varphi_t$.

Since F is C^1, we can find a constant $C > 0$ which only depends on (F and) $\|\varphi\|_{L^\infty(X \times [0,T])}$ such that

$$-C < \frac{\partial F}{\partial s}(\varphi, x) < +C.$$

Consider $H_+(x,t) := e^{-Ct}\dot{\varphi}_t(x)$ and let (x_0, t_0) be a point at which H_+ realizes its maximum on $X \times [0, T]$. If $t_0 = 0$, then $\dot{\varphi}_t(x) \leq e^{CT} \sup_X \dot{\varphi}_0$ for all $(x,t) \in X \times [0, T]$. If $t_0 > 0$, then

$$0 \leq \left(\frac{\partial}{\partial t} - \Delta_t\right)(H_+) = e^{-Ct}\left[\frac{\partial F}{\partial s}(\varphi_t, x) - C\right]\dot{\varphi}$$

hence $\dot{\varphi}_{t_0}(x_0) \leq 0$, since $\frac{\partial F}{\partial s}(\varphi_t, x) - C < 0$. Thus $\dot{\varphi}_t(x) \leq 0$ in this case. All in all, this shows that

$$\dot{\varphi}_t \leq e^{CT} \max\left\{\sup_X \dot{\varphi}_0, 0\right\}.$$

Considering the minimum of $H_-(x,t) := e^{+Ct}\dot{\varphi}_t(x,t)$ yields a similar bound from below and finishes the proof since $\max\{\sup_X \dot{\varphi}_0, -\inf_X \dot{\varphi}_0\} \geq 0$. □

4.2.3 A Priori Estimate on $\Delta\varphi_t$

Recall that we are considering on $X \times [0, T]$

$$\frac{\partial\varphi_t}{\partial t} = \log\left[\frac{(\omega + dd^c\varphi_t)^n}{\omega^n}\right] + F(\varphi_t, x)$$

with initial data $\varphi_0 \in \mathrm{PSH}(X, \omega) \cap C^\infty(X)$. Our aim in this section is to establish an upper bound on $\Delta_\omega\varphi_t$, which is uniform as long as $t > 0$ and is allowed to blow up when t decreases to zero.

4.2.3.1 A Convexity Assumption

To simplify our task, we shall assume that

$$F(s, x) = -G(s) + h(x), \text{ with } s \mapsto G(s) \text{ being convex.}$$

This assumption allows us to bound from above $\Delta_\omega F(\varphi, x)$ as follows:

Lemma 4.2.5. *There exists $C > 0$ which only depends on $\|\varphi_0\|_{L^\infty(X)}$ such that*

$$\Delta_\omega \left(F(\varphi_t, x) \right) \leq C \left[1 + \operatorname{tr}_\omega(\omega_t) \right],$$

where $\omega_t = \omega + dd^c \varphi_t$.

Recall here that for any smooth function h and $(1, 1)$-form β,

$$\Delta_\omega h := n \frac{dd^c h \wedge \omega^{n-1}}{\omega^n} \quad \text{while} \quad \operatorname{tr}_\omega \beta := n \frac{\beta \wedge \omega^{n-1}}{\omega^n}.$$

Proof. Observe that

$$dd^c \left(F(\varphi, x) \right) = -G''(\varphi) d\varphi \wedge d^c \varphi - G'(\varphi) dd^c \varphi \leq -G'(\varphi) dd^c \varphi$$

since G is convex. Now $dd^c \varphi = (\omega + dd^c \varphi) - \omega = \omega_\varphi - \omega = \omega_t - \omega$ is a difference of positive forms and $-C \leq -G'(\varphi) \leq +C$, therefore

$$dd^c \left(F(\varphi, x) \right) \leq C \left(\omega_t + \omega \right),$$

which yields the desired upper bound. $\qquad\qquad\qquad\qquad\qquad\qquad\qquad\qquad\qquad\square$

Our simplifying assumption thus yields a bound from above on $\Delta_\omega \left(F(\varphi, x) \right)$ which depends on $\operatorname{tr}_\omega(\omega_\varphi)$ (and $\|\varphi_0\|_{L^\infty(X)}$) but not on $\|\nabla \varphi_t\|_{L^\infty(X \times [\varepsilon, T])}$. A slightly more involved bound from above is available in full generality, which relies on Blocki's gradient estimate [Bło09]. We refer the reader to the proofs of [SzTo11, Lemmata 2.2 and 2.3] for more details.

4.2.3.2 The Estimate

Proposition 4.2.6. *Assume that $F(s, x) = -G(s) + h(x)$, with $s \mapsto G(s)$ convex. Then*

$$0 \leq \operatorname{tr}_\omega(\omega_t) \leq C \exp\left(C/t \right)$$

where $C > 0$ depends on $\|\varphi_0\|_{L^\infty(X)}$ and $\|\dot\varphi_0\|_{L^\infty(X)}$.

Proof. We set $u(x, t) := \operatorname{tr}_\omega(\omega_t)$ and

$$\alpha(x, t) := t \log u(x, t) - A\varphi_t(x),$$

where $A > 0$ will be specified later. The desired inequality will follow if we can uniformly bound α from above. Our plan is to show that

$$\left(\frac{\partial}{\partial t} - \Delta_t\right)(\alpha) \le C_1 + (Bt + C_2 - A)\mathrm{tr}\,_{\omega_t}(\omega)$$

for uniform constants $C_1, C_2 > 0$ which only depend on $\|\varphi_0\|_{L^\infty(X)}$, $\|\dot\varphi_0\|_{L^\infty(X)}$.
Observe that

$$\left(\frac{\partial}{\partial t} - \Delta_t\right)(\alpha) = \log u + \frac{t}{u}\frac{\partial u}{\partial t} - A\dot\varphi_t - t\Delta_t \log u + A\Delta_t\varphi_t.$$

The last term yields $A\Delta_t\varphi_t = An - A\mathrm{tr}\,_{\omega_t}(\omega)$. The for to last one is estimated thanks
to Proposition 4.1.2,

$$-t\Delta_t \log u \le Bt\,\mathrm{tr}\,_{\omega_t}(\omega) + t\frac{\mathrm{tr}\,_\omega(\mathrm{Ric}(\omega_t))}{\mathrm{tr}\,_\omega(\omega_t)}.$$

It follows from Lemma 4.2.5 that

$$\frac{t}{u}\frac{\partial u}{\partial t} = \frac{t}{u}\Delta_t\left(\log \frac{\omega_t^n}{\omega^n}\right) + \frac{t}{u}\Delta_\omega F(\varphi_t, x)$$

$$= \frac{t}{u}\{-\mathrm{tr}\,_\omega(\mathrm{Ric}\,\omega_t) + \mathrm{tr}\,_\omega(\mathrm{Ric}\,\omega)\} + \frac{t}{u}\Delta_\omega F(\varphi_t, x)$$

$$\le -t\frac{\mathrm{tr}\,_\omega(\mathrm{Ric}\,\omega_t)}{\mathrm{tr}\,_\omega(\omega_t)} + C\frac{(1+u)}{u}.$$

We infer

$$-t\Delta_t \log u + \frac{t}{u}\frac{\partial u}{\partial t} \le Bt\,\mathrm{tr}\,_{\omega_t}(\omega) + C_1,$$

using that u is uniformly bounded below as follows from Proposition 4.1.2 again.
To handle the remaining (first and third) terms, we simply note that $\dot\varphi_t$ is
uniformly bounded below, while

$$\log u \le \log\left[C\mathrm{tr}\,_{\omega_t}(\omega)^{n-1}\right] \le C_2 + C_3\mathrm{tr}\,_{\omega_t}(\omega)$$

by Proposition 4.1.2 and the elementary inequality $\log x < x$. Altogether this yields

$$\left(\frac{\partial}{\partial t} - \Delta_t\right)(\alpha) \le C_4 + (Bt + C_3 - A)\,\mathrm{tr}\,_{\omega_t}(\omega) \le C_4,$$

if we choose $A > 0$ so large that $Bt + C_3 - A < 0$. The desired inequality now
follows from the maximum principle.   □

4.2.4 Proof of Theorem 4.2.1

4.2.4.1 Higher Order Estimates

By Theorem 4.1.4, it follows from our previous estimates that higher order a priori estimates hold as well:

Proposition 4.2.7. *For each fixed $\varepsilon > 0$ and $k \in \mathbb{N}$, there exists $C_k(\varepsilon) > 0$ which only further depends on $\|\varphi_0\|_{L^\infty(X)}$ and $\|\dot{\varphi}_0\|_{L^\infty(X)}$ such that*

$$\|\varphi_t\|_{C^k(X\times[\varepsilon,T])} \leq C_k(\varepsilon).$$

4.2.4.2 A Stability Estimate

Let $0 \leq f, g \in L^2(\omega^n)$ be densities such that

$$\int_X f\omega^n = \int_X g\omega^n = \int_X \omega^n.$$

It follows from the celebrated work of Kolodziej [Kol98] that there exists unique continuous ω-psh functions φ, ψ such that

$$(\omega + dd^c\varphi)^n = f\omega^n, (\omega + dd^c\psi)^n = g\omega^n \quad \text{and} \quad \int_X (\varphi - \psi)\omega^n = 0.$$

We shall need the following stability estimates:

Theorem 4.2.8. *There exists $C > 0$ which only depends on $\|f\|_{L^2}, \|g\|_{L^2}$ such that*

$$\|\varphi - \psi\|_{L^\infty(X)} \leq C\|f - g\|_{L^2(X)}^\gamma,$$

for some uniform exponent $\gamma > 0$.

Such stability estimates go back to the work of Kolodziej [Kol03] and Blocki [Blo03]. Much finer stability results are available by now (see [DZ10, GZ12]). We sketch a proof of this version for the convenience of the reader.

Proof. The proof decomposes in two main steps. We first claim that

$$\|\varphi - \psi\|_{L^2(X)} \leq C\|f - g\|_{L^2(X)}^{\frac{1}{2^{n-1}}}, \tag{4.7}$$

for some appropriate $C > 0$. Indeed we are going to show that

$$\int_X d(\varphi - \psi) \wedge d^c(\varphi - \psi) \wedge \omega^{n-1} \leq C_1 I(\varphi, \psi)^{2^{-(n-1)}}, \tag{4.8}$$

where

$$I(\varphi, \psi) := \int_X (\varphi - \psi) \{(\omega + dd^c \psi)^n - (\omega + dd^c \varphi)^n\} \geq 0$$

is non-negative, as the reader can check that an alternative writing is

$$I(\varphi, \psi) = \sum_{j=0}^{n-1} \int_X d(\varphi - \psi) \wedge d^c(\varphi - \psi) \wedge \omega_\varphi^j \wedge \omega_\psi^{n-1-j}.$$

In our case the Cauchy–Schwarz inequality yields

$$I(\varphi, \psi) = \int_X (\varphi - \psi)(g - f)\omega^n \leq \|\varphi - \psi\|_{L^2} \|f - g\|_{L^2},$$

therefore (4.7) is a consequence of (4.8) and Poincaré's inequality.

To prove (4.8), we write $\omega = \omega_\varphi - dd^c \varphi$ and integrate by parts to obtain,

$$\int d(\varphi - \psi) \wedge d^c(\varphi - \psi) \wedge \omega^{n-1}$$

$$= \int d(\varphi - \psi) \wedge d^c(\varphi - \psi) \wedge \omega_\varphi \wedge \omega^{n-2}$$

$$- \int d(\varphi - \psi) \wedge d^c(\varphi - \psi) \wedge dd^c \varphi \wedge \omega^{n-2}$$

$$= \int d(\varphi - \psi) \wedge d^c(\varphi - \psi) \wedge \omega_{\varphi_1} \wedge \omega^{n-2}$$

$$+ \int d(\varphi - \psi) \wedge d^c \varphi \wedge (\omega_\varphi - \omega_\psi) \wedge \omega^{n-2}$$

We take care of the last term by using Cauchy–Schwarz inequality, which yields

$$\int d(\varphi - \psi) \wedge d^c \varphi \wedge \omega_\varphi \wedge \omega^{n-2} \leq A \left(\int d(\varphi - \psi) \wedge d^c(\varphi - \psi) \wedge \omega_\varphi \wedge \omega^{n-2} \right)^{1/2},$$

where

$$A^2 = \int d\varphi \wedge d^c \varphi \wedge \omega_\varphi \wedge \omega^{n-2}$$

is uniformly bounded from above, since φ is uniformly bounded in terms of $\|f\|_{L^2(X)}$ by the work of Kolodziej [Kol98]. Similarly

$$- \int d(\varphi - \psi) \wedge d^c \varphi \wedge \omega_\psi \wedge \omega^{n-2} \leq B \left(\int d(\varphi - \psi) \wedge d^c(\varphi - \psi) \wedge \omega_\psi \wedge \omega^{n-2} \right)^{1/2},$$

where

$$B^2 = \int d\varphi \wedge d^c\varphi \wedge \omega_\psi \wedge \omega^{n-2}$$

is uniformly bounded from above. Note that both terms can be further bounded from above by the same quantity by bounding from above ω_φ (resp. ω_ψ) by $\omega_\varphi + \omega_\psi$.

Going on this way by induction, replacing at each step ω by $\omega_\varphi + \omega_\psi$, we end up with a control from above of $\int d(\varphi - \psi) \wedge d^c(\varphi - \psi) \wedge \omega^{n-1}$ by a quantity that is bounded from above by $CI(\varphi, \psi)^{2^{-(n-1)}}$ (there are $(n-1)$-induction steps), for some uniform constant $C > 0$. This finishes the proof of the first step.

The second step consists in showing that

$$\|\varphi - \psi\|_{L^\infty(X)} \le C_2 \|\varphi - \psi\|_{L^2(X)}^\gamma$$

for some constants $C_2, \gamma > 0$. We are not going to dwell on this second step here, as it would take us too far. It relies on the comparison techniques between the volume and the Monge–Ampère capacity, as used in [Kol98].   □

4.2.4.3 Conclusion

We are now in position to conclude the proof of Theorem 4.2.1 [at least in case $F(s, x) = -G(s) + h(x)$, with G convex]. Let $\psi_0 \in \mathrm{PSH}(X, \omega)$ be a *continuous* solution to

$$(\omega + dd^c\psi_0)^n = e^{-F(\psi_0, x)}\omega^n.$$

Fix $u_j \in C^\infty(X)$ arbitrary smooth functions which uniformly converge to ψ_0 and let $\psi_j \in \mathrm{PSH}(X, \omega) \cap C^\infty(X)$ be the unique smooth solutions of

$$(\omega + dd^c\psi_j)^n = c_j e^{-F(u_j, x)}\omega^n,$$

normalized by $\int_X (\psi_j - \psi_0)\omega^n = 0$. Here $c_j \in \mathbb{R}$ are normalizing constants wich converge to 1 as $j \to +\infty$, such that

$$c_j \int_X e^{-F(u_j, x)}\omega^n = \int_X \omega^n,$$

and the existence (and uniqueness) of the ψ_j's is provided by Yau's celebrated result [Yau78]. It follows from the stability estimate (Theorem 4.2.8) that

$$\|\psi_j - \psi_0\|_{L^\infty(X)} \longrightarrow 0 \text{ as } j \to +\infty,$$

hence

$$\|\psi_j - u_j\|_{L^\infty(X)} \longrightarrow 0 \text{ as } j \to +\infty.$$

Consider the complex Monge–Ampère flows

$$\frac{\partial \varphi_{t,j}}{\partial t} = \log\left[\frac{(\omega + dd^c \varphi_{t,j})^n}{\omega^n}\right] + F(\varphi_{t,j}, x) - \log c_j,$$

with initial data $\varphi_{0,j} := \psi_j$. It follows from Lemma 4.2.2 that

$$\|\varphi_{t,j} - \varphi_{t,k}\|_{L^\infty(X \times [0,T])} \le e^{\lambda T} \|\psi_j - \psi_k\|_{L^\infty(X)} + \left|\log c_j - \log c_k\right|,$$

thus $(\varphi_{t,j})_j$ is a Cauchy sequence in the Banach space $C^0(X \times [0, T])$. We set

$$\varphi_t := \lim_{j \to +\infty} \varphi_{t,j} \in C^0(X \times [0, T]).$$

Note that $\varphi_t \in \mathrm{PSH}(X, \omega)$ for each $t \in [0, T]$ fixed and $\varphi_0 = \psi_0 = \lim \varphi_{0,j}$ by continuity. Proposition 4.2.7 shows moreover that $(\varphi_{t,j})_j$ is a Cauchy sequence in the Fréchet space $C^\infty(X \times (0, T])$, hence $(x, t) \mapsto \varphi_t(x) \in C^\infty(X \times (0, T])$. Observe that

$$\|\dot{\varphi}_{0,j}\|_{L^\infty(X)} = \|F(\psi_j, x) - F(u_j, x)\|_{L^\infty(X)} \le C \|\psi_j - u_j\|_{L^\infty(X)} \to 0.$$

Lemma 4.2.4 therefore yields for all $t > 0$,

$$\|\dot{\varphi}_t\|_{L^\infty(X)} = \lim_{j \to +\infty} \|\dot{\varphi}_{t,j}\|_{L^\infty(X)} \le C \lim_{j \to +\infty} \|\dot{\varphi}_{0,j}\|_{L^\infty(X)} = 0.$$

This shows that $t \mapsto \varphi_t$ is constant on $(0, T]$, hence constant on $[0, T]$ by continuity. Therefore $\psi_0 \equiv \varphi_t$ is smooth, as claimed.

4.3   Degenerate Parabolic Complex Monge–Ampère Equations

Until further notice, (X, ω_X) denotes a compact Kähler manifold of dimension n endowed with a reference Kähler form.

4.3.1   The Ample Locus

Recall that the *pseudoeffective cone* in $H^{1,1}(X, \mathbb{R})$ is the closed convex cone of classes of closed positive $(1, 1)$-currents in X. A $(1, 1)$-class α in the interior of the pseudoeffective cone is said to be *big*. Equivalently, α is big iff it can be represented by a *Kähler current*, i.e. a closed $(1,1)$-current T which is strictly positive in the sense that $T \geq c\omega_X$ for some $c > 0$. In the special case where the $(1,1)$-form θ is semipositive, it follows from [DemPaun04] that its class is big iff $\int_X \theta^n > 0$, i.e. iff θ is a Kähler form on at least an open subset of X.

The following result is a consequence of Demailly's regularization theorem [Dem92] (cf. [DemPaun04, Theorem 3.4]).

Lemma 4.3.1. *Let θ be a closed real $(1,1)$-form on X, and assume that its class in $H^{1,1}(X, \mathbb{R})$ is big. Then there exists a θ-psh function $\psi_\theta \leq 0$ such that:*

(i) ψ_θ *is of class C^∞ on a Zariski open set $\Omega \subset X$,*
(ii) $\psi_\theta \to -\infty$ *near $\partial\Omega$,*
(iii) $\omega_\Omega := (\theta + dd^c\psi_\theta)|_\Omega$ *is the restriction to Ω of a Kähler form on a compactification $\tilde{X}$ of Ω dominating X.*

More precisely, condition (iii) means that there exists a compact Kähler manifold $(\tilde{X}, \omega_{\tilde{X}})$ and a modification $\pi : \tilde{X} \to X$ such that π is an isomorphism over Ω and $\pi^*\omega_\Omega = \omega_{\tilde{X}}$ on $\pi^{-1}(\Omega)$.

By the Noetherian property of closed analytic subsets, it is easy to see that the set of all Zariski open subsets Ω so obtained admits a largest element, called the *ample locus* of θ and denoted by $\mathrm{Amp}(\theta)$ (see [Bou04, Theorem 3.17]). Note that $\mathrm{Amp}(\theta)$ only depends on the cohomology class of θ.

For later use, we also note:

Lemma 4.3.2. *Let θ be a closed real $(1,1)$-form with big cohomology class, and let $U \subset \mathrm{Amp}(\theta)$ be an arbitrary Zariski open subset. We can then find a θ-psh function τ_U such that τ_U is smooth on U and $\tau_U \to -\infty$ near ∂U.*

Proof. Let ψ_θ be a function as in Lemma 4.3.1, with $\Omega = \mathrm{Amp}(\theta)$. Since $A := X \setminus U$ is a closed analytic subset, it is easy to construct an ω_X-psh function ρ with logarithmic poles along A (see for instance [DemPaun04]). We then set $\tau_U := \psi_\theta + c\rho$ with $c > 0$ small enough to have $\theta + dd^c\psi_\theta \geq \delta\omega_X$ for some $\delta > 0$. $\square$

4.3.2 The Main Result

In the next sections, we will provide a detailed proof of the following result, which is a mild generalization of the technical heart of [ST09]. The assumptions on the measure μ will become more transparent in the context of the Kähler–Ricci flow on varieties with log-terminal singularities, cf. Sect. 4.6.

Theorem 4.3.3. *Let X be a compact Kähler manifold, $T \in (0, +\infty)$, and let $(\theta_t)_{t \in [0,T]}$ be a smooth path of closed semipositive $(1,1)$-forms such that $\theta_t \geq \theta$ for a fixed semipositive $(1,1)$-form θ with big cohomology class. Let also μ be positive measure on X of the form*

$$\mu = e^{\psi^+ - \psi^-} \omega_X^n$$

where

- $\psi^\pm$ *are quasi-psh functions on X (i.e. there exists $C > 0$ such that $\psi^\pm$ are both $C\omega_X$-psh);*
- $e^{-\psi^-} \in L^p$ *for some $p > 1$;*
- $\psi^\pm$ *are smooth on a given Zariski open subset $U \subset \mathrm{Amp}(\theta)$.*

For each continuous θ_0-psh function $\varphi_0 \in C^0(X) \cap \mathrm{PSH}(X, \theta_0)$, there exists a unique bounded continuous function $\varphi \in C_b^0(U \times [0, T))$ with $\varphi|_{U \times \{0\}} = \varphi_0$ and such that on $U \times (0, T)$ φ is smooth and satisfies

$$\frac{\partial \varphi}{\partial t} = \log \left[\frac{(\theta_t + dd^c \varphi)^n}{\mu} \right]. \tag{4.9}$$

Furthermore, φ is in fact smooth up to time T, i.e. $\varphi \in C^\infty(U \times (0, T])$.

Remark 4.3.4. Since $\varphi|_{X \times \{t\}}$ is bounded and θ_t-psh on a Zariski open set of X, it uniquely extends to a bounded θ_t-psh function on X by standard properties of psh functions. We get in this way a natural quasi-psh extension of φ to a bounded function on $X \times [0, T]$, but note that no continuity property is claimed on $X \times [0, T]$ (see however Theorem 4.3.5 below).

As we shall see, uniqueness in Theorem 4.3.3 holds in a strong sense: we have

$$\sup_{U \times [0,T]} |\varphi - \varphi'| = \sup_{U \times \{0\}} |\varphi - \varphi'|$$

for any two

$$\varphi, \varphi' \in C_b^0(U \times [0, T)) \cap C^\infty(U \times (0, T))$$

satisfying (4.9) and such that the restriction to $U \times \{0\}$ of either φ or φ' extends continuously to $X \times \{0\}$.

In the geometric applications to the (unnormalized) Kähler–Ricci flow, the path (θ_t) will be affine as a function of t. In that case, we have a global control on the time derivative:

Theorem 4.3.5. *With the notation of Theorem 4.3.3, assume further that $(\theta_t)_{t \in [0,T]}$ is an* affine *path. For each $\varepsilon > 0$, $\frac{\partial \varphi}{\partial t}$ is then bounded above on $U \times [\varepsilon, T]$, and bounded below on $U \times [\varepsilon, T - \varepsilon]$. In particular, the quasi-psh extension of φ to $X \times [0, T]$ is continuous on $X \times (0, T)$, and on $X \times \{T\}$ as well.*

4.4 A Priori Estimates for Parabolic Complex Monge–Ampère Equations

4.4.1 Setup

Recall that (X, ω_X) is a compact Kähler manifold endowed with a reference Kähler form. In this section, $(\theta_t)_{t \in [0,T]}$ denotes a smooth path of *Kähler forms* on X, and we assume given a *semipositive* $(1, 1)$-form θ with big cohomology class such that

$$\theta_t \geq \theta \text{ for } t \in [0, T].$$

Let also μ be a smooth positive volume form on X, and suppose that $\varphi \in C^\infty(X \times [0, T])$ satisfies

$$\frac{\partial \varphi}{\partial t} = \log \left[\frac{(\theta_t + dd^c \varphi)^n}{\mu} \right]. \tag{4.10}$$

Our goal is to provide a priori estimates on φ that only depend on θ, the sup norm of $\varphi_0 := \varphi|_{X \times \{0\}}$, and the L^p-norm and certain Hessian bounds for the density f of μ. More precisely, we will prove the following result:

Theorem 4.4.1. *With the above notation, suppose that μ is written as*

$$\mu = e^{\psi^+ - \psi^-} \omega_X^n$$

with $\psi^\pm \in C^\infty(X)$, and assume given $C > 0$ and $p > 1$ such that

(i) $-C \leq \sup_X \psi^\pm \leq C$ and $dd^c \psi^\pm \geq -C\omega_X$.
(ii) $\|e^{-\psi^-}\|_{L^p} \leq C$.
(iii) $\|\varphi_0\|_{C^0} \leq C$.

The C^0 norm of φ on $X \times [0, T]$ is then bounded in terms of θ, C, T, p and a bound on the volume $\int_X \theta_t^n$ for $t \in [0, T]$.

Further, φ is bounded in C^∞ topology on $\text{Amp}(\theta) \times (0, T]$, uniformly in terms of θ, C, T, p and C^∞ bounds for (θ_t) on $X \times [0, T]$ and for $\psi^\pm$ on $\text{Amp}(\theta)$. More explicitly, for each compact set $K \Subset \text{Amp}(\theta)$, each $\varepsilon > 0$ and each $k \in \mathbb{N}$, the C^k-norm of φ on $K \times [\varepsilon, T]$ is bounded in terms of θ, C, T, p and C^∞ bounds for (θ_t) on $X \times [0, T]$ and for $\psi^\pm$ in any given neighborhood of K.

During the proof, we shall use the following notation. We introduce the smooth path of Kähler forms

$$\omega_t := \theta_t + dd^c \varphi_t,$$

and denote by $\Delta_t = \text{tr}_{\omega_t} dd^c$ the corresponding time-dependent Laplacian operator on functions. We trivially have

$$\left(\frac{\partial}{\partial t} - \Delta_t\right)\varphi = \dot{\varphi} + \operatorname{tr}_{\theta_t}(\theta_t) - n, \tag{4.11}$$

where $\dot{\varphi}$ is a short-hand for $\frac{\partial\varphi}{\partial t}$. Writing $\dot{\theta}_t$ for the time-derivative of θ_t, it is also immediate to see that

$$\left(\frac{\partial}{\partial t} - \Delta_t\right)\dot{\varphi} = \operatorname{tr}_{\theta_t}(\dot{\theta}_t) \tag{4.12}$$

To simplify the notation, we set $\Omega := \operatorname{Amp}(\theta)$, and choose a θ-psh function ψ_θ as in Lemma 4.3.1, so that $\psi_\theta \to -\infty$ near $\partial\Omega$ and

$$\omega_\Omega := (\theta + dd^c\psi_\theta)|_\Omega$$

is the restriction to Ω of a Kähler form on a compactification of Ω dominating X. Since $\theta_t \geq \theta$ for all t, (4.11) shows that

$$\left(\frac{\partial}{\partial t} - \Delta_t\right)(\varphi - \psi_\theta) \geq \dot{\varphi} + \operatorname{tr}_{\theta_t}(\omega_\Omega) - n \tag{4.13}$$

on $\Omega \times [0, T]$.

As a matter of terminology, we shall say that a quantity is *under control* if it can be bounded by a constant only depending on the desired quantities within the proof of a given lemma.

4.4.2 A Global C^0-Estimate

Lemma 4.4.2. *Suppose that* $\varphi \in C^\infty(X \times [0, T])$ *satisfies (4.10). Assume given* $C > 0$, $p > 1$ *such that*

(i) $\int_X \theta_t^n \leq C$ *for* $t \in [0, T]$;
(ii) $\int_X \mu \geq C^{-1}$ *and* $\|f\|_{L^p} \leq C$ *for the density* $f := \mu/\omega_X^n$.

Then there exists a constant $A > 0$ *only depending on* θ, p, T *and* C *such that*

$$\sup_{X\times[0,T]} |\varphi| \leq \sup_{X\times\{0\}} |\varphi| + A.$$

Proof. **Step 0: an auxiliary construction.** We introduce an auxiliary function, which will also be used in the proof of Lemma 4.4.4 below. For $\varepsilon \in (0, 1/2]$ introduce the Kähler form

$$\eta_\varepsilon := (1 - \varepsilon)\theta + \varepsilon^2\omega_X,$$

and set

$$c_\varepsilon := \log\left(\frac{\int \eta_\varepsilon^n}{\int \mu}\right).$$

Since θ_t is a continuous family of Kähler forms, we can fix $\varepsilon > 0$ small enough such that $\theta_t \geq \varepsilon\omega_X$ for all $t \in [0, T]$. Since $\int_X \theta^n$ is positive, c_ε is under control, even though ε itself is not! Observe also that $\theta_t \geq (1 - \varepsilon)\theta + \varepsilon\theta_t$, and hence

$$\theta_t \geq \eta_\varepsilon \text{ for } t \in [0, T]. \tag{4.14}$$

By [Yau78] there exists a unique smooth η_ε-psh function ρ_ε such that $\sup_X \rho_\varepsilon = 0$ and

$$(\eta_\varepsilon + dd^c\rho_\varepsilon)^n = e^{c_\varepsilon}\mu. \tag{4.15}$$

Since the L^p-norm of the density of $e^{c_\varepsilon}\mu$ is under control and since

$$\tfrac{1}{2}\theta \leq \eta_\varepsilon \leq \theta + \omega_X \leq C_1\omega_X$$

with $C_1 > 0$ only depending on θ, the uniform version of Kolodziej's L^∞-estimates [EGZ09] shows that the C^0 norm of ρ_ε is under control.
Step 1: lower bound. Consider η_ε and ρ_ε as in Step 0, and set

$$H := \varphi - \rho_\varepsilon - c_\varepsilon t.$$

By (4.15) and (4.14) we get

$$\frac{\partial H}{\partial t} = \log\frac{(\theta_t + dd^c\rho_\varepsilon + dd^c H)^n}{(\eta_\varepsilon + dd^c\rho_\varepsilon)^n} \geq \log\frac{(\theta_t + dd^c\rho_\varepsilon + dd^c H)^n}{(\theta_t + dd^c\rho_\varepsilon)^n}$$

on $X \times [0, T]$, and hence $\inf_{X\times[0,T]} H = \inf_{X\times\{0\}} H$ by Proposition 4.1.3. Since c_ε and the C^0 norm of ρ_ε are both under control, we get the desired lower bound for φ.
Step 2: upper bound. By non-negativity of the relative entropy of the probability measure $\mu/\int \mu$ with respect to

$$\frac{(\theta_t + dd^c\varphi)^n}{\int \theta_t^n} = \frac{e^{\dot\varphi}\mu}{\int \theta_t^n}$$

(or, in other words, by concavity of the logarithm and Jensen's inequality), we have

$$\int\left(\log\left(\frac{\int \theta_t^n}{\int \mu}\right) - \dot\varphi\right)\mu \geq 0.$$

It follows that

$$\frac{d}{dt}\left(\int \varphi_t \mu\right) \leq \left(\int \mu\right) \log\left(\int \theta_t^n\right) - \left(\int \mu\right)\log\left(\int \mu\right)$$

$$\leq \|f\|_{L^p}\left(\int \omega_X^n\right)^{1-1/p} \log C + e^{-1} =: A_1$$

is under control, and hence

$$\sup_{t\in[0,T]} \frac{\int \varphi_t \mu}{\int \mu} \leq \frac{\int \varphi_0 \mu}{\int \mu} + A_1 T \leq \sup_X \varphi_0 + A_1 T$$

with $A_1 > 0$ under control. We claim that there exists $B > 0$ under control such that

$$\sup_X \psi \leq \frac{\int \psi \mu}{\int \mu} + B$$

for all θ-psh functions ψ. Applying this with $\psi = \varphi_t$ will yield the desired control on its upper bound. By Skoda's integrability theorem in its uniform version [Zer01], there exist $\delta > 0$ and $B > 0$ only depending on θ such that

$$\int e^{-\delta\psi}\omega_X^n \leq B$$

for all θ-psh functions ψ normalized by $\sup_X \psi = 0$. By Hölder's inequality, it follows that $\int e^{-\delta'\psi}\mu \leq B'$ with δ', B' under control, and the claim follows by Jensen's inequality. □

Remark 4.4.3. The proof given above is directly inspired from that of [ST09, Lemma 3.8]. Let us stress, as a pedagogical note to the non expert reader, that the C^0-estimate thus follows from

- The elementary maximum principle (Proposition 4.1.3);
- Kolodziej's L^∞ estimate for solutions of Monge–Ampère equations [Kol98, EGZ09];
- Skoda's exponential integrability theorem for psh functions (which is in fact also an ingredient in the previous item).

4.4.3 Bounding the Time-Derivative and the Laplacian on the Ample Locus

Lemma 4.4.4. *Suppose that $\varphi \in C^\infty(X \times [0,T])$ satisfies (4.10). Assume that the volume form μ is written as*

$$\mu = e^{\psi^+ - \psi^-} \omega_X^n$$

with $\psi^\pm \in C^\infty(X)$, and let $C > 0$ be a constant such that

(i) $-C\omega_X \le \dot{\theta}_t \le C\omega_X$ for $t \in [0, T]$;
(ii) $\sup_X \psi^\pm \le C$;
(iii) $dd^c \psi^\pm \ge -C\omega_X$.

For each compact set $K \subset \Omega$, we can then find $A > 0$ only depending on K, θ, T and C such that

$$\sup_{K \times [0,T]} t \left(|\dot{\varphi}| + \log |\Delta\varphi| \right) \le A \left(1 + \sup_{X \times [0,T]} |\varphi| - \inf_K \psi^- \right),$$

where Δ denotes the Laplacian with respect to the reference metric ω_X.

Proof. Since ω_Ω extends to a Kähler form on a compactification of Ω dominating X, there exists a constant $c > 0$ under control such that $\omega_\Omega \ge c\omega_X$, and hence

$$\theta_t + dd^c \psi_\theta \ge c\omega_X \tag{4.16}$$

on $\Omega \times [0, T]$ since $\theta_t \ge \theta$.

Step 1: upper bound for $\dot{\varphi}$. We want to apply the maximum principle to

$$H^+ := t\dot{\varphi} + A(\psi_\theta - \varphi),$$

with a constant $A > 0$ to be specified in a moment. Thanks to (4.12) and (4.13), we get

$$\left(\frac{\partial}{\partial t} - \Delta_t \right) H^+ \le -(A - 1)\dot{\varphi} + \mathrm{tr}_{\omega_t} \left(t\dot{\theta}_t - A\omega_\Omega \right) + An.$$

By (i) and (4.16) we have

$$t\dot{\theta}_t - A\omega_\Omega \le TC\omega_X - Ac\omega_X.$$

Choosing

$$A := c^{-1}TC + 1,$$

we obtain

$$\left(\frac{\partial}{\partial t} - \Delta_t \right) H^+ \le -A_1 \dot{\varphi} + A_2 \tag{4.17}$$

with $A_1, A_2 > 0$ under control. We are now in a position to apply the maximum principle. Since $\psi_\theta \to -\infty$ near $\partial\Omega$, H^+ achieves its maximum on $\Omega \times [0, T]$ at some point (x_0, t_0). If $t_0 = 0$ then

$$\sup_{\Omega \times [0,T]} H^+ \le -A \inf_X \varphi,$$

since $\psi_\theta \le 0$. If $t_0 > 0$ then $\left(\frac{\partial}{\partial t} - \Delta_t\right) H^+ \ge 0$ at (x_0, t_0), and (4.17) yields an upper bound for $\dot\varphi$ at (x_0, t_0). It follows that

$$\sup_{\Omega \times [0,T]} H^+ \le C_1 - A \inf_X \varphi.$$

with $C_1 > 0$ under control. Since ψ_θ is bounded below on the given compact set $K \subset \Omega$, we get in particular the desired upper bound on $t\dot\varphi$.

Step 2: lower bound for $\dot\varphi$. We now want to apply the maximum principle to

$$H^- := t(-\dot\varphi + 2\psi^-) + A(\psi_\theta - \varphi),$$

which satisfies by (4.12) and (4.13)

$$\left(\frac{\partial}{\partial t} - \Delta_t\right) H^- \le -(A+1)\dot\varphi + 2\psi^- + \text{tr}_{\omega_t}\left(-t\dot\theta_t - 2tdd^c\psi^- - A\omega_\Omega\right) + An. \tag{4.18}$$

On the one hand, note that

$$-\dot\varphi = \log\left(\frac{\omega_X^n}{\omega_t^n}\right) + \psi^+ - \psi^-,$$

and hence

$$-\dot\varphi \le (\dot\varphi - 2\psi^-) + 2\log\left(\frac{\omega_X^n}{\omega_t^n}\right) + 2C$$

since $\sup_X \psi^+ \le C$. Using $\psi_\theta \le 0$, we also have $t(\dot\varphi - 2\psi^-) \le -(H^- + A\varphi)$, and we get

$$-(A+1)\dot\varphi + 2\psi^- \le -t^{-1}(A+1)(H^- + A\varphi) + 2(A+1)\log\left(\frac{\omega_X^n}{\omega_t^n}\right) + (2A+4)C$$

on $X \times (0, T]$, using this time $\sup_X \psi^- \le C$. On the other hand, (i), (iii) and (4.16) show that

$$-t\dot\theta_t - 2tdd^c\psi^- - A\omega_\Omega \le (3TC - Ac)\omega_X,$$

which is bounded above by $-\omega_X$ if we choose

$$A := c^{-1}(3TC + 1).$$

Plugging these estimates into (4.18), we obtain

$$\left(\frac{\partial}{\partial t} - \Delta_t\right) H^- \leq -t^{-1}(A+1)(H^- + A\varphi) + 2(A+1)\log\left(\frac{\omega_X^n}{\omega_t^n}\right) - \text{tr}_{\omega_t}(\omega_X) + C_2$$

on $\Omega \times (0, T]$, with $C_2 > 0$ under control. By the arithmetico-geometric inequality and the fact that $2(A + 1)\log y - ny^{1/n}$ is bounded above for $y \in (0, +\infty[$, we have

$$2(A + 1)\log\left(\frac{\omega_X^n}{\omega_t^n}\right) - \text{tr}_{\omega_t}(\omega_X) \leq 2(A + 1)\log\left(\frac{\omega_X^n}{\omega_t^n}\right) - n\left(\frac{\omega_X^n}{\omega_t^n}\right)^{1/n} \leq C_3$$

and hence

$$\left(\frac{\partial}{\partial t} - \Delta_t\right) H^- \leq -t^{-1}(A + 1)(H^- + A\varphi) + C_4 \qquad (4.19)$$

with $C_3, C_4 > 0$ under control. We can now apply the maximum principle to obtain as before

$$\sup_{\Omega \times [0,T]} H^- \leq C_5 - A \inf_X \varphi.$$

this yields the desired lower bound on $\dot{\varphi}$.

Step 3: Laplacian bound. We are going to apply the maximum principle to

$$H := t(\log \text{tr}_{\omega_\Omega}(\omega_t) + \psi^-) + A(\psi_\theta - \varphi),$$

with $A > 0$ to be specified below. Since ω_Ω extends to a Kähler metric on some compactification of Ω, its holomorphic bisectional curvature is bounded below by $-C_1$ with $C_1 > 0$ under control, and Proposition 4.1.2 yields

$$-\Delta_t \log \text{tr}_{\omega_\Omega}(\omega_t) \leq \frac{\text{tr}_{\omega_\Omega}\text{Ric}(\omega_t)}{\text{tr}_{\omega_\Omega}(\omega_t)} + C_1\text{tr}_{\omega_t}(\omega_\Omega). \qquad (4.20)$$

On the one hand, we have $\text{Ric}(\omega_t) = \text{Ric}(\mu) - dd^c\dot{\varphi}$ since $\omega_t^n = e^{\dot{\varphi}}\mu$. On the other hand,

$$\frac{\partial}{\partial t}\log \text{tr}_{\omega_\Omega}(\omega_t) = \frac{\text{tr}_{\omega_\Omega}(\dot{\theta}_t + dd^c\dot{\varphi}_t)}{\text{tr}_{\omega_\Omega}(\omega_t)},$$

and hence

$$\left(\frac{\partial}{\partial t} - \Delta_t\right) \log \mathrm{tr}_{\omega_\Omega}(\omega_t) \leq \frac{\mathrm{tr}_{\omega_\Omega}\left(\mathrm{Ric}(\mu) + \dot{\theta}_t\right)}{\mathrm{tr}_{\omega_\Omega}(\omega_t)} + C_1 \mathrm{tr}_{\omega_t}(\omega_\Omega).$$

Now $\dot{\theta}_t \leq C\omega_X$ by assumption, and

$$\mathrm{Ric}(\mu) = -dd^c\psi^+ + dd^c\psi^- + \mathrm{Ric}(\omega_X) \leq C_2\omega_\Omega + dd^c\psi^-$$

for some $C_2 > 0$ under control, using $dd^c\psi^+ \geq -Cc^{-1}\omega_\Omega$ and

$$\mathrm{Ric}(\omega_X) \leq C'\omega_X \leq C'c^{-1}\omega_X.$$

It follows that

$$\left(\frac{\partial}{\partial t} - \Delta_t\right) \log \mathrm{tr}_{\omega_\Omega}(\omega_t) \leq \frac{C_3 + \Delta_{\omega_\Omega}\psi^-}{\mathrm{tr}_{\omega_\Omega}(\omega_t)} + C_1 \mathrm{tr}_{\omega_t}(\omega_\Omega)$$

with $C_3 > 0$ under control. In order to absorb the term involving ψ^- in the left-hand side, we note that $Cc^{-1}\omega_\Omega + dd^c\psi^- \geq 0$, and hence

$$Cc^{-1}\omega_\Omega + dd^c\psi^- \leq \mathrm{tr}_{\omega_t}(Cc^{-1}\omega_\Omega + dd^c\psi^-)\omega_t,$$

which yields after taking the trace with respect to ω_Ω

$$0 \leq \frac{nCc^{-1} + \Delta_{\omega_\Omega}\psi^-}{\mathrm{tr}_{\omega_\Omega}(\omega_t)} \leq Cc^{-1}\mathrm{tr}_{\omega_t}(\omega_\Omega) + \Delta_t\psi^-.$$

Using the trivial inequality $\mathrm{tr}_{\omega_\Omega}(\omega_t)\mathrm{tr}_{\omega_t}(\omega_\Omega) \geq n$ we arrive at

$$\left(\frac{\partial}{\partial t} - \Delta_t\right)(\log \mathrm{tr}_{\omega_\Omega}(\omega_t) + \psi^-) \leq C_4\mathrm{tr}_{\omega_t}(\omega_\Omega) \tag{4.21}$$

with $C_4 > 0$ under control. By (4.21) and (4.11) we thus get

$$\left(\frac{\partial}{\partial t} - \Delta_t\right)H \leq \log \mathrm{tr}_{\omega_\Omega}(\omega_t) + \psi^- - A\dot{\varphi} + (C_4T - A)\mathrm{tr}_{\omega_t}(\omega_\Omega).$$

Lemma 4.1.1 shows that

$$\log \mathrm{tr}_{\omega_\Omega}(\omega_t) + \psi^- \leq \dot{\varphi} + (n-1)\log \mathrm{tr}_{\omega_t}(\omega_\Omega) + C_5, \tag{4.22}$$

since $\sup_X \psi^+ \leq C$ and $\omega_X \leq C_1\omega_\Omega$. If we choose $A := C_4T + 2$, we finally obtain

$$\left(\frac{\partial}{\partial t} - \Delta_t\right) H \leq -\mathrm{tr}_{\omega_t}(\omega_\Omega) - C_6\dot\varphi + C_7, \qquad (4.23)$$

since $(n-1)\log y - 2y \leq -y + O(1)$ for $y \in (0, +\infty)$.

We are now in a position to apply the maximum principle. Since $\psi_\theta \to -\infty$ near $\partial\Omega$, there exists $(x_0, t_0) \in \Omega \times [0, T]$ such that $H(x_0, t_0) = \sup_{\Omega \times [0,T]} H$. If $t_0 = 0$ then

$$\sup_{\Omega \times [0,T]} H \leq A \sup_X |\varphi_0|,$$

using $\psi_\theta \leq 0$. If $t_0 > 0$ then $\left(\frac{\partial}{\partial t} - \Delta_t\right) H \geq 0$ at (x_0, t_0), and (4.23) yields

$$\mathrm{tr}_{\omega_t}(\omega_\Omega) + C_6\dot\varphi \leq C_7,$$

and in particular $\dot\varphi < C_7/C_6$, at (x_0, t_0). Plugging this into (4.22), it follows that

$$\log \mathrm{tr}_{\omega_\Omega}(\omega_t) + \psi^- \leq \dot\varphi + (n-1)\log(-C_6\dot\varphi + C_7) + C_5$$

at (x_0, t_0). Since $y + (n-1)\log(-C_6 y + C_7)$ is bounded above for $y \in]-\infty, C_7/C_6[$, we get $\log \mathrm{tr}_{\omega_\Omega}(\omega_t) + \psi^- \leq C_8$ at (x_0, t_0), and hence

$$\sup_{\Omega \times [0,T]} H = H(x_0, t_0) \leq A \sup_{X \times [0,T]} |\varphi| + TC_8.$$

Finally, there exists a constant $C_K > 0$ such that on the given compact set $K \subset \Omega$ we have $\psi_\theta \geq -C_K$ and $\omega_\Omega \leq C_K\omega$, and the result follows. □

Remark 4.4.5. The arguments used to bound the time-derivative are a combination of the proofs of Lemmas 3.2 and 3.9 in [ST09]. The proof of the Laplacian bound is similar in essence to that of [ST09, Lemma 3.3].

4.4.4 Bounding the Time Derivative in the Affine Case

Lemma 4.4.6. *Under the assumptions of Lemma 4.4.2, suppose that (θ_t) is an affine path, so that $\dot\theta_t$ is independent of t. Then we have*

$$\sup_{X \times [0,T]} (t\dot\varphi) \leq 2 \sup_{X \times [0,T]} |\varphi| + nT,$$

and for each $T' < T$ there exists $A > 0$ only depending on θ, C, p and T' such that

$$\inf_{X \times [0,T']} (t\dot\varphi) \geq -A \left(1 + \sup_{X \times [0,T]} |\varphi| \right).$$

Proof. The upper bound follows directly from the maximum principle applied to

$$H^+ := t\dot\varphi - \varphi - nt,$$

which satisfies

$$\left(\frac{\partial}{\partial t} - \Delta_t \right) H^+ = \mathrm{tr}_{\omega_t}(t\dot\theta_t - \theta_t) = \mathrm{tr}_{\omega_t}(-\theta_0) \leq 0$$

on $X \times [0,T]$. To get the lower bound, take ρ_ε as in Step 0 of the proof of Lemma 4.4.2, and set

$$H^- := -t\dot\varphi - A\varphi + \rho_\varepsilon.$$

where $A > 0$ will be specified in a moment. We then have

$$\left(\frac{\partial}{\partial t} - \Delta_t \right) H^- = -(A+1)\dot\varphi + \mathrm{tr}_{\omega_t}(-t\dot\theta_t - A\theta_t - dd^c\rho_\varepsilon) + An.$$

Since θ_t is affine, we have

$$A\theta_t + t\dot\theta_t = A \left(\theta_0 + t\dot\theta_t + \frac{t}{A}\dot\theta_t \right) = A\theta_{(A+1)t/A},$$

and hence

$$A\theta_t + t\dot\theta_t \geq \eta_\varepsilon$$

for $t \in [0,T']$ by (4.14), if we fix $A \gg 1$ such that $(A+1)T'/A < T$. With this choice of A we get on $X \times [0,T']$

$$\left(\frac{\partial}{\partial t} - \Delta_t \right) H^- \leq (A+1)\log\left(\frac{\mu}{\omega_t^n} \right) - \mathrm{tr}_{\omega_t}(\eta_\varepsilon + dd^c\rho_\varepsilon) + An$$

$$\leq (A+1)\log\left(\frac{\mu}{\omega_t^n} \right) - ne^{c_\varepsilon/n}\left(\frac{\mu}{\omega_t^n} \right)^{1/n} \leq A_1$$

with $A_1 > 0$ under control, using the arithmetico-geometric inequality, (4.15) and the fact that $(A+1)\log y - ne^{c_\varepsilon/n}y^{1/n}$ is bounded above for $y \in (0,+\infty)$. The lower bound for $t\dot\varphi$ follows from the maximum principle, since $\sup_X |\rho_\varepsilon|$ is under control. $\qquad\square$

Remark 4.4.7. This result corresponds to [ST09, Lemma 3.21].

4.4.5 Proof of Theorem 4.4.1

We are now in a position to prove Theorem 4.4.1. Since $\sup_X \psi^+$ is assumed to be bounded below, the mean value inequality for $C\omega_X$-psh functions shows that $\int \psi^+ \omega_X^n$ is also bounded below. By Jensen's inequality and the upper bound on ψ^-, it follows that condition (i) in Lemma 4.4.2 is satisfied. Using the upper bound on ψ^+ and the L^p bound for $e^{-\psi^-}$, we also check condition (ii) of Lemma 4.4.2, which therefore shows that the sup-norm of φ on $X \times [0,T]$ is bounded in terms of C.

By Lemma 4.4.4, on any given neighborhood of $K \times [\varepsilon, T]$ $|\dot{\varphi}|$ and $|\Delta\varphi|$ are bounded in terms of the C, the C^1 norm of (θ_t) on $X \times [0,T]$ and the C^0 norm of ψ^- on the neighborhood in question. We conclude by applying the Evans–Krylov type estimates of Theorem 4.1.4 locally on the ample locus of θ.

4.5 Proof of the Main Theorem

Our goal in this section is to prove Theorems 4.3.3 and 4.3.5.

4.5.1 The Non-degenerate Case

As a first step towards the proof of Theorem 4.3.3, we first consider the non-degenerate case, which amounts to the following result:

Theorem 4.5.1. *Let X be a compact Kähler manifold and $0 < T < +\infty$. Let $(\theta_t)_{t\in[0,T]}$ be a smooth family of Kähler forms on X and let μ be a smooth positive volume form. If $\varphi_0 \in C^\infty(X)$ is strictly θ_0-psh, i.e. $\theta_0 + dd^c\varphi_0 > 0$, then there exists a unique $\varphi \in C^\infty(X \times [0,T])$ such that $\varphi|_{X\times\{0\}} = \varphi_0$ and*

$$\frac{\partial \varphi}{\partial t} = \log\left[\frac{(\theta_t + dd^c\varphi)^n}{\mu}\right]$$

on $X \times [0,T]$.

At least in the case of the Kähler–Ricci flow, this result goes back to [Cao85, Tsu88, Tzha06], see Theorem 3.3.1 in Chap. 3 of the present volume. But since the above statement follows directly from the a priori estimates we have proved so far (Theorem 4.4.1), we may as well provide a proof for completeness.

Proof. Uniqueness follows from the maximum principle (Proposition 4.1.3). By the general theory of non-linear parabolic equations, the solution φ is defined on a maximal half-open interval $[0, T')$ with $T' \leq T$. Since $(\theta_t)_{t \in [0,T]}$ is a smooth path of Kähler metrics, we have $\theta_t \geq c\omega_X$ for all $t \in [0, T]$ if $c > 0$ is small enough, and we may thus apply Theorem 4.4.1 with $\theta = c\omega_X$ and $K = X = \mathrm{Amp}\,(\theta)$ to get that all C^k norms of φ are bounded on $X \times [\varepsilon, T')$ for any fixed $\varepsilon > 0$. It follows that φ extends to a C^∞ function on $X \times [0, T']$. Since $\dot\varphi$ is in particular bounded below, the smooth function $\varphi|_{X \times \{T'\}}$ is strictly $\theta_{T'}$-psh. By the local existence result, we conclude that $T' = T$, since φ could otherwise be extended beyond the maximal existence time T'.   □

4.5.2   A Stability Estimate

If $\varphi, \varphi' \in C^\infty\,(X \times [0, T])$ are two solutions as in Theorem 4.5.1 corresponding to two initial data $\varphi_0, \varphi_0' \in C^\infty(X)$, the maximum principle of Proposition 4.1.3 immediately implies that

$$\|\varphi - \varphi'\|_{C^0(X \times [0,T])} \leq \|\varphi_0 - \varphi_0'\|_{C^0(X)}.$$

In order to treat the general case of Theorem 4.3.3, we need to generalize this estimate when the path of Kähler metrics (θ_t) is allowed to vary as well.

Proposition 4.5.2. *Let* $(\theta_t)_{t \in [0,T]}$ *and* $(\theta_t')_{t \in [0,T]}$ *be two smooth paths of Kähler metrics on* X, *and suppose that* $\varphi, \varphi' \in C^\infty\,(X \times [0, T])$ *satisfy*

$$\frac{\partial \varphi}{\partial t} = \log\left[\frac{(\theta_t + dd^c\varphi)^n}{\mu}\right]$$

and

$$\frac{\partial \varphi'}{\partial t} = \log\left[\frac{(\theta_t' + dd^c\varphi)^n}{\mu}\right],$$

with the same volume form μ *in both cases. As in Theorem 4.4.1, write* μ *as*

$$\mu = e^{\psi^+ - \psi^-}\omega_X^n$$

with $\psi^\pm \in C^\infty(X)$, *and assume given* $C > 0$ *and* $p > 1$ *such that*

(i) $\theta_t \leq C\omega_X$ *and* $\theta_t' \leq C\omega_X$ *for* $t \in [0, T]$;
(ii) $-C \leq \sup_X \psi^\pm \leq C$;
(iii) $dd^c\psi^\pm \geq -C\omega_X$;
(iv) $\|e^{-\psi^-}\|_{L^p} \leq C$.

Finally, let θ be a semipositive $(1, 1)$-form with big cohomology class such that

(v) $\theta_t \geq \theta$ and $\theta_t' \geq \theta$ for $t \in [0, T]$.

For each compact subset $K \Subset \mathrm{Amp}(\theta)$, we can then find a constant $A > 0$ only depending on K, θ, C, p and T such that

$$\|\varphi - \varphi'\|_{C^0(K \times [0,T])} \leq \|\varphi_0 - \varphi_0'\|_{C^0(X)}$$

$$+ A \left(1 + \|\varphi_0\|_{C^0(X)} + \|\varphi_0'\|_{C^0(X)} - \inf_K \psi^- \right)$$

$$\|\theta_t - \theta_t'\|_{C^0(X \times [0,T])}.$$

Proof. Set $N := \|\varphi_0 - \varphi_0'\|_{C^0(X)}$ and $M = \|\theta_t - \theta_t'\|_{C^0(X \times [0,T])}$. We may assume that $M > 0$, and we set for $\lambda \in [0, M]$

$$\theta_t^\lambda := \left(1 - \tfrac{\lambda}{M}\right) \theta_t + \tfrac{\lambda}{M} \theta_t'.$$

For each λ fixed, $(\theta_t^\lambda)_{t \in [0,T]}$ is a smooth path of Kähler forms, and Theorem 4.5.1 yields a unique solution $\varphi^\lambda \in C^\infty (X \times [0, T])$ to the parabolic complex Monge–Ampère equation

$$\begin{cases} \dfrac{\partial \varphi^\lambda}{\partial t} = \log \left[\dfrac{(\theta_t^\lambda + dd^c \varphi^\lambda)^n}{\mu} \right] \\ \varphi^\lambda|_{X \times \{0\}} = \left(1 - \tfrac{\lambda}{M}\right) \varphi_0 + \tfrac{\lambda}{M} \varphi_0' \end{cases} \tag{4.24}$$

By the local existence theory, φ^λ depends smoothly on the parameter λ. If we denote by Δ_t^λ the Laplacian with respect to the Kähler form

$$\omega_t^\lambda := \theta_t^\lambda + dd^c \varphi_t^\lambda,$$

then we have

$$\left(\frac{\partial}{\partial t} - \Delta_t^\lambda \right) \left(\frac{\partial \varphi^\lambda}{\partial \lambda} \right) = \mathrm{tr}_{\omega_t^\lambda} \left(\frac{\partial \theta_t^\lambda}{\partial \lambda} \right),$$

and hence

$$\left(\frac{\partial}{\partial t} - \Delta_t^\lambda \right) \left(\frac{\partial \varphi^\lambda}{\partial \lambda} \right) = M^{-1} \mathrm{tr}_{\omega_t^\lambda} \left(\theta_t' - \theta_t \right) \leq \mathrm{tr}_{\omega_t^\lambda}(\omega_X), \tag{4.25}$$

by definition of M.

Using the notation of Sect. 4.4.1, we are going to apply the maximum principle to the function $H \in C^\infty (\Omega \times [0, T])$ defined by

$$H := e^{-At}\left(\frac{\partial\varphi^\lambda}{\partial\lambda}\right) + A\psi^- + A^2(\psi_\theta - \varphi^\lambda),$$

where $A > 0$ will be specified below. Using (4.13), (4.25) and $dd^c\psi^- \geq -C\omega_X$, we compute

$$\left(\frac{\partial}{\partial t} - \Delta_t^\lambda\right)H \leq -Ae^{-At}\left(\frac{\partial\varphi^\lambda}{\partial\lambda}\right) + \operatorname{tr}_{\omega_t^\lambda}\left(e^{-At}\omega_X + AC\omega_X\right)$$

$$+ A^2\log\left[\frac{\mu}{(\omega_t^\lambda)^n}\right] - A^2\operatorname{tr}_{\omega_t^\lambda}(\omega_\Omega) + A^2 n$$

$$= -AH + A^2\psi^- + A^3(\psi_\theta - \varphi^\lambda) + (1 + AC)\operatorname{tr}_{\omega_t^\lambda}(\omega_X)$$

$$+ A^2\psi^+ - A^2\psi^- + A^2\log\left[\frac{\omega_X^n}{(\omega_t^\lambda)^n}\right] - A^2\operatorname{tr}_{\omega_t^\lambda}(\omega_\Omega) + A^2 n.$$

Since $\omega_\Omega \geq c\omega_X$ for some $c > 0$ under control, the arithmetico-geometric inequality allows us, just as before, to choose $A \gg 1$ under control such that

$$(1 + AC)\operatorname{tr}_{\omega_t^\lambda}(\omega_X) + A^2\log\left[\frac{\omega_X^n}{(\omega_t^\lambda)^n}\right] - A^2\operatorname{tr}_{\omega_t^\lambda}(\omega_\Omega) \leq A_1$$

with $A_1 > 0$ under control. By Lemma 4.4.2, there exists $A_2 > 0$ under control such that

$$\sup_{X\times[0,T]} |\varphi^\lambda| \leq \|\varphi_0\|_{C^0(X)} + \|\varphi_0'\|_{C^0(X)} + A_2. \tag{4.26}$$

Using $\psi^+ \leq C$ and $\psi_\theta \leq 0$, we finally get

$$\left(\frac{\partial}{\partial t} - \Delta_t^\lambda\right)H \leq -AH + A^3\left(\|\varphi_0\|_{C^0(X)} + \|\varphi_0'\|_{C^0(X)}\right) + A_3$$

with $A_3 > 0$ under control. Now

$$H|_{\Omega\times\{0\}} \leq M^{-1}N + A\psi^- + A^2\left(\|\varphi_0\|_{C^0(X)} + \|\varphi_0'\|_{C^0(X)}\right),$$

and the maximum principle thus yields

$$\sup_{\Omega\times[0,T]} H \leq M^{-1}N + A_4\left(1 + \|\varphi_0\|_{C^0(X)} + \|\varphi_0'\|_{C^0(X)}\right).$$

Since ψ_θ is bounded below on the given compact set K, we get using again (4.26)

$$\sup_{K \times [0,T]} \frac{\partial \varphi^\lambda}{\partial \lambda} \leq M^{-1} N + A_5 \left(1 + \|\varphi_0\|_{C^0(X)} + \|\varphi'_0\|_{C^0(X)} - \inf_K \psi^- \right).$$

Integrating with respect to $\lambda \in [0, M]$ and exchanging the roles of φ and φ' yields the desired result.   □

Remark 4.5.3. The proof of Proposition 4.5.2 is directly adapted from that of [ST09, Lemma 3.14].

4.5.3 The General Case

We now consider as in Theorem 4.3.3 a smooth path $(\theta_t)_{t \in [0,T]}$ of closed semipositive $(1,1)$-forms such that $\theta_t \geq \theta$ for a fixed closed semipositive $(1,1)$-form θ with big cohomology class. Let μ be a positive measure on X of the form

$$\mu = e^{\psi^+ - \psi^-} \omega_X^n$$

where

- $\psi^\pm$ are quasi-psh functions on X (i.e. there exists $C > 0$ such that $\psi^\pm$ are both $C\omega_X$-psh);
- $e^{-\psi^-} \in L^p$ for some $p > 1$;
- $\psi^\pm$ are smooth on a given Zariski open subset $U \subset \mathrm{Amp}\,(\theta)$.

Given $\varphi_0 \in C^0(X) \cap \mathrm{PSH}(X, \theta_0)$, our goal is to prove the existence and uniqueness of

$$\varphi \in C_b^0 (U \times [0, T]) \cap C^\infty (U \times (0, T])$$

such that $\varphi|_{U \times \{0\}} = \varphi_0$ and

$$\frac{\partial \varphi}{\partial t} = \log \left[\frac{(\theta_t + dd^c \varphi)^n}{\mu} \right] \tag{4.27}$$

on $U \times (0, T)$.

4.5.3.1 Existence

We regularize the data. By Demailly [Dem92], there exist two sequences $\psi_k^\pm \in C^\infty(X)$ such that

- $\psi_k^\pm$ decreases pointwise to $\psi^\pm$ on X, and the convergence is in C^∞ topology on U;

- $dd^c \psi_k^{\pm} \geq -C\omega_X$ for a fixed constant $C > 0$.

Note that $\left| \sup_X \psi_k^{\pm} \right|$ is bounded independently of k, while we have for all k

$$\left\| e^{-\psi_k^-} \right\|_{L^p} \leq \left\| e^{-\psi^-} \right\|_{L^p}.$$

By Richberg's theorem, we similarly get a decreasing sequence $\varphi_0^j \in C^\infty(X)$ such that $\delta_j := \sup_X \left| \varphi_0^j - \varphi_0 \right| \to 0$ and $\theta_0 + dd^c \varphi_0^j > -\varepsilon_j \omega$ with $\varepsilon_j \to 0$. We then set

- $\theta_t^j := \theta_t + \varepsilon_j \omega_X;$
- $\mu_{k,l} = e^{\psi_k^+ - \psi_l^-} \omega_X^n.$

Since $(\theta_t^j)_{t \in [0,T]}$ is a smooth path of Kähler forms, $\mu_{k,l}$ is a smooth positive volume form and φ_0^j is smooth and strictly θ_0^j-psh, Theorem 4.5.1 shows that there exists a unique function $\varphi^{j,k,l} \in C^\infty (X \times [0, T])$ such that

$$\begin{cases} \dfrac{\partial \varphi^{j,k,l}}{\partial t} = \log \left[\dfrac{\left(\theta_t^j + dd^c \varphi^{j,k,l} \right)^n}{\mu_{k,l}} \right] \\[2em] \varphi^{j,k,l}|_{X \times \{0\}} = \varphi_0^j. \end{cases} \tag{4.28}$$

By Theorem 4.4.1, $\varphi^{j,k,l}$ is uniformly bounded on $X \times [0, T]$, and bounded in C^∞ topology on $U \times (0, T]$.

Furthermore, the maximum principle (Proposition 4.1.3) shows that for each j fixed the sequence $\varphi^{j,k,l}$ is increasing (resp. decreasing) with respect to k (resp. l). As a consequence,

$$\varphi^{j,k} = \lim_{l \to \infty} \varphi^{j,k,l}, \quad \varphi^j = \lim_{k \to \infty} \varphi^{j,k}$$

define bounded functions on $X \times [0, T]$ that are uniformly bounded in C^∞ topology on $U \times (0, T]$. Note also that $\varphi^j|_{X \times \{0\}} = \varphi_0^j$ by construction. The stability estimate of Proposition 4.5.2 shows that for each compact $K \subset U$ there exists $A_K > 0$ such that

$$\sup_{K \times [0,T]} \left| \varphi^{i,k,l} - \varphi^{j,k,l} \right| \leq A_K \left(\delta_i + \delta_j + \varepsilon_i + \varepsilon_j \right) \tag{4.29}$$

for all i, j, k, l, and hence

$$\sup_{K \times [0,T]} \left| \varphi^i - \varphi^j \right| \leq A_K \left(\delta_i + \delta_j + \varepsilon_i + \varepsilon_j \right)$$

for all i, j. As a consequence, (φ^i) is a Cauchy sequence in the Fréchet space $C^0 (U \times [0, T])$, and hence converges uniformly on compact sets of $U \times [0, T]$ to a bounded function $\varphi \in C^0 (U \times [0, T])$, the convergence being in C^∞ topology on $U \times (0, T]$. Passing to the limit in (4.28) shows that φ satisfies (4.27) on $U \times (0, T)$, and φ coincides with $\varphi_0 = \lim_j \varphi_0^j$ on $U \times \{0\}$.

4.5.3.2 Uniqueness

Let φ be the function just constructed, and suppose that

$$\varphi' \in C_b^0 (U \times [0, T]) \cap C^\infty (U \times (0, T))$$

satisfies (4.27). We are going to show that

$$\sup_{U \times [0,T]} |\varphi - \varphi'| = \sup_{U \times \{0\}} |\varphi - \varphi'|,$$

which will in particular imply the desired uniqueness statement.

By Lemma 4.3.2, we can choose a θ-psh function $\tau_U \leq 0$ that is smooth on U and tends to $-\infty$ near ∂U. Fix $0 < c \ll 1$ with $c \theta \leq \omega_X$, so that $\omega_X + c \, dd^c \tau_U \geq 0$.

For a given index j define $H^j \in C^0 (U \times [0, T)) \cap C^\infty (U \times (0, T))$ by

$$H^j := \varphi^j - \varphi' - c\varepsilon_j \tau_U,$$

using the same notation as in the proof of the existence of φ. On $U \times (0, T)$ we have

$$\frac{\partial H^j}{\partial t} = \log \left[\frac{(\theta_t + \varepsilon_j \omega_X + dd^c \varphi^j)^n}{(\theta_t + dd^c \varphi')^n} \right]$$

$$= \log \left[\frac{(\theta_t + dd^c \varphi' + dd^c H^j + \varepsilon_j (\omega_X + c \, dd^c \tau_U))^n}{(\theta_t + dd^c \varphi')^n} \right]$$

$$\geq \log \left[\frac{(\theta_t + dd^c \varphi' + dd^c H^j)^n}{(\theta_t + dd^c \varphi')^n} \right],$$

and hence

$$\inf_{U \times [0,T)} H^j = \inf_{U \times \{0\}} H^j$$

by Proposition 4.1.3. Since $\varphi_0^j \to \varphi_0$ uniformly on $U \times \{0\}$ and $\tau_U \leq 0$, we get in the limit as $j \to \infty$

$$\inf_{U \times [0,T)} (\varphi - \varphi') \geq \inf_{U \times \{0\}} (\varphi - \varphi').$$

In order to prove the similar inequality with the roles of φ and φ' exchanged, we need to introduce yet another parameter in the construction of φ, in order to allow more flexibility. For each $\delta \in [0, 1)$, $(1 - \delta)\varphi^j$ is smooth and strictly $(1 - \delta)\theta_0^j$-psh, and Theorem 4.5.1 thus yields a unique function $\varphi^{\delta,j,k,l} \in C^\infty (X \times [0, T])$ such that

$$
\begin{cases}
\dfrac{\partial \varphi^{\delta,j,k,l}}{\partial t} = \log \left[\dfrac{\left((1 - \delta)\theta_t^j + dd^c \varphi^{\delta,j,k,l} \right)^n}{\mu_{k,l}} \right] \\[3mm]
\varphi^{\delta,j,k,l}\big|_{X \times \{0\}} = (1 - \delta)\varphi^j .
\end{cases}
\tag{4.30}
$$

If we further require that $\delta \in [0, 1/2]$, then $(1 - \delta)\theta_t^j \geq \frac{1}{2}\theta$ for all j and t, and we thus see just as before that $\varphi^{\delta,j,k,l}$ is monotonic with respect to k and l, uniformly bounded on $X \times [0, T]$ and bounded in C^∞ topology on $U \times (0, T]$, and that for each compact $K \Subset U$ we have an estimate

$$
\sup_{K \times [0,T)} \left| \varphi^{\delta,i,k,l} - \varphi^{\delta,j,k,l} \right| \leq A_K(\varepsilon_i + \varepsilon_j + \delta_i + \delta_j).
$$

We may thus consider

$$
\varphi^\delta = \lim_{j \to \infty} \lim_{k \to \infty} \lim_{l \to \infty} \varphi^{\delta,j,k,l},
$$

which belongs to $C^0 (U \times [0, T]) \cap C^\infty (U \times (0, T))$ and satisfies

$$
\frac{\partial \varphi^\delta}{\partial t} = \log \left[\frac{((1 - \delta)\theta_t + dd^c \varphi^\delta)^n}{\mu} \right].
$$

Since $\|\varphi_0^{j,k,l} - \varphi_0^{\delta,j,k,l}\|_{C^0(X)}$ and $\|\theta_t^j - (1 - \delta)\theta_t^j\|_{C^0(X \times [0,T])}$ are both $O(\delta)$ uniformly with respect to j, k, l, Proposition 4.5.2 shows that

$$
\sup_{K \times [0,T]} \left| \varphi^{\delta,j,k,l} - \varphi^{j,k,l} \right| \leq C_K \delta
$$

for each compact $K \Subset U$, with $C_K > 0$ independent of δ, j, k, l, and hence in the limit

$$
\sup_{K \times [0,T]} \left| \varphi^\delta - \varphi \right| \leq C_K \delta
\tag{4.31}
$$

for all $\delta \in [0, 1/2]$. Now define for each $\delta \in [0, 1/2]$ a function $H^\delta \in C^0 (U \times [0, T]) \cap C^\infty (U \times (0, T])$ by

$$
H^\delta := \varphi' - \varphi^\delta - \delta \tau_U .
$$

We have

$$
\frac{\partial H^\delta}{\partial t} = \log \left[\frac{\left((1-\delta)\theta_t + dd^c\varphi^\delta + \delta(\theta_t + dd^c\tau_U) + dd^c H^\delta\right)^n}{\left((1-\delta)\theta_t + dd^c\varphi^\delta\right)^n} \right]
$$

$$
\geq \log \left[\frac{\left((1-\delta)\theta_t + dd^c\varphi^\delta + dd^c H^\delta\right)^n}{\left((1-\delta)\theta_t + dd^c\varphi^\delta\right)^n} \right],
$$

and hence

$$
\inf_{U\times[0,T]} H^\delta = \inf_{U\times\{0\}} H^\delta.
$$

Since $\tau_U \leq 0$, we get

$$
\varphi' - \varphi^\delta \geq \delta\tau_U + \inf_{U\times\{0\}} (\varphi' - (1-\delta)\varphi)
$$

on $U \times [0, T]$, and hence

$$
\inf_{U\times[0,T]} (\varphi' - \varphi) \geq \inf_{U\times\{0\}} (\varphi' - \varphi)
$$

in the limit as $\delta \to 0$, using (4.31) and the fact that φ is bounded.

4.5.4 The Affine Case: Proof of Theorem 4.3.5

We use the notation of the existence proof above. If (θ_t) is an affine path, then so is $\theta_t^j = \theta_t + \varepsilon_j\omega_X$. We may thus apply Lemma 4.4.6 to conclude that for each $\varepsilon > 0$, $\frac{\partial\varphi^{j,k,l}}{\partial t}$ is uniformly bounded above on $X \times [\varepsilon, T]$, and uniformly bounded below on $X \times [\varepsilon, T - \varepsilon]$. Since φ is a limit of $\varphi^{j,k,l}$ in C^∞-topology on $U \times (0, T]$, we conclude as desired that $\frac{\partial\varphi}{\partial t}$ is bounded above on $U \times [\varepsilon, T]$ and bounded below on $U \times [\varepsilon, T - \varepsilon]$.

Denote also by φ the quasi-psh extension to $X \times [0, T]$ and let $\varepsilon > 0$. Since the time derivative is bounded above on $X \times [\varepsilon, T]$, there exists $C_\varepsilon > 0$ such that

$$
(\theta_t + dd^c\varphi)^n \leq C_\varepsilon\mu
$$

on $X \times \{t\}$ for each $t \in [\varepsilon, T]$. By the results of [EGZ11] (which rely on viscosity techniques), if follows that φ is continuous on $X \times \{t\}$ for each $t \in [\varepsilon, T]$. Since the time derivative is bounded on $U \times [\varepsilon, T - \varepsilon]$, φ is also uniformly Lipschitz continuous in the time variable on $X \times [\varepsilon, T - \varepsilon]$, and it follows as desired that φ is continuous on $X \times (0, T)$.

Remark 4.5.4. It is of course reasonable to expect that φ is in fact continuous on the whole of $X \times [0, T]$.

4.6 The Kähler–Ricci Flow on a Log Terminal Variety

4.6.1 Forms and Currents with Local Potentials

Let X be a complex analytic space with normal singularities, and denote by n its dimension. Since closed $(1, 1)$-forms and currents on X are not necessarily locally dd^c-exact in general, we need to rely on a specific terminology (compare [EGZ09, Sect. 5.2]). We refer for instance to [Dem85] for the basic facts on smooth functions, distributions and psh functions on a complex analytic space. The main point for us is that any psh function on X_{reg} uniquely extends to a psh function on X by normality, see [GR56]. For lack of a proper reference, we include:

Lemma 4.6.1. *Any pluriharmonic distribution on X is locally the real part of a holomorphic function, i.e. the kernel of the dd^c operator on the sheaf $\mathcal{D}'_X$ of germs of distributions coincides with the sheaf $\Re\mathcal{O}_X$ of real parts of holomorphic germs.*

Proof. If $u \in \mathcal{D}'_X$ satisfies $dd^c u = 0$, then $\pm u$ is psh on X_{reg}, and hence extends to a psh function on X by Grauert and Remmert [GR56]. In particular, $\pm u$ is usc and bounded above, which means that u is the germ of a continuous (finite valued) function.

From there on, the proof is basically the same as [FS90, Proposition 1.1]. Let $\pi : X' \to X$ be a proper bimeromorphic morphism with X' smooth. Since $\pm(u \circ \pi)$ is psh on the complex manifold X', we have $u \in \pi_* (\Re\mathcal{O}_{X'})$. We will thus be done if we prove that

$$\pi_* (\Re\mathcal{O}_{X'}) = \Re\mathcal{O}_X.$$

Since X is normal, Zariski's main theorem implies that $\pi_*\mathcal{O}_{X'} = \mathcal{O}_X$, and hence π has connected fibers. The claim is thus that the coboundary morphism

$$\pi_* (\Re\mathcal{O}_{X'}) \to R^1\pi_*(i\mathbb{R}) \tag{4.32}$$

associated to the short exact sequence

$$0 \to i\mathbb{R} \to \mathcal{O}_{X'} \to \Re\mathcal{O}_{X'} \to 0$$

is zero. For each $x \in X$, the composition of (4.32) with the restriction morphism

$$R^1\pi_* (i\mathbb{R})_x \to H^1 \left(\pi^{-1}(x), i\mathbb{R}\right) \tag{4.33}$$

is zero, because it factors through

$$H^0 \left(\pi^{-1}(x), \Re\mathcal{O}_{\pi^{-1}(x)} \right) = \mathbb{R}$$

by the maximum principle, $\pi^{-1}(x)$ being compact and connected. But (4.33) is in fact an isomorphism, just because π is a proper map between locally compact spaces (cf. for instance [Dem09, Theorem 9.10, p. 223]), and it follows as desired that (4.32) is zero.   □

Definition 4.6.2. A $(1, 1)$-*form (resp.* $(1, 1)$-*current) with local potentials* on X is defined to be a section of the quotient sheaf $\mathcal{C}_X^\infty/\Re\mathcal{O}_X$ (resp. $\mathcal{D}_X'/\Re\mathcal{O}_X$). We also introduce the *Bott–Chern cohomology space*

$$H_{\mathrm{BC}}^{1,1}(X) := H^1(X, \Re\mathcal{O}_X).$$

Thanks to Lemma 4.6.1, a $(1, 1)$-form with local potentials can be more concretely described as a closed $(1, 1)$-form θ on X that is locally of the form $\theta = dd^c u$ for a smooth function u. We say that θ is a *Kähler form* if u is strictly psh. Similarly, a closed $(1, 1)$-current T with local potentials is locally of the form $dd^c \varphi$ where φ is a distribution. Since X is normal, and hence locally irreducible, $dd^c \varphi$ is a positive current iff φ is a psh function.

The sheaves $\mathcal{C}_X^\infty$ and $\mathcal{D}_X'$ being soft, hence acyclic, the cohomology long exact sequence shows that $H_{\mathrm{BC}}^{1,1}(X)$ is isomorphic to the quotient of the space of $(1, 1)$-forms (resp. currents) with local potentials by $dd^c C^\infty(X)$ (resp. $dd^c \mathcal{D}'(X)$). In particular, any $(1, 1)$-current T with local potentials can be (globally) written as

$$T = \theta + dd^c \varphi$$

where θ is a $(1, 1)$-form with local potentials and φ is a distribution.

Note also that $H_{\mathrm{BC}}^{1,1}(X)$ is finite dimensional when X is compact, as follows from the cohomology long exact sequence associated to

$$0 \to i\mathbb{R} \to \mathcal{O}_X \to \Re\mathcal{O}_X \to 0$$

and the finite dimensionality of $H^1(X, \mathcal{O}_X)$ and $H^2(X, \mathbb{R})$.

Proposition 4.6.3. *Let* $\alpha \in H_{\mathrm{BC}}^{1,1}(X)$ *be a* $(1, 1)$-*class on a normal complex space* X, *and let* T *be a closed positive* $(1, 1)$-*current on* X_{reg} *representing the restriction* $\alpha|_{X_{\mathrm{reg}}}$ *to the regular part of* X. *Then:*

 (i) T uniquely extends as a positive $(1, 1)$-current with local potentials on X, and the dd^c-class of the extension coincides with α;

 (ii) if X is compact Kähler and if T has locally bounded potentials on an open subset U of X_{reg}, then the positive measure T^n, defined on U in the sense of Bedford–Taylor, has finite total mass on U.

Proof. Let θ be a $(1, 1)$-form with local potentials representing α. On X_{reg} we then have $T = \theta|_{X_{\text{reg}}} + dd^c \varphi$, where φ is a θ-psh function on X_{reg}. If U is a small enough neighborhood of a given point of X, then $\theta = dd^c u$ for some smooth function u on U, and $u + \varphi$ is a psh function on U_{reg}. By the Riemann-type extension property for psh functions [GR56], $u + \varphi$ uniquely extends to a psh function on U, and (i) easily follows.

Point (ii) follows from [BEGZ10, Proposition 1.16] (which is in turn an easy consequence of Demailly's regularization theorem [Dem92]). More precisely, choose a resolution of singularities $\pi : X' \to X$, where X' can be taken to be a compact Kähler manifold and π is an isomorphism above X_{reg}. Denoting by $\langle \pi^* T^n \rangle$ the top-degree non-pluripolar product of $\pi^* T$ on X' (in the sense of [BEGZ10]), we then have

$$\int_U T^n = \int_{\pi^{-1}(U)} \pi^* T^n \leq \int_{X'} \langle \pi^* T^n \rangle < +\infty. \qquad \square$$

We will also use the following simple fact.

Lemma 4.6.4. *Let* $\pi : X' \to X$ *be a bimeromorphic morphism between normal compact complex spaces, let* $A \subset X$ *and* $A' \subset X'$ *be closed analytic subsets of codimension at least 2, and let u be a psh function on* $(X' \setminus A') \cap \pi^{-1}(X \setminus A)$. *Then u is constant.*

Proof. Since A' has codimension at least 2, u extends to a psh function on $\pi^{-1}(X \setminus A)$ by Grauert and Remmert [GR56]. By Zariski's main theorem, π has connected fibers, and u therefore descends to a psh function v on $X \setminus A$, which extends to a psh function on X since A has codimension at least 2. It follows that v is constant. $\qquad \square$

4.6.2 Log Terminal Singularities

Recall that a complex space X is $\mathbb{Q}$-*Gorenstein* if it has normal singularities and if its canonical bundle K_X exists as a $\mathbb{Q}$-line bundle, which means that there exists $r \in \mathbb{N}$ and a line bundle L on X such that $L|_{X_{\text{reg}}} = rK_{X_{\text{reg}}}$.

Let X be a $\mathbb{Q}$-Gorenstein space, and choose a *log resolution* of X, i.e. a projective bimeromorphic morphism $\pi : X' \to X$ which is an isomorphism over X_{reg} and whose exceptional divisor $E = \sum_i E_i$ has simple normal crossings. There is a unique collection of rational numbers a_i, called the *discrepancies* of X (with respect to the chosen log resolution) such that

$$K_{X'} \sim_{\mathbb{Q}} \pi^* K_X + \sum_i a_i E_i,$$

where $\sim_{\mathbb{Q}}$ denotes $\mathbb{Q}$-linear equivalence. By definition, X has *log terminal singularities* iff $a_i > -1$ for all i. This definition is independent of the choice of a log resolution; this will be a consequence of the following analytic interpretation of log terminal singularities as a *finite volume* condition. As an example, quotient singularities are log terminal, and conversely every two-dimensional log terminal singularity is a quotient singularity (see for instance [KolMori98] for more information on log terminal singularities).

After replacing X with a small open subset, we may choose a local generator σ of the line bundle rK_X for some $r \in \mathbb{N}^*$. Restricting to X_{reg}, we define a smooth positive volume form by setting

$$\mu_\sigma := \left(i^{rn^2} \sigma \wedge \bar{\sigma} \right)^{1/r}. \tag{4.34}$$

Such measures are called *adapted measures* in [EGZ09]. The key fact is then the following analytic interpretation of the discrepancies:

Lemma 4.6.5. *Let z_i be a local equation of E_i, defined on a neighborhood $U \subset X'$ of a given point of E. Then we have*

$$\left(\pi^* \mu_\sigma \right)_{U \setminus E} = \prod_i |z_i|^{2a_i} \, dV$$

for some smooth volume form dV on U.

This result is a straightforward consequence of the change of variable formula. As a consequence, a $\mathbb{Q}$-Gorenstein variety X has log terminal singularities iff every (locally defined) adapted measure μ_σ has locally finite mass near each singular point of X. The construction of adapted measures can be globalized as follows: let ϕ be a smooth metric on the $\mathbb{Q}$-line bundle K_X. Then

$$\mu_\phi := \left(\frac{i^{rn^2} \sigma \wedge \bar{\sigma}}{|\sigma|_{r\phi}} \right)^{1/r} \tag{4.35}$$

becomes independent of the choice of a local generator σ of rK_X, and hence defines a smooth positive volume form on X_{reg}, which has locally finite mass near points of X_{sing} iff X is log terminal.

Remark 4.6.6. In [ST09], an adapted measure of the form μ_ϕ for a smooth metric ϕ on K_X is called a *smooth volume form*. We prefer to avoid this terminology, which has the drawback that ω^n is in general not smooth in this sense even when ω is a smooth positive $(1,1)$-form on X. This is in fact already the case for quotient singularities.

The following result illustrates why log terminal singularities are natural in the context of Kähler–Einstein geometry.

Proposition 4.6.7. *Let X be a $\mathbb{Q}$-Gorenstein compact Kähler space, and let ω be a Kähler form on X_{reg} with non-negative Ricci curvature. Assume also that $[\omega] \in H^{1,1}_{\mathrm{BC}}(X_{\mathrm{reg}})$ extends to X. Then X necessarily has log terminal singularities.*

Recall that $[\omega]$ extends to X iff ω extends as a positive $(1,1)$-current with local potentials, by Proposition 4.6.3.

Proof. The volume form ω^n, defined on X_{reg}, induces a Hermitian metric on $K_{X_{\mathrm{reg}}}$. If σ is a local generator of the line bundle rK_X near a given singular point of X, we can consider its pointwise norm $|\sigma|$ on X_{reg}, and it is easy to check from the definitions that

$$\mu_\sigma = |\sigma|^{2/r} \omega^n.$$

Since σ is local generator, $dd^c \log |\sigma|$ is equal to minus the curvature form of the metric on $rK_{X_{\mathrm{reg}}}$, i.e.

$$dd^c \log |\sigma| = r \mathrm{Ric}(\omega).$$

The assumption therefore implies that $\log |\sigma|$ is a local psh function on X_{reg}, and hence extends to a local psh function on X by Grauert and Remmert [GR56]. In particular, $\log |\sigma|$ is locally bounded above on X, and we thus get near each singular point of X $\mu_\sigma \leq C\omega^n$ for some constant $C > 0$. Since $[\omega]$ extends to X, ω^n has finite total mass on X_{reg} by Proposition 4.6.3, and it follows as desired that μ_σ has locally finite mass on X. $\qquad\square$

4.6.3 The Kähler–Ricci Flow on a Log Terminal Variety

Given an initial projective variety X_0 with log terminal singularities and K_{X_0} pseudoeffective, each step of the Minimal Model Program produces a birational morphism $f : X \to Y$ with X, Y projective and normal, X log terminal and K_X f-ample. The following result, due to Song and Tian [ST09], shows that it is then possible to run the (unnormalized) Kähler–Ricci flow on X, starting from an initial positive current with continuous local potentials coming from Y (the actual assumption on the initial current in [ST09] being in fact slightly more demanding).

Theorem 4.6.8. *Let $f : X \to Y$ be a bimeromorphic morphism between two normal compact Kähler spaces such that X is log terminal and K_X is f-ample. Let also $\alpha \in H^{1,1}_{\mathrm{BC}}(Y)$ be a Kähler class on Y, so that $f^*\alpha + t[K_X]$ is a Kähler class in $H^{1,1}_{\mathrm{BC}}(X)$ for $0 < t \ll 1$, and set*

$$T_0 := \sup \{t \in (0, +\infty) \mid f^*\alpha + t[K_X] \text{ is Kähler on } X\}.$$

Given a positive $(1, 1)$*-current* ω_0 *with continuous local potentials on* X *and* $[\omega_0] = f^*\alpha$, *there is a unique way to include* ω_0 *in a family* $(\omega_t)_{t \in [0,T_0)}$ *of positive* $(1, 1)$*-currents with continuous local potentials on* X *such that*

(i) $[\omega_t] = f^*\alpha + t[K_X]$ *for all* $t \in [0, T_0)$;
(ii) *setting* $\Omega := X_{\mathrm{reg}} \setminus \mathrm{Exc}(f)$, *the local potentials of* ω_t *are continuous on* $\Omega \times [0, T_0)$, *and locally bounded on* $X \times [0, T]$ *for each* $T < T_0$;
(iii) $(\omega_t)_{t \in (0,T_0)}$ *restricts to a smooth path of Kähler forms on* Ω *satisfying*

$$\frac{\partial \omega_t}{\partial t} = -\mathrm{Ric}(\omega_t).$$

Moreover, the measures ω_t^n *are uniformly comparable to any given adapted measure as long as* t *stays in a compact subset of* $(0, T)$.

This result of course applies in particular when f is the identity map and α is any Kähler class on X. This special case of Theorem 4.6.8 yields the following result for the normalized Kähler–Ricci flow:

Corollary 4.6.9. *Let* X *be a projective complex variety with log terminal singularities such that* $\pm K_X$ *ample. Then each positive* $(1, 1)$*-current with continuous local potentials* ω_0 *such that* $[\omega_0] = [\pm K_X]$ *extends in a unique way to a family* $(\omega_t)_{t \in [0,+\infty)}$ *of positive* $(1, 1)$*-currents with continuous local potentials such that*

(i) $[\omega_t] = [\pm K_X]$ *for all* $t \in [0, +\infty)$;
(ii) *the local potentials of* ω_t *are continuous on* $X_{\mathrm{reg}} \times [0, +\infty)$, *and bounded on* $X \times [0, T]$ *for each* $T \in (0, +\infty)$;
(ii) $(\omega_t)_{t \in (0,+\infty)}$ *restricts to a smooth path of Kähler forms on* X_{reg} *satisfying*

$$\frac{\partial \omega_t}{\partial t} = -\mathrm{Ric}(\omega_t) \mp \omega_t.$$

Moreover, the volume forms ω_t^n *are uniformly comparable to any given adapted measure as long as* t *stays in a compact subset of* $(0, T)$.

Indeed, as is well-known, setting

$$\omega_s' := (1 \pm s)\omega_{\pm \log(1 \pm s)}$$

defines a bijection between the solutions of

$$\frac{\partial \omega_t}{\partial t} = -\mathrm{Ric}(\omega_t) \mp \omega_t$$

on $X_{\mathrm{reg}} \times (0, +\infty)$ and those of

$$\frac{\partial \omega_s'}{\partial s} = -\mathrm{Ric}(\omega_s')$$

on $X_{\text{reg}} \times (0, T_0)$, with $[\omega'_s] = [\pm K_X] + s[K_X]$ and $T_0 = +\infty$ when $+K_X$ is ample (resp. $T_0 = 1$ when $-K_X$ is ample).

Proof of Theorem 4.6.8. Since α can be represented by a Kähler form on Y, we may choose a closed semipositive $(1, 1)$-form θ_0 with local potentials on X such that $[\theta_0] = f^*\alpha$. We thus have $\omega_0 = \theta_0 + dd^c\varphi_0$ with φ_0 a continuous θ_0-psh function on X. Given $T \in (0, T_0)$, we can choose a Kähler form θ_T representing $f^*\alpha + T[K_X]$, by definition of T_0. For $t \in [0, T]$ set

$$\theta_t := \theta_0 + t\chi$$

with $\chi := T^{-1}(\theta_T - \theta_0)$, which defines an affine path of semipositive $(1, 1)$-forms with local potentials. For $t \in [0, T]$, the path of currents we are looking is of the form $\omega_t = \theta_t + dd^c\varphi_t$ with

$$\varphi \in C_b^0 (\Omega \times [0, T]) \cap C^\infty (\Omega \times (0, T])$$

and $\varphi|_{\Omega \times \{0\}} = \varphi_0$. Since $\chi = T^{-1}(\theta_T - \theta_0)$ is a representative of $[K_X]$, we can find a smooth metric ϕ on the $\mathbb{Q}$-line bundle K_X having χ as its curvature form. If we denote by $\mu := \mu_\phi$ the corresponding adapted measure, it follows from the definitions that for any Kähler form ω on an open subset U of X_{reg} we have

$$- dd^c \log \left[\frac{\omega^n}{\mu} \right] = \chi + \text{Ric}(\omega) \tag{4.36}$$

On $\Omega \times (0, T]$, the equation $\frac{\partial \omega_t}{\partial t} = -\text{Ric}(\omega_t)$ is thus equivalent to

$$dd^c \left(\frac{\partial \varphi}{\partial t} \right) = dd^c \log \left[\frac{(\theta_t + dd^c\varphi)^n}{\mu} \right].$$

By Lemma 4.6.4, this amounts to

$$\frac{\partial \varphi}{\partial t} = \log \left[\frac{(\theta_t + dd^c\varphi)^n}{\mu} \right] + c(t)$$

for some smooth function $c : [0, T] \to \mathbb{R}$, since

$$\Omega = f^{-1}(Y \setminus Y_{\text{sing}}) \cap (X \setminus X_{\text{sing}})$$

and X, Y each have a singular locus of codimension at least 2 by normality. After choosing a primitive of $c(t)$, we can absorb it in the left-hand side, and we end up with showing the existence and uniqueness of

$$\varphi \in C_b^0 (\Omega \times [0, T]) \cap C^\infty (\Omega \times (0, T])$$

such that $\varphi|_{\Omega \times \{0\}} = \varphi_0$ and

$$\frac{\partial \varphi}{\partial t} = \log \left[\frac{(\theta_t + dd^c \varphi)^n}{\mu} \right]$$

on $\Omega \times (0, T]$. Since θ_T is a Kähler form, we have

$$\theta_T \geq \theta := c \, \theta_0$$

for $0 < c \ll 1$, and hence $\theta_t \geq \theta$ for all $t \in [0, T]$. Now let $\pi : X' \to X$ be a log resolution, which is thus in particular an isomorphism above X_{reg}, and pick a Kähler form $\omega_{X'}$ on X'. Since X has log terminal singularities, by Lemma 4.6.5 the measure $\mu' := \pi^* \mu$ is of the form

$$\mu' := e^{\psi^+ - \psi^-} \omega_{X'}^n$$

where $\psi^\pm$ are quasi-psh functions on X' with logarithmic poles along the exceptional divisor E, smooth on $X' \setminus E = \pi^{-1}(X_{\text{reg}})$, and such that $e^{-\psi^-} \in L^p$ for some $p > 1$. We also have $\theta_t' := \pi^* \theta_t \geq \theta' := \pi^* \theta$. Finally, since $[\theta']$ is the pull-back by $f \circ \pi$ of a Kähler class on Y, we have

$$\text{Amp}(\theta') = X' \setminus \text{Exc}(f \circ \pi) = \pi^{-1}(\Omega) \simeq \Omega.$$

Using Theorems 4.3.3 and 4.3.5, it is now easy to conclude the proof of Theorem 4.6.8.   □

References

[Aub78] T. Aubin, Equation de type Monge-Ampère sur les variétés kählériennes compactes. Bull. Sci. Math. **102**, 63–95 (1978)

[BM87] S. Bando, T. Mabuchi, Uniqueness of Einstein Kähler metrics modulo connected group actions, in *Algebraic Geometry* (Sendai, 1985), ed. by T. Oda. Advanced Studies in Pure Mathematics, vol. 10 (Kinokuniya, 1987), pp. 11–40 (North-Holland, Amsterdam, 1987)

[BBGZ13] R. Berman, S. Boucksom, V. Guedj, A. Zeriahi, A variational approach to complex Monge-Ampère equations. Publ. Math. I.H.E.S. **117**, 179–245 (2013)

[Blo03] Z. Błocki, Uniqueness and stability for the complex Monge-Ampère equation on compact Kähler manifolds. Indiana Univ. Math. J. **52**(6), 1697–1701 (2003)

[Bło09] Z. Błocki, A gradient estimate in the Calabi-Yau theorem. Math. Ann. **344**, 317–327 (2009)

[Bou04] S. Boucksom, Divisorial Zariski decompositions on compact complex manifolds. Ann. Sci. Ecole Norm. Sup. (4) **37**(1), 45–76 (2004)

[BEGZ10] S. Boucksom, P. Eyssidieux, V. Guedj, A. Zeriahi, Monge-Ampère equations in big cohomology classes. Acta Math. **205**, 199–262 (2010)

[Cao85] H.D. Cao, Deformation of Kähler metrics to Kähler-Einstein metrics on compact Kähler manifolds. Invent. Math. **81**(2), 359–372 (1985)

[Dem85] J.P. Demailly, Mesures de Monge-Ampère et caractérisation géométrique des variétés algébriques affines. Mém. Soc. Math. Fr. **19**, 124 p. (1985)

[Dem92] J.P. Demailly, Regularization of closed positive currents and intersection theory. J. Algebr. Geom. **1**(3), 361–409 (1992)

[Dem09] J.P. Demailly, Complex analytic and differential geometry (2009), OpenContent book available at http://www-fourier.ujf-grenoble.fr/~demailly/manuscripts/agbook.pdf

[DemPaun04] J.-P. Demailly, M. Paun, Numerical characterization of the Kähler cone of a compact Kähler manifold. Ann. Math. (2) **159**(3), 1247–1274 (2004)

[DZ10] S. Dinew, Z. Zhang, On stability and continuity of bounded solutions of degenerate complex Monge-Ampère equations over compact Kähler manifolds. Adv. Math. **225**(1), 367–388 (2010)

[EGZ09] P. Eyssidieux, V. Guedj, A. Zeriahi, Singular Kähler-Einstein metrics. J. Am. Math. Soc. **22**, 607–639 (2009)

[EGZ11] P. Eyssidieux, V. Guedj, A. Zeriahi, Viscosity solutions to degenerate complex Monge-Ampère equations. Comm. Pure Appl. Math. **64**, 1059–1094 (2011)

[FS90] A. Fujiki, G. Schumacher, The moduli space of extremal compact Kähler manifolds and generalized Weil-Petersson metrics. Publ. Res. Inst. Math. Sci. **26**(1), 101–183 (1990)

[Gill11] M. Gill, Convergence of the parabolic complex Monge-Ampère equation on compact Hermitian manifolds. Comm. Anal. Geom. **19**(2), 277–303 (2011)

[GR56] H. Grauert, R. Remmert, Plurisubharmonische Funktionen in komplexen Räumen. Math. Z. **65**, 175–194 (1956)

[GZ12] V. Guedj, A. Zeriahi, Stability of solutions to complex Monge-Ampère equations in big cohomology classes. Math. Res. Lett. **19**(5), 1025–1042 (2012)

[KolMori98] J. Kollár, S. Mori, in *Birational Geometry of Algebraic Varieties*. Cambridge Tracts in Mathematics, vol. 134 (Cambridge University Press, Cambridge, 1998)

[Kol98] S. Kołodziej, The complex Monge-Ampère equation. Acta Math. **180**(1), 69–117 (1998)

[Kol03] S. Kołodziej, The Monge-Ampère equation on compact Kähler manifolds. Indiana Univ. Math. J. **52**(3), 667–686 (2003)

[Lieb96] G.M. Lieberman, *Second Order Parabolic Differential Equations* (World Scientific, River Edge, 1996)

[Siu87] Y.T. Siu, in *Lectures on Hermitian-Einstein Metrics for Stable Bundles and Kähler-Einstein Metrics*. DMV Seminar, vol. 8 (Birkhäuser, Basel, 1987)

[ST09] J. Song, G. Tian, The Kähler–Ricci flow through singularities (2009). Preprint [arXiv:0909.4898]

[SzTo11] G. Székelyhidi, V. Tosatti, Regularity of weak solutions of a complex Monge-Ampère equation. Anal. PDE **4**, 369–378 (2011)

[Tzha06] G. Tian, Z. Zhang, On the Kähler–Ricci flow on projective manifolds of general type. Chin. Ann. Math. Ser. B **27**(2), 179–192 (2006)

[Tsu88] H. Tsuji, Existence and degeneration of Kähler-Einstein metrics on minimal algebraic varieties of general type. Math. Ann. **281**(1), 123–133 (1988)

[Yau78] S.T. Yau, On the Ricci curvature of a compact Kähler manifold and the complex Monge-Ampère equation, I. Comm. Pure Appl. Math. **31**(3), 339–411 (1978)

[Zer01] A. Zeriahi, Volume and capacity of sublevel sets of a Lelong class of psh functions. Indiana Univ. Math. J. **50**(1), 671–703 (2001)

Chapter 5
The Kähler–Ricci Flow on Fano Manifolds

Huai-Dong Cao

Abstract In these lecture notes, we aim at giving an introduction to the Kähler–Ricci flow (KRF) on Fano manifolds. It covers mostly the developments of the KRF in its first 20 years (1984–2003), especially an essentially self-contained exposition of Perelman's uniform estimates on the scalar curvature, the diameter, and the Ricci potential function for the normalized Kähler–Ricci flow (NKRF), including the monotonicity of Perelman's μ-entropy and κ-noncollapsing theorems for the Ricci flow on compact manifolds. The lecture notes is based on a mini-course on KRF delivered at University of Toulouse III in February 2010, a talk on Perelman's uniform estimates for NKRF at Columbia University's Geometry and Analysis Seminar in Fall 2005, and several conference talks, including "Einstein Manifolds and Beyond" at CIRM (Marseille—Luminy, fall 2007), "Program on Extremal Kähler Metrics and Kähler–Ricci Flow" at the De Giorgi Center (Pisa, spring 2008), and "Analytic Aspects of Algebraic and Complex Geometry" at CIRM (Marseille— Luminy, spring 2011).

Introduction

In these lecture notes, we aim at giving an introduction to the Kähler–Ricci flow (KRF) on Fano manifolds, i.e., compact Kähler manifolds with positive first Chern class. It will cover some of the developments of the KRF in its first 20 years (1984–2003), especially an essentially self-contained exposition of Perelman's uniform estimates on the scalar curvature, the diameter, and the Ricci potential function (in C^1-norm) for the normalized Kähler–Ricci flow (NKRF), including the monotonicity of Perelman's μ-entropy and κ-noncollapsing theorems for the Ricci flow

H.-D. Cao (✉)
Department of Mathematics, University of Macau, Macau, China

Department of Mathematics, Lehigh University, Bethlehem, PA 18015, USA
e-mail: huc2@lehigh.edu

S. Boucksom et al. (eds.), *An Introduction to the Kähler–Ricci Flow*, Lecture Notes in Mathematics 2086, DOI 10.1007/978-3-319-00819-6_5, © Springer International Publishing Switzerland 2013

on compact manifolds. Except in the last section where we shall briefly discuss the formation of singularities of the KRF in Fano case, much of the recent progress since Perelman's uniform estimates are not touched here, especially those by Phong–Sturm [PS06] and Phong–Song–Sturm–Weinkove [PSSW09, PSSW08b, PSSW11] (see also [Pal08, CZ09, Sz10, Tos10a, MSz09, Zha11] etc.) tying the convergence of the NKRF to a notion of GIT stability for the diffeomorphism group, in the spirit of the conjecture of Yau [Yau93] (see also [Tian97, Don02]). We hope to discuss these developments, as well as many works related to Kähler Ricci solitons, on another occasion. We also refer the readers to the lecture notes by J. Song and B. Weinkove in Chap. 3 of the present volume for some of the other significant developments in KRF.

In spring 1982, Yau invited Richard Hamilton to give a talk at the Institute for Advanced Study (IAS) on his newly completed seminal work "Three-manifolds with positive Ricci curvature" [Ham82]. Shortly after, Yau asked me, Ben Chow and Ngaiming Mok to present Hamilton's work on the Ricci flow in details at Yau's IAS geometry seminar. At the time, Ben Chow and I were first year graduate students, and Mok was an instructor at Princeton University. There was another fellow first year graduate student, S. Bando, working with Yau. It was clear to us that Yau was very excited about Hamilton's work and saw its great potential. He encouraged us to study and pursue Hamilton's Ricci flow.

Besides attending courses at Princeton and Yau's lecture series in geometric analysis at IAS, I spent most of 1982 preparing for Princeton's General Examination, a 3-hour oral exam covering two basic subjects (Real and Complex Analysis and Algebra) plus two additional advanced topics. But I also continued to study Hamilton's paper. After I passed the General Exam in January 1983, I went to see Yau and asked for his suggestion for a thesis problem. Yau immediately gave me the problem to study the Ricci flow on Kähler manifolds, especially the long time existence and convergence on Fano manifolds. At the time I hardly knew any complex geometry (but I did not dare to tell Yau so). In the following months, I spent a lot of time reading and trying to understand Yau's seminal paper on the Calabi conjecture [Yau78], and also Calabi's paper on extremal Kähler metrics [Cal82] suggested by Yau. In the mean time, it happened that Yau invited Calabi to visit IAS in spring 1983 and I benefited a great deal from Calabi's lecture series on "Vanishing theorems in Kähler geometry" at IAS that spring.

By spring 1984 I had managed to prove the long time existence of the canonical Kähler–Ricci flow by adopting Yau's celebrated a priori estimates for the Calabi conjecture to the parabolic case, as well as the convergence to Kähler–Einstein metrics when the first Chern class c_1 is either negative or zero. The convergence proof when $c_1 = 0$ used a version of the Li-Yau type estimate for positive solutions to the heat equation with evolving metrics and an argument of J. Moser. But little progress was made toward long time behavior when $c_1 > 0$. Without being fully aware of the significance and the difficulties of the problem in the Fano case, I felt kind of uneasy that I did not meet my adviser's high expectation. But to my relief, Yau seemed quite pleased and encouraged me to write up the work. That resulted in my 1985 paper [Cao85]. In fall of 1984, several of Yau's Princeton graduate students, including me and B. Chow, followed him to San Diego where both Richard

Hamilton and Rick Schoen also arrived. By then Bando had used the short time property of the flow to classify three-dimensional compact Kähler manifolds of nonnegative bisectional curvature [Bando84] and graduated from Princeton. Shortly after our arrival in San Diego, following Hamilton's work in [Ham86], Ben Chow and I also used the short time property of the flow to classify compact Kähler manifolds with nonnegative curvature operator in all dimensions [CaoChow86]. In 1988, Mok's work [Mok88] was published in which he was able to show (in 1986) nonnegative bisectional curvature is preserved in all dimensions. By combining the short time property of the flow and the existence of special rational curves by Mori [Mori79], Mok proved the generalized Frankel conjecture in its full generality (see also a recent new proof by H. Gu [Gu09]). Around the same time, Tsuji [Tsu88] extended my work on the KRF for the negative Chern class case to compact complex manifolds of general type (see also the related later work of Tian–Zhang [Tzha06]). This is a brief history of the KRF in its early years.

Late 1980s and 1990s saw great advances in the Ricci flow by Hamilton [Ham88, Ham93a, Ham93b, Ham95b, Ham95a, Ham97, Ham99] which laid the foundation for his program to use the Ricci flow to attack the Poincaré and geometrization conjectures. In particular, the works of Hamilton [Ham88] and Ben Chow [Chow91] imply that every metric on a compact Riemann surface can be deformed to a metric of constant curvature under the Ricci flow. During the same period, there were several developments in the KRF, including the constructions of $U(n)$-invariant Kähler–Ricci soliton examples by Koiso [Koiso90] and the author [Cao94][1]; the Li–Yau–Hamilton inequalities and the Harnack inequality for the KRF [Cao92, Cao97]; the important work of W.-X. Shi [Shi90, Shi97], another former student of Yau, using the noncompact KRF to approach Yau's conjecture that a complete noncompact Kähler manifold with positive bisectional curvature is biholomorphic to the complex Euclidean space $\mathbb{C}^n$ (see [ChauT08] for a recent survey on the subject), etc. In addition, in 1991 at Columbia University, I observed that Mabuchi's K-energy [Mab86] and the functional defined in Ding–Tian [DT92] are monotone decreasing under the KRF (Cao, 1991, unpublished work on the Kähler–Ricci flow). The fact that the K-energy is monotone under the KRF turned out to be quite useful, and was applied in the work of Chen–Tian [CheT02] 10 years later.

In November 2002 and spring 2003, Perelman [Per02, Per03q, Per03b] made astounding breakthroughs in the Ricci flow. In April 2003, in a private lecture at MIT, Perelman presented in detail his uniform scalar curvature and diameter estimates for the NKRF based on the monotonicity of his $\mathcal{W}$-functional and μ-entropy, and the powerful ideas in his κ-noncollapsing results. We remark that prior to Perelman's lecture at MIT, such uniform estimates had appeared only in the special case when NKRF has positive bisectional curvature, in the work of Chen and Tian [CheT02] for the Kähler surface case (see also [CheT06] for the higher dimensional case) assuming in addition the existence of K–E metrics; and also in the work of B.-L Chen, X.-P. Zhu and the author [CCZ03] in all dimensions and without assuming the existence of K–E metrics.

---

[1]My work was carried out at Columbia University in early 1990s.

From Hamilton and Perelman's works to the recent proof of the 1/4-pinching differentiable sphere theorem by Brendle–Schoen [BS09], we have seen spectacular applications of the Ricci flow and its sheer power of flowing to canonical metrics/structures without a priori knowing their existence. Let us hope to see similar phenomena happen to the KRF.

5.1 Preliminaries

In this section, we fix our notations and recall some basic facts and formulas in Kähler Geometry.

5.1.1 Kähler Metrics and Kähler Forms

Let (X^n, g) be a compact Kähler manifold of complex dimension n with the Kähler metric g. In local holomorphic coordinates $(z^1, \cdots, z^n)$, denote its Kähler form by

$$\omega = \frac{\sqrt{-1}}{2} \sum_{i,j} g_{i\bar{j}} dz^i \wedge d\bar{z}^j. \tag{5.1}$$

By definition, g is Kähler means that its Kähler form ω is a *closed* real (1,1) form, or equivalently,

$$\partial_k g_{i\bar{j}} = \partial_i g_{k\bar{j}} \quad \text{and} \quad \partial_{\bar{k}} g_{i\bar{j}} = \partial_{\bar{j}} g_{i\bar{k}} \tag{5.2}$$

for all $i, j, k = 1, \cdots n$. Here $\partial_k = \partial/\partial z^k$ and $\partial_{\bar{k}} = \partial/\partial \bar{z}^k$.

The cohomology class $[\omega]$ represented by ω in $H^2(X, \mathbb{R})$ is called the Kähler class of the metric $g_{i\bar{j}}$. By the Hodge theory, two Kähler metrics $g_{i\bar{j}}$ and $\tilde{g}_{i\bar{j}}$ belong to the same Kähler class if and only if $g_{i\bar{j}} = \tilde{g}_{i\bar{j}} + \partial_i \partial_{\bar{j}} \varphi$, or equivalently,

$$\omega = \tilde{\omega} + \frac{\sqrt{-1}}{2\pi} \partial \bar{\partial} \varphi \tag{5.3}$$

for some real valued smooth function φ on X.

The volume of (X, g) is given by

$$\text{Vol}(X, g) = \int_X \omega^{[n]}, \tag{5.4}$$

where we have followed the convention of Calabi [Cal82] to denote $\omega^{[n]} = \omega^n/n!$ so that the volume form is given by

$$dV = \det(g_{i\bar{j}}) \wedge_{i=1}^n \left(\frac{\sqrt{-1}}{2} dz^i \wedge d\bar{z}^i \right) = \omega^{[n]}. \tag{5.5}$$

Clearly, by Stokes' theorem, if g and $\tilde{g}$ are in the same Kähler class then we have

$$\text{Vol}(X, g) = \text{Vol}(X, \tilde{g}).$$

5.1.2 Curvatures and the First Chern Class

The Christoffel symbols of the metric $g_{i\bar{j}}$ are given by

$$\Gamma_{ij}^k = g^{k\bar{\ell}} \partial_i g_{j\bar{\ell}} \quad \text{and} \quad \Gamma_{\bar{i}\bar{j}}^{\bar{k}} = g^{\ell\bar{k}} \partial_{\bar{i}} g_{\ell\bar{j}}, \tag{5.6}$$

where $(g^{i\bar{j}}) = ((g_{i\bar{j}})^{-1})^T$. It is a basic fact in Kähler geometry that, for each point $x_0 \in X^n$, there exists a system of holomorphic normal coordinates $(z^1, \cdots, z^n)$ at x_0 such that

$$g_{i\bar{j}}(x_0) = \delta_{i\bar{j}} \quad \text{and} \quad \partial_k g_{i\bar{j}}(x_0) = 0, \qquad \forall i, j, k = 1, \cdots n. \tag{5.7}$$

The curvature tensor of the metric $g_{i\bar{j}}$ is defined as $R_{i\ k\bar{\ell}}^{\ j} = -\partial_{\bar{\ell}} \Gamma_{ik}^j$, or by lowering j to the second index,

$$R_{i\bar{j}k\bar{\ell}} = g_{p\bar{j}} R_{i\ k\bar{\ell}}^{\ p} = -\partial_k \partial_{\bar{\ell}} g_{i\bar{j}} + g^{p\bar{q}} \partial_k g_{i\bar{q}} \partial_{\bar{\ell}} g_{p\bar{j}}. \tag{5.8}$$

From (5.2) and (5.8), we immediately see that $R_{i\bar{j}k\bar{\ell}}$ is symmetric in i and k, in $\bar{j}$ and $\bar{\ell}$, and in the pairs $\{i\bar{j}\}$ and $\{k\bar{\ell}\}$.

We say that (X^n, g) has positive (holomorphic) bisectional curvature, or positive holomorphic sectional curvature, if

$$R_{i\bar{j}k\bar{\ell}} v^i v^{\bar{j}} w^k w^{\bar{\ell}} > 0, \quad \text{or} \quad R_{i\bar{j}k\bar{\ell}} v^i v^{\bar{j}} v^k v^{\bar{\ell}} > 0$$

respectively, for all nonzero vectors v and w in the holomorphic tangent bundle $T_x X$ of X at x for all $x \in X$.

The Ricci tensor of the metric $g_{i\bar{j}}$ is obtained by taking the trace of $R_{i\bar{j}k\bar{\ell}}$:

$$R_{i\bar{j}} = g^{k\bar{\ell}} R_{i\bar{j}k\bar{\ell}} = -\partial_i \partial_{\bar{j}} \log \det(g). \tag{5.9}$$

From (5.9), it is clear that the Ricci form

$$\text{Ric} = \frac{\sqrt{-1}}{2} \sum_{i,j} R_{i\bar{j}} dz^i \wedge d\bar{z}^j \tag{5.10}$$

is real and closed. It is well known that the first Chern class $c_1(X) \in H^2(X, Z)$ of X is represented by the Ricci form:

$$[\text{Ric}] = \pi c_1(X). \tag{5.11}$$

Finally, the scalar curvature of the metric $g_{i\bar{j}}$ is

$$R = g^{i\bar{j}} R_{i\bar{j}}. \tag{5.12}$$

Hence, the total scalar curvature

$$\int_X R dV = \int_X \text{Ric} \wedge \omega^{[n-1]}, \tag{5.13}$$

depends only on the Kähler class of ω and the first Chern class $c_1(X)$.

5.1.3 Covariant Derivatives

Given any smooth function f, we denote by

$$\nabla_i f = \partial_i f, \qquad \nabla_{\bar{i}} f = \partial_{\bar{i}} f.$$

For any (1,0)-form v_i, its covariant derivatives are defined as

$$\nabla_j v_i = \partial_j v_i - \Gamma_{ij}^k v_k \quad \text{and} \quad \nabla_{\bar{j}} v_i = \partial_{\bar{j}} v_i. \tag{5.14}$$

Similarly, for covariant two-tensors, we have

$$\nabla_k v_{i\bar{j}} = \partial_k v_{i\bar{j}} - \Gamma_{ik}^p v_{p\bar{j}}, \quad \nabla_{\bar{k}} v_{i\bar{j}} = \partial_{\bar{k}} v_{i\bar{j}} - \Gamma_{\bar{j}\bar{k}}^{\bar{p}} v_{i\bar{p}},$$

$$\nabla_k v_{ij} = \partial_k v_{ij} - \Gamma_{ik}^p v_{pj} - \Gamma_{jk}^p v_{ip}, \quad \text{and} \quad \nabla_{\bar{k}} v_{ij} = \partial_{\bar{k}} v_{ij}.$$

Now, in the Kähler case, the second Bianchi identity in Riemannian geometry translates into the relations

$$\nabla_p R_{i\bar{j}k\bar{\ell}} = \nabla_k R_{i\bar{j}p\bar{\ell}} \quad \text{and} \quad \nabla_{\bar{p}} R_{i\bar{j}k\bar{\ell}} = \nabla_{\bar{\ell}} R_{i\bar{j}k\bar{p}}. \tag{5.15}$$

Covariant differentiations of the same type can be commuted freely, e.g.,

$$\nabla_k \nabla_j v_i = \nabla_j \nabla_k v_i, \qquad \nabla_{\bar{k}} \nabla_{\bar{j}} v_i = \nabla_{\bar{j}} \nabla_{\bar{k}} v_i, \tag{5.16}$$

etc. But we shall need the following formulas when commuting covariant derivatives of different types:

$$\nabla_k \nabla_{\bar{j}} v_i - \nabla_{\bar{j}} \nabla_k v_i = -R_{i\bar{j}k\bar{\ell}} v_\ell, \tag{5.17}$$

$$\nabla_k \nabla_{\bar{\ell}} v_{i\bar{j}} - \nabla_{\bar{\ell}} \nabla_k v_{i\bar{j}} = -R_{i\bar{p}k\bar{\ell}} v_{p\bar{j}} + R_{p\bar{j}k\bar{\ell}} v_{i\bar{p}}, \tag{5.18}$$

etc.

We define

$$|\nabla f|^2 = g^{i\bar{j}} \partial_i f \partial_{\bar{j}} f, \tag{5.19}$$

$$|\text{Ric}|^2 = g^{i\bar{\ell}} g^{k\bar{j}} R_{i\bar{j}} R_{k\bar{\ell}}, \tag{5.20}$$

and

$$|\mathrm{Rm}|^2 = g^{i\bar{q}} g^{p\bar{j}} g^{k\bar{s}} g^{r\bar{\ell}} R_{i\bar{j}k\bar{\ell}} R_{p\bar{q}r\bar{s}}. \tag{5.21}$$

The norm square $|S|^2$ of any other type of covariant tensor S is defined similarly. Finally, the Laplace operator on a tensor S is, in normal coordinates, defined as

$$\Delta S = \frac{1}{2} \sum_k (\nabla_k \nabla_{\bar{k}} + \nabla_{\bar{k}} \nabla_k) S. \tag{5.22}$$

5.1.4 Kähler–Einstein Metrics and Kähler–Ricci Solitons

It is well known that a Kähler metric $g_{i\bar{j}}$ is Kähler–Einstein if

$$R_{i\bar{j}} = \lambda g_{i\bar{j}}$$

for some real number $\lambda \in \mathbb{R}$. Kähler–Ricci solitons are extensions of K–E metrics: a Kähler metric $g_{i\bar{j}}$ is called a gradient Kähler–Ricci (K–R) soliton if there exists a real-valued smooth function f on X such that

$$R_{i\bar{j}} = \lambda g_{i\bar{j}} - \partial_i \partial_{\bar{j}} f \quad \text{and} \quad \nabla_i \nabla_j f = 0. \tag{5.23}$$

It is called *shrinking* if $\lambda > 0$, *steady* if $\lambda = 0$, and *expanding* if $\lambda < 0$. The function f is called a potential function.

Note that the second equation in (5.23) is equivalent to saying the gradient vector field

$$\nabla f = (g^{i\bar{j}} \partial_{\bar{j}} f) \frac{\partial}{\partial z^i}$$

is holomorphic. By scaling, we can normalize $\lambda = 1, 0, -1$ in (5.23). The concept of Ricci soliton was introduced by Hamilton [Ham88] in mid 1980s. It has since played a significant role in Hamilton's Ricci flow as Ricci solitons often arise as singularity models (see, e.g., [Cao10] for a survey). Note that when f is a constant function, K–R solitons are simply K–E metrics.

Clearly, if X^n admits a K–E metric or K–R soliton g then the first Chern class is necessarily definite, as

$$\pi c_1(X) = \lambda[\omega_g].$$

When $c_1(X) = 0$ it follows from Yau's solution to the Calabi conjecture that in each Kähler class there exists a unique Calabi–Yau metric (i.e., Ricci-flat Kähler metric) g in that class. Moreover, when $c_1(X) < 0$, Aubin [Aub78] and Yau [Yau78] proved independently that there exists a unique Kähler–Einstein metric in the class $-\pi c_1(X)$.

However, in the Fano case (i.e., $c_1(X) > 0$), it is well known that there exist obstructions to the existence of a K–E metric g in the class of $\omega \in \pi c_1(X)$ with $R_{i\bar{j}} = g_{i\bar{j}}$. One of the obstructions is the Futaki invariant defined as follows: take any Kähler metric g with $\omega \in \pi c_1(X)$. Then its Kähler class $[\omega]$ agrees with its Ricci class [Ric]. Hence, by the Hodge theory, there exists a real-valued smooth function f, called the Ricci potential of the metric g, such that

$$R_{i\bar{j}} = g_{i\bar{j}} - \partial_i \partial_{\bar{j}} f. \tag{5.24}$$

In [Fut83], Futaki proved that the functional $F : \eta(X) \to C$ defined by

$$F(V) = \int_X \nabla_V f \, \omega^{[n]} = \int_X (V \cdot \nabla f) \omega^{[n]} \tag{5.25}$$

on the space $\eta(X)$ of holomorphic vector fields depends only on the class $\pi c_1(X)$, but not the metric g. In particular, if a Fano manifold X^n admits a positive K–E metric, then the Futaki invariant F defined above must be zero.

On the other hand, it turns out that compact stead and expanding K–R solitons are necessarily K–E (cf.). If g is a *shrinking* K–R soliton satisfying

$$R_{i\bar{j}} = g_{i\bar{j}} - \partial_i \partial_{\bar{j}} f \quad \text{and} \quad \nabla_i \nabla_j f = 0 \tag{5.26}$$

with non-constant function f then, taking $V = \nabla f$, we have

$$F(\nabla f) = \int_X |\nabla f|^2 \omega^{[n]} \neq 0. \tag{5.27}$$

The existence of compact (shrinking) K–R solitons were shown independently by Koiso [Koiso90] and the author [Cao94], and later by X. Wang and X. Zhu [WZ04]. The noncompact example was first found by Feldman–Ilmanen–Knopf [FIK03], see also A. Dancer-Wang [DW11] and Futaki–Wang [FutW11] for further examples.

We remark that Bando and Mabuchi [BM87] proved that positive K–E metrics are unique in the sense that any two positive K–E metrics on X^n only differ by an automorphism of X^n. Moreover, Tian and Zhu [TZ02] extended the definition of the Futaki invariant by introducing a corresponding obstruction to the existence of (shrinking) K–R solitons on Fano manifolds. They also proved the Bando–Mabuchi type uniqueness result for K–R solitons [TZ00].

5.2 The (Normalized) Kähler–Ricci Flow

In this section we introduce the Kähler–Ricci flow (KRF) and the normalized Kähler–Ricci flow (NKRF) on Fano manifolds, i.e., compact Kähler manifolds with positive first Chern class. We state the basic long time existence of solutions to the NKRF proved by the author in [Cao85], derive the evolution equations of various curvature tensors, and present Mok's result on preserving the non-negativity of the holomorphic bisectional curvature under the KRF.

5.2.1 Kähler–Ricci Flow and Normalized Kähler–Ricci Flow

On any given Kähler manifold $(X^n, \tilde{g}_{i\bar{j}})$, the Kähler–Ricci flow deforms the initial metric $\tilde{g}$ by the equation

$$\frac{\partial}{\partial t} g_{i\bar{j}}(t) = -R_{i\bar{j}}(t), \quad g(0) = \tilde{g}, \tag{5.28}$$

or equivalently, in terms of the Kähler forms, by

$$\frac{\partial}{\partial t} \omega(t) = -\mathrm{Ric}(\omega(t)), \quad \omega(0) = \omega_0. \tag{5.29}$$

Note that, by (5.29), the Kähler class $[\omega(t)]$ of the evolving metric $g_{i\bar{j}}(t)$ satisfies the ODE

$$\frac{d}{dt}[\omega(t)] = -\pi c_1(X),$$

from which it follows that

$$[\omega(t)] = [\omega_0] - t\pi c_1(X). \tag{5.30}$$

Proposition 5.2.1. *Given any initial Kähler metric $\tilde{g}$ on a compact Kähler manifold X^n, KRF (5.28) admits a unique solution $g(t)$ for a short time.*

Proof. We consider the nonlinear, strictly parabolic, scalar equation of Monge–Ampéré type

$$\frac{\partial}{\partial t}\varphi = \log\frac{\det(\tilde{g}_{i\bar{j}} - t\tilde{R}_{i\bar{j}} + \partial_i\partial_{\bar{j}}\varphi)}{\det(\tilde{g}_{i\bar{j}})}, \quad \varphi(0) = 0$$

as in [Bando84]. Then, this parabolic equation admits a unique solution φ for a short time, and it is easy to verify that

$$g_{i\bar{j}}(t) =: \tilde{g}_{i\bar{j}} - t\tilde{R}_{i\bar{j}} + \partial_i\partial_{\bar{j}}\varphi$$

gives rise to a short time solution to KRF (5.28) for small $t > 0$. This proves the existence. For the uniqueness, suppose $h_{i\bar{j}}$ is another solution to KRF (5.28). Then, by (5.30), we have

$$h_{i\bar{j}} = \tilde{g}_{i\bar{j}} - t\tilde{R}_{i\bar{j}} + \partial_i\partial_{\bar{j}}\psi$$

for some real-valued function ψ. But then we must have

$$\partial_i\partial_{\bar{j}}\left(\frac{\partial}{\partial t}\psi\right) = -R_{i\bar{j}} + \tilde{R}_{i\bar{j}}.$$

Hence, by (5.9) and by adjusting with an additive function in t only, we have

$$\frac{\partial}{\partial t} \psi = \log \frac{\det(\tilde{g}_{i\bar{j}} - t\tilde{R}_{i\bar{j}} + \partial_i \partial_{\bar{j}} \psi)}{\det(\tilde{g}_{i\bar{j}})}.$$

Note that $h_{i\bar{j}}(0) = \tilde{g}_{i\bar{j}}$ forces $\psi(0)$ to be a constant function. Therefore φ and ψ differ by a function in t only which in turn implies that $g = h$.

Alternatively, by the work of Hamilton [Ham82] (see also De Turck [Det83]), there exists a unique solution $g(t)$ to (5.28), regarded as the Ricci flow for Riemannian metric, for a short time with $\tilde{g}$ as the initial metric. Moreover, Hamilton [Ham95a] observed that the holonomy group does not change under the Ricci flow for a short time. Thus, the solution $g(t)$ we obtained remains Kähler for $t > 0$. □

Lemma 5.2.2. *Under the Kähler–Ricci flow (5.28), the volume of $(X, g_{i\bar{j}}(t))$ changes by*

$$\frac{d}{dt} \text{Vol}(X, g(t)) = -\int_X R(t)\, \omega^{[n]}(t).$$

Proof. Under KRF (5.28), we have

$$\frac{\partial}{\partial t} \omega^{[n]} = (\frac{\partial}{\partial t} \log \det(g_{i\bar{j}})) \omega^{[n]}$$

and

$$\frac{\partial}{\partial t} \log \det(g_{i\bar{j}}) = g^{i\bar{j}} \frac{\partial}{\partial t} g_{i\bar{j}} = -g^{i\bar{j}} R_{i\bar{j}} = -R.$$

Therefore, the volume element $dV = \omega^{[n]}$ changes by

$$\frac{\partial}{\partial t} \omega^{[n]} = -R\omega^{[n]}. \tag{5.31}$$

□

From now on, we consider a Fano manifold $(X^n, \tilde{g}_{i\bar{j}})$ such that

$$[\omega_0] = [\tilde{\omega}] = \pi c_1(X), \tag{5.32}$$

and we deform the initial metric $\tilde{g}$ by the KRF (5.28).

To keep the volume unchanged, we consider the normalized Kähler–Ricci flow

$$\frac{\partial}{\partial t} g_{i\bar{j}} = -R_{i\bar{j}} + g_{i\bar{j}}, \quad g(0) = \tilde{g} \tag{5.33}$$

or equivalently

$$\frac{\partial}{\partial t}\omega = -\text{Ric}(\omega) + \omega, \quad \omega(0) = \omega_0. \tag{5.34}$$

From the proof of Lemma 5.2.2, it is easy to see that the following holds (in fact, under NKRF (5.33) the solution $g(t)$ has the same Kähler class):

Lemma 5.2.3. *Under the normalized Kähler–Ricci flow (5.33), we have*

$$\frac{\partial}{\partial t}(dV) = (n - R)dV.$$

By (5.30) and (5.32), it follows that

$$[\omega(t)] = \pi(1 - t)c_1(X),$$

showing that $[\omega(t)]$ shrinks homothetically and would become degenerate at $t = 1$. This suggests that if $[0, T)$ is the maximal existence time interval of solution $\hat{g}(t)$ to KRF (5.28), then T cannot exceed 1. We shall see that the NKRF (5.33) has solution $g(t)$ exists for all time $0 \le t < \infty$, which in turn implies that $T = 1$ for KRF (5.28).

By direct calculations, one can easily verify the following relations between the solutions to KRF (5.28) and NKRF (5.33).

Lemma 5.2.4. *Let $\hat{g}_{i\bar{j}}(s), 0 \le s < 1$, and $g_{i\bar{j}}(t), 0 \le t < \infty$, be solutions to the KRF (5.28) and the NKRF (5.33) respectively. Then, $\hat{g}_{i\bar{j}}(s)$ and $g_{i\bar{j}}(t)$ are related by*

$$\hat{g}_{i\bar{j}}(s) = (1 - s)g_{i\bar{j}}(t(s)), \quad t = -\log(1 - s)$$

and

$$g_{i\bar{j}}(t) = e^t \hat{g}_{i\bar{j}}(s(t)), \quad s = 1 - e^{-t}.$$

Corollary 5.2.5. *Let $\hat{g}_{i\bar{j}}(s)$ and $g_{i\bar{j}}(t)$ be as in Lemma 5.2.4. Then, their scalar curvatures and the norm square of their curvature tensors are related respectively by*

$$(1 - s)\hat{R}(s) = R(t(s)),$$

and

$$(1 - s)|\widehat{\text{Rm}}|_{\hat{g}(s)} = |\text{Rm}|_{g(t(s))}.$$

5.2.2 The Long Time Existence of the NKRF

First of all, it is well known that the NKRF (5.33) is equivalent to a parabolic scalar equation of complex Monge–Ampère type on the Kähler potential. For any given initial metric $g_0 = \tilde{g}$ satisfying (5.32), consider

$$g_{i\bar{j}}(t) = \tilde{g}_{i\bar{j}} + \partial_i \partial_{\bar{j}} \varphi, \tag{5.35}$$

where $\varphi = \varphi(t)$ is a time-dependent, real-valued, smooth unknown function on X. Then,

$$\frac{\partial}{\partial t} g_{i\bar{j}} = \partial_i \partial_{\bar{j}} \varphi_t$$

and

$$-R_{i\bar{j}} + g_{i\bar{j}} = -R_{i\bar{j}} + \tilde{g}_{i\bar{j}} + \partial_i \partial_{\bar{j}} \varphi = -R_{i\bar{j}} + \tilde{R}_{i\bar{j}} + \partial_i \partial_{\bar{j}} (\tilde{f} + \varphi)$$

$$= \partial_i \partial_{\bar{j}} \log \frac{\omega^n}{\tilde{\omega}^n} + \partial_i \partial_{\bar{j}} (\tilde{f} + \varphi).$$

Here $\tilde{f}$ is the Ricci potential of $\tilde{g}_{i\bar{j}}$ as defined in (5.24). Thus, the NKRF (5.33) reduces to

$$\partial_i \partial_{\bar{j}} \varphi_t = \partial_i \partial_{\bar{j}} \log \frac{\omega^n}{\tilde{\omega}^n} + \partial_i \partial_{\bar{j}} (\tilde{f} + \varphi),$$

or equivalently,

$$\frac{\partial}{\partial t} \varphi = \log \frac{\det(\tilde{g}_{i\bar{j}} + \partial_i \partial_{\bar{j}} \varphi)}{\det(\tilde{g}_{i\bar{j}})} + \tilde{f} + \varphi + b(t) \tag{5.36}$$

for some function $b(t)$ of t only.

Note that (5.36) is strictly parabolic, so standard PDE theory implies its short time existence (cf. [Baker10]). Clearly, we have

Lemma 5.2.6. *If φ solves the parabolic scalar equation (5.36), then $g_{i\bar{j}}(t)$, as defined in (5.35), is a solution to the NKRF (5.33).*

Now we can state the following long time existence result shown by the author [Cao85], based on the parabolic version of Yau's a priori estimates in [Yau78]. We refer the readers to [Cao85], or the lecture notes by Song and Weinkove [SW] in this volume, for a proof.

Theorem 5.2.7 ([Cao85]). *The solution $\varphi(t)$ to (5.36) exists for all time $0 \le t < \infty$. Consequently, the solution $g_{i\bar{j}}(t)$ to the normalized Kähler–Ricci flow (5.33) exists for all time $0 \le t < \infty$.*

5.2.3 Preserving Positivity of the Bisectional Curvature

To derive the curvature evolution equations for both KRF and NKRF, we consider

$$\frac{\partial}{\partial t} g_{i\bar{j}} = -R_{i\bar{j}} + \lambda g_{i\bar{j}}. \tag{5.37}$$

Lemma 5.2.8. *Under (5.37), we have*

$$\frac{\partial}{\partial t} R_{i\bar{j}} = \Delta R_{i\bar{j}} + R_{i\bar{j}k\bar{\ell}} R_{\ell\bar{k}} - R_{i\bar{k}} R_{k\bar{j}}, \tag{5.38}$$

and

$$\frac{\partial}{\partial t} R = \Delta R + |\mathrm{Ric}|^2 - \lambda R. \tag{5.39}$$

Proof. First of all, from (5.9), we get

$$\frac{\partial}{\partial t} R_{i\bar{j}} = -\nabla_i \nabla_{\bar{j}} (g^{k\bar{\ell}} \frac{\partial}{\partial t} g_{k\bar{\ell}}) = \nabla_i \nabla_{\bar{j}} R. \tag{5.40}$$

On the other hand, by using the commuting formulas (5.16)–(5.18) for covariant differentiations, we have

$$\nabla_k \nabla_{\bar{k}} R_{i\bar{j}} = \nabla_k \nabla_{\bar{j}} R_{i\bar{k}} = \nabla_{\bar{j}} \nabla_k R_{i\bar{k}} - R_{k\bar{j}i\bar{\ell}} R_{\ell\bar{k}} + R_{k\bar{j}\ell\bar{k}} R_{i\bar{\ell}}$$

$$= \nabla_{\bar{j}} \nabla_i R - R_{i\bar{j}k\bar{\ell}} R_{\ell\bar{k}} + R_{i\bar{\ell}} R_{l\bar{j}},$$

and

$$\nabla_k \nabla_{\bar{k}} R_{i\bar{j}} = \nabla_{\bar{k}} \nabla_k R_{i\bar{j}}.$$

Hence,

$$\Delta R_{i\bar{j}} = \frac{1}{2} \left(\nabla_k \nabla_{\bar{k}} + \nabla_{\bar{k}} \nabla_k \right) R_{i\bar{j}} = \nabla_i \nabla_{\bar{j}} R - R_{i\bar{j}k\bar{\ell}} R_{\ell\bar{k}} + R_{i\bar{\ell}} R_{l\bar{j}}. \tag{5.41}$$

Therefore, (5.38) follows from (5.40) and (5.41)

Next, using the evolution equation of $R_{i\bar{j}}$, we have

$$\frac{\partial}{\partial t} R = \frac{\partial}{\partial t} (g^{i\bar{j}} R_{i\bar{j}}) = g^{i\bar{j}} (\Delta R_{i\bar{j}} + R_{i\bar{j}k\bar{\ell}} R_{\ell\bar{k}} - R_{i\bar{k}} R_{k\bar{j}}) + R_{i\bar{j}} (R_{j\bar{i}} - \lambda g_{j\bar{i}})$$

$$= \Delta R + |\mathrm{Ric}|^2 - \lambda R. \qquad \square$$

Lemma 5.2.9. *Under (5.37), we have*

$$\frac{\partial}{\partial t} R_{i\bar{j}k\bar{l}} = \Delta R_{i\bar{j}k\bar{l}} + R_{i\bar{j}p\bar{q}} R_{q\bar{p}k\bar{l}} + R_{i\bar{l}p\bar{q}} R_{q\bar{p}k\bar{j}} - R_{i\bar{p}k\bar{q}} R_{p\bar{j}q\bar{l}} + \lambda R_{i\bar{j}k\bar{l}}$$

$$- \frac{1}{2} (R_{i\bar{p}} R_{p\bar{j}k\bar{l}} + R_{p\bar{j}} R_{i\bar{p}k\bar{l}} + R_{k\bar{p}} R_{i\bar{j}p\bar{l}} + R_{p\bar{l}} R_{i\bar{j}k\bar{p}}).$$

Proof. From (5.8) and by using normal coordinates, we have

$$\frac{\partial}{\partial t} R_{i\bar{j}k\bar{l}} = \partial_k \partial_{\bar{l}} R_{i\bar{j}} + \lambda R_{i\bar{j}k\bar{l}} = \partial_k (\nabla_{\bar{l}} R_{i\bar{j}} + \Gamma^{\bar{p}}_{\bar{j}\bar{l}} R_{i\bar{p}}) + \lambda R_{i\bar{j}k\bar{l}}$$

$$= \nabla_k \nabla_{\bar{l}} R_{i\bar{j}} - R_{i\bar{p}} R_{p\bar{j}k\bar{l}} + \lambda R_{i\bar{j}k\bar{l}}.$$

On the other hand, by (5.15) and covariant differentiation commuting formulas (5.16)–(5.18), we obtain

$$\nabla_{\bar{p}} \nabla_p R_{i\bar{j}k\bar{l}} = \nabla_k \nabla_{\bar{l}} R_{i\bar{j}} - R_{i\bar{j}p\bar{q}} R_{q\bar{p}k\bar{l}} + R_{i\bar{p}k\bar{q}} R_{p\bar{j}q\bar{l}} - R_{i\bar{l}p\bar{q}} R_{q\bar{p}k\bar{j}} + R_{i\bar{j}p\bar{l}} R_{k\bar{p}},$$

and

$$\nabla_p \nabla_{\bar{p}} R_{i\bar{j}k\bar{l}} = \nabla_{\bar{p}} \nabla_p R_{i\bar{j}k\bar{l}} - R_{i\bar{q}} R_{q\bar{j}k\bar{l}} + R_{q\bar{j}} R_{i\bar{q}k\bar{l}} - R_{k\bar{q}} R_{i\bar{j}q\bar{l}} + R_{q\bar{l}} R_{i\bar{j}k\bar{q}}.$$

Hence,

$$\Delta R_{i\bar{j}k\bar{l}} = \frac{1}{2} \left(\nabla_p \nabla_{\bar{p}} + \nabla_{\bar{p}} \nabla_p \right) R_{i\bar{j}k\bar{l}}$$

$$= \nabla_k \nabla_{\bar{l}} R_{i\bar{j}} - R_{i\bar{j}p\bar{q}} R_{q\bar{p}k\bar{l}} + R_{i\bar{p}k\bar{q}} R_{p\bar{j}q\bar{l}} - R_{i\bar{l}p\bar{q}} R_{q\bar{p}k\bar{j}}$$

$$+ \frac{1}{2} (-R_{i\bar{p}} R_{p\bar{j}k\bar{l}} + R_{p\bar{j}} R_{i\bar{p}k\bar{l}} + R_{k\bar{p}} R_{i\bar{j}p\bar{l}} + R_{p\bar{l}} R_{i\bar{j}k\bar{p}}),$$

and Lemma 5.2.9 follows.   □

Remark 5.2.1. Clearly, the Ricci evolution equation (5.38) is also a consequence of Lemma 5.2.9, but the proof in Lemma 5.2.8 is more direct and easier.

The Ricci flow in general seems to prefer positive curvatures: positive Ricci curvature is preserved in three-dimension [Ham82]; positive scalar curvature, positive curvature operator [Ham86] and positive isotropic curvature [BS09, Ng07] are preserved in all dimensions. Here we present a proof of Mok's theorem that positive bisectional curvature is preserved under KRF.

Theorem 5.2.10 ([Mok88]). *Let* $(X^n, \tilde{g})$ *be a compact Kähler manifold of nonnegative holomorphic bisectional curvature, and let* $g_{i\bar{j}}(t)$ *be a solution to the KRF (5.28) or NKRF (5.33) on* $X^n \times [0, T)$. *Then, for* $t > 0$, $g_{i\bar{j}}(t)$ *also has nonnegative holomorphic bisectional curvature. Furthermore, if the holomorphic* *bisectional curvature is positive at one point at $t = 0$, then $g_{i\bar{j}}(t)$ has positive holomorphic bisectional curvature at all points for $t > 0$.*

Proof. Let us denote by

$$F_{i\bar{j}k\bar{l}} =: R_{i\bar{j}p\bar{q}}R_{q\bar{p}k\bar{l}} - R_{i\bar{p}k\bar{q}}R_{p\bar{j}q\bar{l}} + R_{i\bar{l}p\bar{q}}R_{q\bar{p}k\bar{j}} + \lambda R_{i\bar{j}k\bar{l}}$$

$$- \frac{1}{2}(R_{i\bar{p}}R_{p\bar{j}k\bar{l}} + R_{p\bar{j}}R_{i\bar{p}k\bar{l}} + R_{k\bar{p}}R_{i\bar{j}p\bar{l}} + R_{p\bar{l}}R_{i\bar{j}k\bar{p}})$$

so that by Lemma 5.2.9

$$\frac{\partial}{\partial t}R_{i\bar{j}k\bar{l}} = \Delta R_{i\bar{j}k\bar{l}} + F_{i\bar{j}k\bar{l}}.$$

By a version of Hamilton's strong tensor maximum principle (cf. [Bando84]), it suffices to show that the following "null-vector condition" holds: for any $(1,0)$ vectors $V = (v^i)$ and $W = (w^i)$, we have

$$F_{i\bar{j}k\bar{l}}v^i v^{\bar{j}}w^k w^{\bar{l}} \geq 0 \quad \text{whenever} \quad R_{i\bar{j}k\bar{l}}v^i v^{\bar{j}}w^k w^{\bar{l}} = 0, \qquad \text{(NVC)}$$

or simply,

$$F_{V\bar{V}W\bar{W}} =: F(V, \bar{V}, W, \bar{W}) \geq 0 \quad \text{whenever} \quad R_{V\bar{V}W\bar{W}} =: \text{Rm}(V, \bar{V}, W, \bar{W}) = 0.$$

Claim 2.1. If $R_{V\bar{V}W\bar{W}} = 0$, then for any $(1,0)$ vector Z, we have

$$R_{V\bar{Z}W\bar{W}} = R_{V\bar{V}W\bar{Z}} = 0.$$

For real parameter $s \in \mathbb{R}$, consider

$$G(s) = \text{Rm}(V + sZ, \bar{V} + s\bar{Z}, W, \bar{W}).$$

Since the bisectional curvature is nonnegative and $R_{V\bar{V}W\bar{W}} = 0$, it follows that $G'(0) = 0$ which implies that

$$\text{Re}\,(R_{V\bar{Z}W\bar{W}}) = 0.$$

Suppose $R_{V\bar{Z}W\bar{W}} \neq 0$, and let $R_{V\bar{Z}W\bar{W}} = |R_{V\bar{Z}W\bar{W}}|e^{\sqrt{-1}\theta}$. Then, replacing Z by $e^{-\sqrt{-1}\theta}Z$ in the above, we get

$$0 = \text{Re}\,(e^{-\sqrt{-1}\theta}R_{V\bar{Z}W\bar{W}}) = |R_{V\bar{Z}W\bar{W}}|,$$

a contradiction. Thus, we must have

$$R_{V\bar{Z}W\bar{W}} = 0.$$

Similarly, we have $R_{V\bar{V}W\bar{Z}} = 0$.

By Claim 2.1, we see that if $R_{V\bar{V}W\bar{W}} = 0$ then

$$F_{V\bar{V}W\bar{W}} = R_{V\bar{V}Y\bar{Z}}R_{Z\bar{Y}W\bar{W}} - |R_{V\bar{Y}W\bar{Z}}|^2 + |R_{V\bar{W}Y\bar{Z}}|^2.$$

Therefore, (NVC) follows immediately from the following

Claim 2.2. Suppose $R_{V\bar{V}W\bar{W}} = 0$. Then, for any $(1,0)$ vectors Y and Z,

$$R_{V\bar{V}Y\bar{Z}}R_{Z\bar{Y}W\bar{W}} \geq |R_{V\bar{Y}W\bar{Z}}|^2.$$

First of all, consider

$$H(s) = \text{Rm}(V + sY, \bar{V} + s\bar{Y}, W + sZ, \bar{W} + s\bar{Z})$$
$$= s^2 \left(R_{V\bar{V}Z\bar{Z}} + R_{Y\bar{Y}W\bar{W}} + R_{V\bar{Y}W\bar{Z}} + R_{Y\bar{V}ZW} + R_{V\bar{Y}Z\bar{W}} + R_{Y\bar{V}W\bar{Z}} \right)$$
$$+ O(s^3).$$

Here we have used Claim 2.1.

Since $H(s) \geq 0$ and $H(0) = 0$, we have $H''(0) \geq 0$. Hence, by taking $Y = \zeta^k e_k$ and $Z = \eta^\ell e_\ell$ with respect to any basis $\{e_1, \cdots e_n\}$, we obtain a real semi-positive definite bilinear form $Q(Y, Z)$:

$$0 \leq Q(Y, Z) =: R_{V\bar{V}Z\bar{Z}} + R_{Y\bar{Y}W\bar{W}} + R_{V\bar{Y}W\bar{Z}} + R_{Y\bar{V}ZW} + R_{V\bar{Y}Z\bar{W}} + R_{Y\bar{V}W\bar{Z}}$$
$$= R_{V\bar{V}e_k\bar{e}_\ell}\eta^k\eta^{\bar{\ell}} + R_{e_k\bar{e}_\ell W\bar{W}}\zeta^k\zeta^{\bar{\ell}} + R_{V\bar{e}_k W\bar{e}_\ell}\zeta^{\bar{k}}\eta^{\bar{\ell}} + R_{e_k\bar{V}e_\ell\bar{W}}\zeta^k\eta^\ell$$
$$+ R_{V\bar{e}_k e_\ell\bar{W}}\zeta^{\bar{k}}\eta^\ell + R_{e_k\bar{V}W\bar{e}_\ell}\zeta^k\eta^{\bar{\ell}} \qquad\qquad \square$$

Next, we need a useful linear algebra fact (cf. Lemma 4.1 in [Cao92]):

Lemma 5.2.11. *Let A and C be two $m \times m$ real symmetric semi-positive definite matrices, and let B be a real $m \times m$ matrix such that the $2m \times 2m$ real symmetric matrix*

$$G_1 = \begin{pmatrix} A & B \\ B^T & C \end{pmatrix}$$

is semi-positive definite. Then, we have

$$\text{Tr}(AC) \geq |B|^2.$$

Proof. Consider the associated matrix

$$G_2 = \begin{pmatrix} C & -B \\ -B^T & A \end{pmatrix}.$$

It is clear that G_2 is also symmetric and semi-positive definite. Hence, we get

$$\text{Tr}(G_1 G_2) \geq 0.$$

However,

$$G_1 G_2 = \begin{pmatrix} AC - BB^T & BA - AB \\ B^T C - CB^T & CA - B^T B \end{pmatrix}.$$

Therefore,

$$\text{Tr}(AC) - |B|^2 = \frac{1}{2}\text{Tr}(G_1 G_2) \geq 0. \qquad \square$$

As a special case, by taking

$$G_1 = \begin{pmatrix} ReA & -ImA & Re(B+D)^T & -Im(B+D)^T \\ ImA & ReA & Im(B-D)^T & Re(B-D)^T \\ Re(B+D) & Im(B-D) & ReC & -ImC \\ -Im(B+D) & Re(B-D) & ImC & ReC \end{pmatrix},$$

we immediately obtain the following (see [Cao92, Lemma 4.2])

Corollary 5.2.12. *Let A, B, C, D be complex matrices with A and C being Hermitian. Suppose that the (real) quadratic form*

$$\sum A_{k\bar{l}}\eta^k\overline{\eta^l} + C_{k\bar{l}}\zeta^k\overline{\zeta^l} + 2Re(B_{k\bar{l}}\eta^k\overline{\zeta^l}) + 2Re(D_{kl}\eta^k\zeta^l), \quad \eta, \zeta \in \mathbb{C}^n,$$

is semi-positive definite. Then we have

$$Tr(AC) \geq |B|^2 + |D|^2,$$

i.e.

$$\sum A_{k\bar{l}}C_{l\bar{k}} \geq \sum |B_{k\bar{l}}|^2 + |D_{kl}|^2.$$

Now, by applying Corollary 5.2.12 to the above semi-positive definite real bi-linear form Q, one gets

$$R_{V\bar{V}Y\bar{Z}}R_{Z\bar{Y}W\bar{W}} \geq |R_{V\bar{Y}W\bar{Z}}|^2 + |R_{V\bar{W}Y\bar{Z}}|^2.$$

We have thus proved (NVC), which concludes the proof of Theorem 5.2.10.

Remark 5.2.13. S. Bando [Bando84] first proved Theorem 5.2.10 for $n = 3$, and W.-X. Shi [Shi97] extended Theorem 5.2.10 to the complete noncompact case with bounded curvature tensor.

Furthermore, by slightly modifying the above proof of Theorem 5.2.10, R. Hamilton and the author (Cao and Hamilton, 1992, unpublished work) observed in 1992 at IAS that nonnegative *holomorphic orthogonal bisectional curvature*, $\mathrm{Rm}(V, \bar{V}, W, \bar{W}) \geq 0$ whenever $V \perp W$, is also preserved under KRF. For the reader's convenience, we provide the proof below.

Theorem 5.2.14 (Cao-Hamilton). *Let $g_{i\bar{j}}(t)$ be a solution to the KRF (5.28) on a complete Kähler manifold with bounded curvature. If $g_{i\bar{j}}(0)$ has nonnegative holomorphic orthogonal bisectional curvature, then it remains so for $g_{i\bar{j}}(t)$ for $t > 0$.*

Proof. First of all, by using a certain special evolving orthonormal frame $\{e_\alpha\}$ under KRF (5.28) similarly as in [Ham86] (see also [Shi97, Sect. 5]), one obtains the simplified evolution equation

$$\frac{\partial}{\partial t} R_{\alpha\bar{\beta}\gamma\bar{\delta}} = \Delta R_{\alpha\bar{\beta}\gamma\bar{\delta}} + R_{\alpha\bar{\beta}\mu\bar{\nu}} R_{\nu\bar{\mu}\gamma\bar{\delta}} + R_{\alpha\bar{\delta}\mu\bar{\nu}} R_{\nu\bar{\mu}\gamma\bar{\beta}} - R_{\alpha\bar{\mu}\gamma\bar{\nu}} R_{\mu\bar{\beta}\nu\bar{\delta}}, \qquad (5.42)$$

where $R_{\alpha\bar{\beta}\gamma\bar{\delta}}$ is the Riemannian curvature tensor components with respect to the evolving frame $\{e_\alpha\}$.

Again, by Hamilton's tensor maximal principle, it suffices to check the corresponding null-vector condition:

$$G_{\alpha\bar{\alpha}\beta\bar{\beta}} \geq 0, \text{ whenever } R_{\alpha\bar{\alpha}\beta\bar{\beta}} = 0 \text{ for any } e_\alpha \perp e_\beta, \qquad (\text{NVC}')$$

where

$$G_{\alpha\bar{\beta}\gamma\bar{\delta}} = R_{\alpha\bar{\beta}\mu\bar{\nu}} R_{\nu\bar{\mu}\gamma\bar{\delta}} + R_{\alpha\bar{\delta}\mu\bar{\nu}} R_{\nu\bar{\mu}\gamma\bar{\beta}} - R_{\alpha\bar{\mu}\gamma\bar{\nu}} R_{\mu\bar{\beta}\nu\bar{\delta}}.$$

Now, without loss of generality, we assume $R_{1\bar{1}2\bar{2}} = 0$ for a pair of unit $(1, 0)$-vectors $e_1 \perp e_2$. Then we need to show $G_{1\bar{1}2\bar{2}} \geq 0$.

Claim 2.3. If $e_i \perp e_1$, then $R_{1\bar{1}2\bar{i}} = 0$, similarly, if $e_i \perp e_2$, then $R_{2\bar{2}1\bar{i}} = 0$.

The first statement in Claim 2.3 follows from the simple fact that if $e_i \perp e_1$, then $\mathrm{Rm}(e_1, \overline{e_1}, e_2 + se_i, \overline{e_2 + se_i}) \geq 0$ for arbitrary complex number s. The proof of second statement is similar.

Claim 2.4. $R_{1\bar{2}1\bar{1}} = R_{1\bar{2}2\bar{2}}$.

Note that $(e_1 + se_2) \perp (e_2 - \bar{s}e_1)$ for any complex number s, hence

$$\mathrm{Rm}(e_1 + se_2, \overline{e_1 + se_2}, e_2 - \bar{s}e_1, \overline{e_2 - s\overline{e_1}}) \geq 0.$$

Again its first order derivative vanishes at point $s = 0$, and Claim 2.4 follows.

Claim 2.5. $G_{1\bar{1}2\bar{2}} = R_{1\bar{1}i\bar{j}}R_{j\bar{i}2\bar{2}} - |R_{1\bar{1}2\bar{j}}|^2 + |R_{1\bar{2}\mu\bar{v}}|^2$, where $3 \le i, j \le n$ and $1 \le \mu, v \le n$.

From the definition of $G_{1\bar{1}2\bar{2}}$, the assumption that $R_{1\bar{1}2\bar{2}} = 0$ and the above two claims, we have:

$$
\begin{aligned}
G_{1\bar{1}2\bar{2}} &= R_{1\bar{2}\mu\bar{v}}R_{v\bar{\mu}2\bar{1}} + R_{1\bar{1}\mu\bar{v}}R_{v\bar{\mu}2\bar{2}} - R_{1\bar{\mu}2\bar{v}}R_{\mu\bar{2}v\bar{1}} \\
&= R_{1\bar{2}\mu\bar{v}}R_{v\bar{\mu}2\bar{1}} \\
&\quad + R_{1\bar{1}i\bar{j}}R_{j\bar{i}2\bar{2}} + R_{1\bar{1}1\bar{2}}R_{2\bar{1}2\bar{2}} + R_{1\bar{1}2\bar{1}}R_{1\bar{2}2\bar{2}} \\
&\quad - R_{1\bar{1}2\bar{j}}R_{i\bar{2}j\bar{1}} - R_{1\bar{1}2\bar{1}}R_{1\bar{2}1\bar{1}} - R_{1\bar{2}2\bar{2}}R_{2\bar{2}2\bar{1}} \\
&= R_{1\bar{1}i\bar{j}}R_{j\bar{i}2\bar{2}} - |R_{1\bar{1}2\bar{j}}|^2 + |R_{1\bar{2}\mu\bar{v}}|^2.
\end{aligned}
$$

Now for arbitrary $(1,0)$-vectors $X, Y \perp e_1, e_2$ and real number s, we have the following:

$$
(e_\alpha + sX) \perp \left(e_\beta + sY - s^2 e_\alpha < \bar{X}, Y >\right).
$$

Thus using Claim 2.3, we have

$$
\begin{aligned}
0 &\le \mathrm{Rm}(e_1 + sX, \overline{e_1} + s\bar{X}, e_2 + sY - s^2 e_1 < \bar{X}, Y >, \overline{e_2} + s\bar{Y} - s^2 \overline{e_1} < X, \bar{Y} >) \\
&= s^2 \left(R_{2\bar{2}X\bar{X}} + R_{1\bar{1}Y\bar{Y}} + 2\mathrm{Re}R_{X\bar{1}Y\bar{2}} + 2\mathrm{Re}(R_{X\bar{Y}2\bar{1}} - R_{1\bar{1}2\bar{1}} < X, \bar{Y} >)\right) \\
&\quad + O(s^3)
\end{aligned}
$$

Hence, for all s, X and Y,

$$
\left(R_{2\bar{2}X\bar{X}} + R_{1\bar{1}Y\bar{Y}} + 2\mathrm{Re}R_{X\bar{1}Y\bar{2}} + 2\mathrm{Re}(R_{X\bar{Y}2\bar{1}} - R_{1\bar{1}2\bar{1}} < X, \bar{Y} >)\right) \ge 0
$$

By using Corollary 5.2.12 again, we obtain

$$
R_{1\bar{1}i\bar{j}}R_{j\bar{i}2\bar{2}} \ge |R_{i\bar{1}j\bar{2}}|^2 + |R_{j\bar{2}2\bar{1}} - R_{1\bar{1}2\bar{1}}g_{i\bar{j}}|^2.
$$

This together with Claim 2.5 implies that $G_{1\bar{1}2\bar{2}} \ge 0$. The proof of Theorem 5.2.14 is completed. $\qquad\square$

Remark 5.2.2. Wilking [Wil10] has provided a nice uniform Lie Algebra approach treating all known nonnegativity curvature conditions preserved under the Ricci flow and KRF so far, including nonnegative bisectional curvature and nonnegative orthogonal bisectional curvature.

5.3   The Li–Yau–Hamilton Inequalities for KRF

In [LiYau86], Li–Yau developed a fundamental gradient estimate, now called
Li–Yau estimate (aka differential Harnack inequality), for positive solutions to the
heat equation on a complete Riemannian manifold with nonnegative Ricci curvature.
They used it to derive the Harnack inequality for such solutions by a path integration.
Shortly after, based on a suggestion of Yau, Hamilton [Ham88] derived a similar
estimate for the scalar curvature of solutions to the Ricci flow on Riemann surfaces
with positive curvature. Hamilton subsequently obtained a matrix version of the
Li–Yau estimate [Ham93a] for solutions to the Ricci flow with positive curvature
operator in all dimensions. This matrix version of the Li–Yau estimate is now called
Li–Yau–Hamilton estimate, and it played a central role in the analysis of formation
of singularities and the application of the Ricci flow to three-manifold topology.
Around the same time, the author derived the (matrix) Li–Yau–Hamilton estimate
for the KRF with nonnegative bisectional curvature and obtained the Harnack
inequality for the evolving scalar curvature by a similar path integration argument.
We remark that our Li–Yau–Hamilton estimate for the KRF in the non compact
case played a crucial role in the works of Chen–Tang–Zhu [CTZ04], Ni [Ni05] and
Chau–Tam [ChauT06]. The presentation here essentially follows the original papers
of Hamilton [Ham88, Ham93a, Ham93b] and Cao [Cao92, Cao97].

   We shall start by recalling the well-known Li–Yau inequality for positive
solutions to the heat equation on complete Riemannian manifolds with nonnegative
Ricci curvature, and the important observation that Li–Yau inequality becomes
equality on the standard heat kernel on the Euclidean space. Then, following
Hamilton, we show how one could derive the matrix Li–Yau–Hamilton quadratic
for the KRF from the equation of expanding Kahler–Ricci solitons. Finally we state
and sketch the matrix Li–Yau–Hamilton inequality for the KRF with nonnegative
bisectional curvature.

5.3.1   The Li–Yau Estimate for the Two-Dimensional
Ricci Flow

Let us begin by describing the Li–Yau estimate [LiYau86] for positive solutions
to the heat equation on a complete Riemannian manifold with nonnegative Ricci
curvature.

Theorem 5.3.1 ([LiYau86]). *Let* (M, g_{ij}) *be an n-dimensional complete
Riemannian manifold with nonnegative Ricci curvature. Let* $u(x, t)$ *be any positive
solution to the heat equation*

$$\frac{\partial u}{\partial t} = \Delta u \qquad on \ M \times [0, \infty).$$

Then, for all $t > 0$, we have

$$\frac{\partial u}{\partial t} - \frac{|\nabla u|^2}{u} + \frac{n}{2t}u \geq 0 \qquad on \ \ M \times (0,\infty). \tag{5.44}$$

We remark that, as observed by Hamilton (cf. [Ham93a]), one can in fact prove that for any vector field V^i on M,

$$\frac{\partial u}{\partial t} + 2\nabla u \cdot V + u|V|^2 + \frac{n}{2t}u \geq 0. \tag{5.45}$$

If we take the optimal vector field $V = -\nabla u/u$, then we recover the inequality (5.44).

Now we consider the Ricci flow on a Riemann surface. Since in (real) dimension two the Ricci curvature is given by

$$R_{ij} = \frac{1}{2}Rg_{ij},$$

the Ricci flow becomes

$$\frac{\partial g_{ij}}{\partial t} = -Rg_{ij}. \tag{5.46}$$

Now let $g_{ij}(t)$ be a complete solution of the Ricci flow (5.46) on a Riemann surface M and $0 \leq t < T$. Then the scalar curvature R evolves by the semilinear equation

$$\frac{\partial R}{\partial t} = \triangle R + R^2$$

on $M \times [0, T)$. Suppose the scalar curvature of the initial metric is bounded, nonnegative everywhere and positive somewhere. Then it follows from the maximum principle that the scalar curvature $R(x,t)$ of the evolving metric remains nonnegative. Moreover, from the standard strong maximum principle (which works in each local coordinate neighborhood), the scalar curvature is positive everywhere for $t > 0$. In [Ham88], Hamilton obtained the following Li–Yau estimate for the scalar curvature $R(x,t)$.

Theorem 5.3.2 ([Ham88]). *Let $g_{ij}(t)$ be a complete solution to the Ricci flow on a surface M. Assume the scalar curvature of the initial metric is bounded, nonnegative everywhere and positive somewhere. Then the scalar curvature $R(x,t)$ satisfies the Li–Yau estimate*

$$\frac{\partial R}{\partial t} - \frac{|\nabla R|^2}{R} + \frac{R}{t} \geq 0. \tag{5.47}$$

Proof. By the above discussion, we know $R(x,t) > 0$ for $t > 0$. If we set

$$L = \log R(x,t) \quad \text{for} \quad t > 0,$$

then

$$\frac{\partial}{\partial t} L = \frac{1}{R}(\Delta R + R^2)$$

$$= \Delta L + |\nabla L|^2 + R$$

and (5.47) is equivalent to

$$\frac{\partial L}{\partial t} - |\nabla L|^2 + \frac{1}{t} = \Delta L + R + \frac{1}{t} \geq 0.$$

Following Li–Yau [LiYau86] in the linear heat equation case, we consider the quantity

$$Q = \frac{\partial L}{\partial t} - |\nabla L|^2 = \Delta L + R.$$

Then by a direct computation,

$$\frac{\partial Q}{\partial t} = \frac{\partial}{\partial t}(\Delta L + R)$$

$$= \Delta \left(\frac{\partial L}{\partial t} \right) + R\Delta L + \frac{\partial R}{\partial t}$$

$$= \Delta Q + 2\nabla L \cdot \nabla Q + 2|\nabla^2 L|^2 + 2R(\Delta L) + R^2$$

$$\geq \Delta Q + 2\nabla L \cdot \nabla Q + Q^2.$$

So we get

$$\frac{\partial}{\partial t}\left(Q + \frac{1}{t} \right) \geq \Delta \left(Q + \frac{1}{t} \right) + 2\nabla L \cdot \nabla \left(Q + \frac{1}{t} \right) + \left(Q - \frac{1}{t} \right)\left(Q + \frac{1}{t} \right).$$

Hence by the maximum principle argument, we obtain

$$Q + \frac{1}{t} \geq 0.$$

This proves the theorem.   □

5.3.2 Li–Yau Estimate and Expanding Solitons

To prove inequality (5.47) for the scalar curvature of solutions to the Ricci flow in higher dimensions is not so simple. It turns out that one does not get inequality (5.47) directly, but rather indirectly as the trace of certain matrix estimate when either curvature operator (in the Riemannian case) or bisectional curvature (in the Kähler case) is nonnegative. The key ingredient in formulating this matrix version is an important observation by Hamilton that the Li–Yau inequality, as well as its matrix version, becomes equality on the expanding solitons which he first discovered for the case of the heat equation on $\mathbb{R}^n$. This led him and the author to formulate and prove the matrix differential Harnack inequality, now called Li–Yau–Hamilton estimates, for the Ricci flow in higher dimensions [Ham93a, Ham93b] and the Kähler–Ricci flow [Cao92, Cao97] respectively.

To illustrate, let us examine the heat equation case first. Consider the heat kernel

$$u(x,t) = (4\pi t)^{-n/2} e^{-|x|^2/4t}, \tag{5.48}$$

which can be considered as an expanding soliton solution for the standard heat equation on $\mathbb{R}^n$.

Differentiating the function u once, we get

$$\nabla_j u = -u\frac{x_j}{2t} \quad \text{or} \quad \nabla_j u + uV_j = 0, \tag{5.49}$$

where

$$V_j = \frac{x_j}{2t} = -\frac{\nabla_j u}{u}.$$

Differentiating (5.49) again, we have

$$\nabla_i \nabla_j u + \nabla_i u V_j + \frac{u}{2t}\delta_{ij} = 0. \tag{5.50}$$

To make the expression in (5.50) symmetric in i, j, we multiply V_i to (5.49) and add to (5.50) and obtain

$$\nabla_i \nabla_j u + \nabla_i u V_j + \nabla_j u V_i + uV_i V_j + \frac{u}{2t}\delta_{ij} = 0. \tag{5.51}$$

Taking the trace in (5.51) and using the equation $\partial u/\partial t = \Delta u$, we arrive at

$$\frac{\partial u}{\partial t} + 2\nabla u \cdot V + u|V|^2 + \frac{n}{2t}u = 0,$$

which shows that the Li–Yau inequality (5.44) becomes an equality on our expanding soliton solution u! Moreover, we even have the matrix identity (5.51).

Based on the above observation and in a similar process, Hamilton [Ham93a] found a matrix quantity, which vanishes on expanding gradient Ricci solitons and is nonnegative for any solution to the Ricci flow with nonnegative curvature operator. At the same time, the author [Cao92] (see also [Cao97]) proved the Li–Yau–Hamilton estimate for the Kähler–Ricci flow with nonnegative bisectional curvature, see below.

To formulate the Li–Yau–Hamilton quadric, let us consider a homothetically expanding gradient Kähler–Ricci soliton g satisfying

$$R_{i\bar{j}} + \frac{1}{t}g_{i\bar{j}} = \nabla_i V_{\bar{j}}, \qquad \nabla_i V_j = 0 \tag{5.52}$$

with $V_i = \nabla_i f$ for some real-valued smooth function f on X. Differentiating (5.52) and commuting give the first order relations

$$\nabla_k R_{i\bar{j}} = \nabla_k \nabla_{\bar{j}} V_i - \nabla_{\bar{j}} \nabla_k V_i = -R_{k\bar{j}i\bar{p}} V_p,$$

or

$$\nabla_k R_{i\bar{j}} + R_{i\bar{j}k\bar{p}} V_p = 0, \tag{5.53}$$

and

$$\nabla_k R_{i\bar{j}} V_{\bar{k}} + R_{i\bar{j}k\bar{p}} V_p V_{\bar{k}} = 0. \tag{5.54}$$

Differentiating (5.53) again and using the first equation in (5.52), we get

$$\nabla_{\bar{l}} \nabla_k R_{i\bar{j}} + \nabla_{\bar{p}} R_{i\bar{j}k\bar{l}} V_p + R_{i\bar{j}k\bar{p}} R_{p\bar{l}} + \frac{1}{t} R_{i\bar{j}k\bar{l}} = 0. \tag{5.55}$$

Taking the trace in (5.55), we get

$$\Delta R_{i\bar{j}} + \nabla_{\bar{k}} R_{i\bar{j}} V_k + R_{i\bar{j}k\bar{l}} R_{l\bar{k}} + \frac{1}{t} R_{i\bar{j}} = 0. \tag{5.56}$$

Symmetrizing by adding (5.54) to (5.56), we arrive at

$$\Delta R_{i\bar{j}} + \nabla_k R_{i\bar{j}} V_{\bar{k}} + \nabla_{\bar{k}} R_{i\bar{j}} V_k + R_{i\bar{j}k\bar{l}} R_{l\bar{k}} + R_{i\bar{j}k\bar{l}} V_l V_{\bar{k}} + \frac{1}{t} R_{i\bar{j}} = 0,$$

or, by (5.38), equivalently

$$\frac{\partial}{\partial t} R_{i\bar{j}} + \nabla_k R_{i\bar{j}} V_{\bar{k}} + \nabla_{\bar{k}} R_{i\bar{j}} V_k + R_{i\bar{k}} R_{k\bar{j}} + R_{i\bar{j}k\bar{l}} V_l V_{\bar{k}} + \frac{1}{t} R_{i\bar{j}} = 0. \tag{5.57}$$

5.3.3 Li–Yau–Hamilton Estimates and Harnack's Inequalities

We now state the Li–Yau–Hamilton estimates and the Harnack inequalities for KRF and NKRF with nonnegative holomorphic bisectional curvature.

Theorem 5.3.3 ([Cao92, Cao97]). *Let $g_{i\bar{j}}(t)$ be a complete solution to the Kähler–Ricci flow on X^n with bounded curvature and nonnegative bisectional curvature and $0 \le t < T$. For any point $x \in X$ and any vector V in the holomorphic tangent space $T_x^{1,0}X$, let*

$$Z_{i\bar{j}} = \frac{\partial}{\partial t}R_{i\bar{j}} + R_{i\bar{k}}R_{k\bar{j}} + \nabla_k R_{i\bar{j}}V^k + \nabla_{\bar{k}} R_{i\bar{j}}V^{\bar{k}} + R_{i\bar{j}k\bar{\ell}}V^k V^{\bar{\ell}} + \frac{1}{t}R_{i\bar{j}}.$$

Then we have

$$Z_{i\bar{j}}W^i W^{\bar{j}} \ge 0$$

for all $x \in X$, $V, W \in T_x^{1,0}X$, and $t > 0$.

The proof of Theorem 5.3.3 is based on Hamilton's strong tensor maximum principle and involves a large amount of calculations. We refer the interested reader to the original papers [Cao92, Cao97] for details.

Corollary 5.3.4 ([Cao92, Cao97]). *Under the assumptions of Theorem 5.3.3, the scalar curvature R satisfies the estimate*

$$\frac{\partial R}{\partial t} + \nabla_i RV^i + \nabla_{\bar{i}} RV^{\bar{i}} + R_{i\bar{j}}V^i V^{\bar{j}} + \frac{R}{t} \ge 0 \tag{5.58}$$

for all $x \in X$ and $t > 0$. In particular,

$$\frac{\partial R}{\partial t} - \frac{|\nabla R|^2}{R} + \frac{R}{t} \ge 0. \tag{5.59}$$

Proof. The first inequality (5.58) follows by taking the trace of $Z_{i\bar{j}}$ in Theorem 5.3.3. By taking $V = -\nabla \log R$ in (5.58) and observing $R_{i\bar{j}} \le Rg_{i\bar{j}}$ (because $R_{i\bar{j}} \ge 0$), we obtain the second inequality (5.59). □

As a consequence of Corollary 5.3.4, we obtain the following Harnack inequality for the scalar curvature R by taking the Li–Yau type path integral as in [LiYau86].

Corollary 5.3.5 ([Cao92, Cao97]). *Let $g_{i\bar{j}}(t)$ be a complete solution to the Kähler–Ricci flow on X^n with bounded and nonnegative bisectional curvature. Then for any points $x_1, x_2 \in X$, and $0 < t_1 < t_2$, we have*

$$R(x_1, t_1) \le \frac{t_2}{t_1}e^{d_{t_1}(x_1, x_2)^2/4(t_2-t_1)} R(x_2, t_2).$$

Here $d_{t_1}(x_1, x_2)$ denotes the distance between x_1 and x_2 with respect to $g_{i\bar{j}}(t_1)$.

Proof. Take the geodesic path $\gamma(\tau)$, $\tau \in [t_1, t_2]$, from x_1 to x_2 at time t_1 with constant velocity $d_{t_1}(x_1, x_2)/(t_2 - t_1)$. Consider the space-time path $\eta(\tau) = (\gamma(\tau), \tau)$, $\tau \in [t_1, t_2]$. We compute

$$
\log \frac{R(x_2, t_2)}{R(x_1, t_1)} = \int_{t_1}^{t_2} \frac{d}{d\tau} \log R(\gamma(\tau), \tau) d\tau
$$

$$
= \int_{t_1}^{t_2} \frac{1}{R} \left(\frac{\partial R}{\partial \tau} + \nabla R \cdot \frac{d\gamma}{d\tau} \right) d\tau
$$

$$
\geq \int_{t_1}^{t_2} \left(\frac{\partial \log R}{\partial \tau} - |\nabla \log R|^2_{g(\tau)} - \frac{1}{4} \left| \frac{d\gamma}{d\tau} \right|^2_{g(\tau)} \right) d\tau.
$$

Then, by the Li–Yau estimate (5.59) for R in Corollary 5.3.4 and the fact that the metric is shrinking (since the Ricci curvature is nonnegative), we have

$$
\log \frac{R(x_2, t_2)}{R(x_1, t_1)} \geq \int_{t_1}^{t_2} \left(-\frac{1}{\tau} - \frac{1}{4} \left| \frac{d\gamma}{d\tau} \right|^2_{g(t_1)} \right) d\tau
$$

$$
= \log \frac{t_1}{t_2} - \frac{d_{t_1}(x_1, x_2)^2}{4(t_2 - t_1)}.
$$

Now the desired Harnack inequality follows by exponentiating.   □

Finally, we can convert Corollaries 5.3.4 and 5.3.5 to the NKRF case and yield the following Li–Yau type estimate and Harnack's inequality.

Theorem 5.3.6 ([Cao92]). *Let $g_{i\bar{j}}(t)$ be a solution to NKRF on $X^n \times [0, \infty)$ with nonnegative bisectional curvature. Then, the scalar curvature R satisfies*

(a) the Li–Yau type estimate: for any $t > 0$ and $x \in X$,

$$
\frac{\partial R}{\partial t} - \frac{|\nabla R|^2}{R} + \frac{R}{1 - e^{-t}} \geq 0; \tag{5.60}
$$

(b) the Harnack inequality: for any $0 < t_1 < t_2$ and any $x, y \in X$,

$$
R(x, t_1) \leq \frac{e^{t_2} - 1}{e^{t_1} - 1} \exp\{e^{t_2 - t_1} \frac{d^2_{t_1}(x, y)}{4(t_2 - t_1)}\} R(y, t_2), \tag{5.61}
$$

Proof. Part (a): Let $\hat{g}_{i\bar{j}}(s)$ be the associated solution to KRF on $X \times [0, 1)$. By Lemma 5.2.4, Corollaries 5.2.5 and 5.3.4, we have

$$
R = (1 - s)\hat{R}, \qquad 1 - e^{-t} = s,
$$

and

$$\frac{\partial \hat{R}}{\partial s} - \frac{|\nabla \hat{R}|_{\hat{g}}^2}{\hat{R}} + \frac{\hat{R}}{s} \geq 0.$$

It is then easy to check that they are translated into (5.60).

Part (b): By the Li–Yau path integration argument as in the proof of Corollary 5.3.5 but use (5.60) instead, we get

$$\log \frac{R(y, t_2)}{R(x, t_1)} \geq \int_{t_1}^{t_2} \left(-\frac{1}{1 - e^{-\tau}} - \frac{1}{4} \left| \frac{d\gamma}{d\tau} \right|_{g(\tau)}^2 \right) d\tau$$

$$= \log \frac{e^{t_1} - 1}{e^{t_2} - 1} - \frac{1}{4} \Delta(x, t_1; y, t_2).$$

where

$$\Delta(x, t_1; y, t_2) = \inf_{\gamma} \int_{t_1}^{t_2} |\gamma'(\tau)|_{g(\tau)}^2 d\tau. \tag{5.62}$$

But, the NKRF equation and the assumption of $\mathrm{Ric}_g \geq 0$ imply that, for $t_1 < t_2$,

$$g(t_2) \leq e^{t_2 - t_1} g(t_1).$$

Hence,

$$\Delta(x, t_1; y, t_2) \leq e^{t_2 - t_1} \frac{d_{t_1}^2(x, y)}{(t_2 - t_1)}.$$

Therefore,

$$\log \frac{R(y, t_2)}{R(x, t_1)} \geq \log \frac{e^{t_1} - 1}{e^{t_2} - 1} - e^{t_2 - t_1} \frac{d_{t_1}^2(x, y)}{4(t_2 - t_1)}. \qquad \square$$

5.4   Perelman's Entropy and Noncollapsing Theorems

In this section, we review Perelman's $\mathcal{W}$-functional and the associated μ-entropy. We show that the μ-entropy is monotone under the Ricci flow and use this important fact to prove a strong κ-noncollapsing theorem for the Ricci flow on compact Riemannian manifolds. These results and the ideas in the proof play a crucial role in the next two sections when we discuss the uniform estimates on the diameter and the scalar curvature of the NKFR.

5.4.1   Perelman's $\mathcal{W}$-Functional and μ-Entropy for the Ricci Flow

Let M be a compact n-dimensional manifold. Consider the following functional, due to Perelman [Per02],

$$\mathcal{W}(g_{ij}, f, \tau) = \int_M [\tau(R + |\nabla f|^2) + f - n](4\pi\tau)^{-\frac{n}{2}} e^{-f} dV \qquad (5.63)$$

under the constraint

$$(4\pi\tau)^{-\frac{n}{2}} \int_M e^{-f} dV = 1. \qquad (5.64)$$

Here g_{ij} is any given Riemannian metric, f is any smooth function on M, and τ is a positive scale parameter. Clearly the functional $\mathcal{W}$ is invariant under simultaneous scaling of τ and g_{ij} (or equivalently the parabolic scaling), and invariant under diffeomorphism. Namely, for any positive number $a > 0$ and any diffeomorphism $\varphi \in \text{Diff}(M^n)$,

$$\mathcal{W}(\varphi^* g_{ij}, \varphi^* f, \tau) = \mathcal{W}(g_{ij}, f, \tau) \quad \text{and} \quad \mathcal{W}(a g_{ij}, f, a\tau) = \mathcal{W}(g_{ij}, f, \tau). \qquad (5.65)$$

In [Per02] Perelman derived the following first variation formula (see also [CZ06])

Lemma 5.4.1 ([Per02]).   *If* $v_{ij} = \delta g_{ij}$, $h = \delta f$, *and* $\eta = \delta\tau$, *then*

$$\delta\mathcal{W}(v_{ij}, h, \eta)$$

$$= \int_M -\tau v_{ij} \left(R_{ij} + \nabla_i \nabla_j f - \frac{1}{2\tau} g_{ij} \right) (4\pi\tau)^{-\frac{n}{2}} e^{-f} dV$$

$$+ \int_M \left(\frac{v}{2} - h - \frac{n}{2\tau}\eta \right) [\tau(R + 2\Delta f - |\nabla f|^2) + f - n - 1](4\pi\tau)^{-\frac{n}{2}} e^{-f} dV$$

$$+ \int_M \eta \left(R + |\nabla f|^2 - \frac{n}{2\tau} \right) (4\pi\tau)^{-\frac{n}{2}} e^{-f} dV.$$

Here $v = g^{ij} v_{ij}$.

By Lemma 5.4.1 and direct computations (cf. [Per02, CZ06]), one obtains

Lemma 5.4.2 ([Per02]).   *If* $g_{ij}(t)$, $f(t)$ *and* $\tau(t)$ *evolve according to the system*

$$\begin{cases} \dfrac{\partial g_{ij}}{\partial t} = -2R_{ij}, \\[2mm] \dfrac{\partial f}{\partial t} = -\Delta f + |\nabla f|^2 - R + \dfrac{n}{2\tau}, \\[2mm] \dfrac{\partial \tau}{\partial t} = -1, \end{cases}$$

then

$$\frac{d}{dt}\mathcal{W}(g_{ij}(t), f(t), \tau(t)) = \int_M 2\tau \left| R_{ij} + \nabla_i \nabla_j f - \frac{1}{2\tau} g_{ij} \right|^2 (4\pi\tau)^{-\frac{n}{2}} e^{-f} dV,$$

and $\int_M (4\pi\tau)^{-\frac{n}{2}} e^{-f} dV$ is constant. In particular $\mathcal{W}(g_{ij}(t), f(t), \tau(t))$ is nondecreasing in time and the monotonicity is strict unless we are on a shrinking gradient soliton.

Now we define

$$\mu(g_{ij}, \tau) = \inf \left\{ \mathcal{W}(g_{ij}, f, \tau) \mid f \in C^\infty(M), \frac{1}{(4\pi\tau)^{n/2}} \int_M e^{-f} dV = 1 \right\}.$$
(5.66)

Note that if we set $u = e^{-f/2}$, then the functional $\mathcal{W}$ can be expressed as

$$\mathcal{W} = \mathcal{W}(g_{ij}, u, \tau) = (4\pi\tau)^{-\frac{n}{2}} \int_M [\tau(Ru^2 + 4|\nabla u|^2) - u^2 \log u^2 - nu^2] dV$$
(5.67)

and the constraint (5.64) becomes

$$(4\pi\tau)^{-\frac{n}{2}} \int_M u^2 dV = 1.$$
(5.68)

Thus $\mu(g_{ij}, \tau)$ corresponds to the best constant of a logarithmic Sobolev inequality. Since the non-quadratic term is subcritical (in view of Sobolev exponent), it is rather straightforward to show that

$$\inf\left\{ (4\pi\tau)^{-\frac{n}{2}} \int_M [\tau(4|\nabla u|^2 + Ru^2) - u^2 \log u^2 - nu^2] dV : (4\pi\tau)^{-\frac{n}{2}} \int_M u^2 dV = 1 \right\}$$

is achieved by some nonnegative function $u \in H^1(M)$ which satisfies the Euler–Lagrange equation

$$\tau(-4\Delta u + Ru) - 2u \log u - nu = \mu(g_{ij}, \tau)u.$$

One can further show that u is positive (see [Rot81]). Then the standard regularity theory of elliptic PDEs shows that u is smooth. We refer the reader to Rothaus [Rot81] for more details. It follows that $\mu(g_{ij}, \tau)$ is achieved by a minimizer f satisfying the nonlinear equation

$$\tau(2\Delta f - |\nabla f|^2 + R) + f - n = \mu(g_{ij}, \tau).$$
(5.69)

It turns out that the μ-entropy has the following important monotonicity property under the Ricci flow:

Proposition 5.4.3 ([Per02]). *Let $g_{ij}(t)$ be a solution to the Ricci flow*

$$\frac{\partial g_{ij}}{\partial t} = -2R_{ij}$$

on $M^n \times [0, T)$ with $0 < T < \infty$, then $\mu(g_{ij}(t), T_0 - t)$ is nondecreasing along the Ricci flow for any $T_0 \geq T$; moreover, the monotonicity is strict unless we are on a shrinking gradient soliton.

Proof. Fix any time t_0, let f_0 be a minimizer of $\mu(g_{ij}(t_0), T_0 - t_0)$. Note that the backward heat equation

$$\frac{\partial f}{\partial t} = -\Delta f + |\nabla f|^2 - R + \frac{n}{2\tau}$$

is equivalent to the linear equation

$$\frac{\partial}{\partial t}((4\pi\tau)^{-\frac{n}{2}}e^{-f}) = -\Delta((4\pi\tau)^{-\frac{n}{2}}e^{-f}) + R((4\pi\tau)^{-\frac{n}{2}}e^{-f}).$$

Thus we can solve the backward heat equation of f with $f|_{t=t_0} = f_0$ to obtain $f(t)$ for $t \in [0, t_0]$, satisfying constraint (5.64). Then, for $t \leq t_0$, it follows from Lemma 5.4.2 that

$$\mu(g_{ij}(t), T_0 - t) \leq \mathcal{W}(g_{ij}(t), f(t), T_0 - t)$$
$$\leq \mathcal{W}(g_{ij}(t_0), f(t_0), T_0 - t_0)$$
$$= \mu(g_{ij}(t_0), T_0 - t_0),$$

and the second inequality is strict unless we are on a shrinking gradient soliton. $\quad\square$

5.4.2 Strong κ-Noncollapsing of the Ricci Flow

We now apply the monotonicity of the μ-entropy in Proposition 5.4.3 to prove a strong version of Perelman's **no local collapsing theorem**, which is extremely important because it gives a local injectivity radius estimate in terms of the local curvature bound.

Definition 5.4.4. Let $g_{ij}(t), 0 \leq t < T$, be a solution to the Ricci flow on an n-dimensional manifold M, and let κ, r be two positive constants. We say that the solution $g_{ij}(t)$ is κ-**noncollapsed** at $(x_0, t_0) \in M \times [0, T)$ on the scale r if we have

$$V_{t_0}(x_0, r) \geq \kappa r^n,$$

whenever

$$|\mathrm{Rm}|(x, t_0) \le r^{-2}$$

for all $x \in B_{t_0}(x_0, r)$. Here $V_{t_0}(x_0, r)$ denotes the volume with respect to $g_{ij}(t_0)$ of the geodesic ball $B_{t_0}(x_0, r)$ centered at $x_0 \in M$ and of radius r with respect to the metric $g_{ij}(t_0)$.

Remark 5.4.1. In [Per02], Perelman also defined κ-noncollapsing by requiring the curvature bound $|\mathrm{Rm}|(x, t) \le r^{-2}$ on the (backward) parabolic cylinder $B_{t_0}(x_0, r) \times [t_0 - r^2, t_0]$.

The following result was proved in [CZ06] (cf. Theorem 3.3.3 in [CZ06])).

Theorem 5.4.5 (Strong no local collapsing theorem). *Let M be a compact Riemannian manifold, and let $g_{ij}(t)$ be a solution to the Ricci flow on $M^n \times [0, T)$ with $0 < T < +\infty$. Then there exists a positive constant κ, depending only the initial metric g_0 and T, such that $g_{ij}(t)$ is κ-noncollapsed at very point $(x_0, t_0) \in M \times [0, T)$ on all scales less than $\sqrt{T}$. In fact, for any $(x_0, t_0) \in M \times [0, T)$ and $0 < r \le \sqrt{T}$ we have*

$$V_{t_0}(x_0, r) \ge \kappa r^n,$$

whenever

$$R(\cdot, t_0) \le r^{-2} \quad on \ B_{t_0}(x_0, r).$$

Proof. Recall that

$$\mu(g_{ij}, \tau) = \inf \left\{ \mathcal{W}(g_{ij}, u, \tau) \ \middle| \ \int_M (4\pi\tau)^{-\frac{n}{2}} u^2 dV = 1 \right\}.$$

where,

$$\mathcal{W}(g_{ij}, u, \tau) = (4\pi\tau)^{-\frac{n}{2}} \int_M [\tau(Ru^2 + 4|\nabla u|^2) - u^2 \log u^2 - nu^2] dV.$$

Set

$$\mu_0 = \inf_{0 \le \tau \le 2T} \mu(g_{ij}(0), \tau) > -\infty. \tag{5.70}$$

By the monotonicity of $\mu(g_{ij}(t), \tau - t)$ in Proposition 5.4.3, we have

$$\mu_0 \le \mu(g_{ij}(0), t_0 + r^2) \le \mu(g_{ij}(t_0), r^2) \tag{5.71}$$

for $t_0 < T$ and $r^2 \le T$.

Take a smooth cut-off function $\zeta(s)$, $0 \leq \zeta \leq 1$, such that

$$\zeta(s) = \begin{cases} 1, & |s| \leq 1/2, \\ 0, & |s| \geq 1 \end{cases}$$

and $|\zeta'| \leq 2$ everywhere. Define a test function $u(x)$ on M by

$$u(x) = e^{L/2}\zeta\left(\frac{d_{t_0}(x_0, x)}{r}\right),$$

where the constant L is chosen so that

$$(4\pi r^2)^{-\frac{n}{2}} \int_M u^2 dV_{t_0} = 1$$

Note that

$$|\nabla u|^2 = e^L r^{-2}|\zeta'(\frac{d_{t_0}(x_0, x)}{r})|^2 \quad \text{and} \quad u^2 \log u^2 = Lu^2 + e^L\zeta^2 \log \zeta^2.$$

Also, by the definition of $u(x)$, we have

$$(4\pi r^2)^{-\frac{n}{2}} e^L V_{t_0}(x_0, r/2) \leq 1, \tag{5.72}$$

and

$$(4\pi)^{-\frac{n}{2}} r^{-n} e^L V_{t_0}(x_0, r) \geq 1. \tag{5.73}$$

Now it follows from (5.71) and the upper bound assumption on R that

$$\mu_0 \leq \mathcal{W}(g_{ij}(t_0), u, r^2)$$

$$= (4\pi r^2)^{-\frac{n}{2}} \int_M [r^2(Ru^2 + 4|\nabla u|^2) - u^2 \log u^2 - nu^2]$$

$$\leq 1 - L - n + (4\pi r^2)^{-\frac{n}{2}} e^L \int_M (4|\zeta'|^2 - \zeta^2 \log \zeta^2)$$

$$\leq 1 - L - n + (4\pi r^2)^{-\frac{n}{2}} e^L (16 + e^{-1}) V_{t_0}(x_0, r).$$

Here, in the last inequality, we have used the elementary fact that $-s \log s \leq e^{-1}$ for $0 \leq s \leq 1$. Combining the above with (5.72), we arrive at

$$\mu_0 \leq 1 - L - n + (16 + e^{-1})\frac{V_{t_0}(x_0, r)}{V_{t_0}(x_0, r/2)}. \tag{5.74}$$

Notice that if we have the volume doubling property

$$V_{t_0}(x_0, r) \leq C V_{t_0}(x_0, r/2)$$

for some universal constant $C > 0$, then (5.73) and (5.74) together would imply

$$V_{t_0}(x_0, r) \geq \exp\{\mu_0 + n - 1 - (16 + e^{-1})C\}r^n, \tag{5.75}$$

thus proving the theorem. We now describe how to bypass such a volume doubling property by a clever argument[2] pointed out by B.-L. Chen back in 2003.

Notice that the above argument is also valid if we replace r by any positive number $0 < a \leq r$. Thus, at least we have shown the following

Assertion: Set

$$\kappa = \min\left\{\exp[\mu_0 + n - 1 - (16 + e^{-1})3^n], \frac{1}{2}\alpha_n\right\},$$

where α_n is the volume of the unit ball in $\mathbb{R}^n$. Then, for any $0 < a \leq r$, we have

$$V_{t_0}(x_0, a) \geq \kappa a^n, \tag{$*$}_a$$

whenever the volume doubling property,

$$V_{t_0}(x_0, a) \leq 3^n V_{t_0}(x_0, a/2),$$

holds.

Now we finish the proof by contradiction. Suppose $(*)_a$ fails for $a = r$. Then we must have

$$V_{t_0}(x_0, \frac{r}{2}) < 3^{-n} V_{t_0}(x_0, r)$$

$$< 3^{-n} \kappa r^n$$

$$< \kappa \left(\frac{r}{2}\right)^n.$$

This says that $(*)_{r/2}$ would also fail. By induction, we deduce that

$$V_{t_0}(x_0, \frac{r}{2^k}) < \kappa \left(\frac{r}{2^k}\right)^n \quad \text{for all } k \geq 1.$$

---

[2]Perelman also used a similar argument in proving his uniform diameter estimate for the NKRF, see the proof of Claim 6.1 in Sect. 5.6.

But this contradicts the fact that

$$\lim_{k \to \infty} \frac{V_{t_0}(x_0, \frac{r}{2^k})}{\left(\frac{r}{2^k}\right)^n} = \alpha_n.$$ □

5.4.3 The μ-Entropy and the Strong Noncollapsing Estimate for KRF and NKRF

To convert the κ-noncollapsing theorem for the Ricci flow to the KRF and NKRF, first note that for any local holomorphic coordinates $(z^1, \cdots, z^n)$ with $z^i = x^i + \sqrt{-1}y^i$, $(x^1, \cdots, x^n, y^1, \cdots, y^n)$ form a preferred smooth local coordinates with

$$\frac{\partial}{\partial z^i} = \frac{1}{2}\left(\frac{\partial}{\partial x^i} - \sqrt{-1}\frac{\partial}{\partial y^i}\right) \quad \text{and} \quad \frac{\partial}{\partial \bar{z}^i} = \frac{1}{2}\left(\frac{\partial}{\partial x^i} + \sqrt{-1}\frac{\partial}{\partial y^i}\right).$$

Thus, in terms of the corresponding Riemannian metric ds^2, we have

$$ds^2\left(\frac{\partial}{\partial x^i}, \frac{\partial}{\partial x^j}\right) = ds^2\left(\frac{\partial}{\partial y^i}, \frac{\partial}{\partial y^j}\right) = 2\Re(g_{i\bar{j}})$$

while

$$ds^2\left(\frac{\partial}{\partial x^i}, \frac{\partial}{\partial y^j}\right) = 2\Im(g_{i\bar{j}}).$$

In particular, for any $(z^1, \cdots, z^n)$ with $g_{i\bar{j}} = \delta_{i\bar{j}}$ (e.g., under normal coordinates), then

$$ds^2\left(\frac{\partial}{\partial x^i}, \frac{\partial}{\partial x^j}\right) = ds^2\left(\frac{\partial}{\partial y^i}, \frac{\partial}{\partial y^j}\right) = 2\delta_{ij} \quad \text{and} \quad ds^2\left(\frac{\partial}{\partial x^i}, \frac{\partial}{\partial y^j}\right) = 0.$$

(Thus, we can symbolically express the Riemannian metric $g_{\mathbb{R}} = ds^2 = 2g_{i\bar{j}}$.)

On the other hand, if $R_{i\bar{j}} = \lambda\delta_{i\bar{j}}$ under the normal holomorphic coordinates $(z^1, \cdots, z^n)$ then, for the Riemannian Ricci tensor Ric_{ds^2}, we have

$$\text{Ric}_{ds^2}\left(\frac{\partial}{\partial x^i}, \frac{\partial}{\partial x^j}\right) = \text{Ric}_{ds^2}\left(\frac{\partial}{\partial y^i}, \frac{\partial}{\partial y^j}\right) = 2\lambda\delta_{ij} \quad \text{and} \quad \text{Ric}_{ds^2}\left(\frac{\partial}{\partial x^i}, \frac{\partial}{\partial y^j}\right) = 0.$$

That is,

$$\text{Ric}_{ds^2} = \lambda ds^2,$$

so we have the same Einstein constant λ.

Note that we also have the following relations:

- The scalar curvature: $R_{ds^2} = 2R$
- The Laplace operator: $\Delta_{ds^2} = 2\Delta$
- The norm square of the gradient of a function: $|\nabla f|^2_{ds^2} = 2|\nabla f|^2$, etc.

In particular, we have

$$R_{ds^2} + |\nabla f|^2_{ds^2} = 2(R + |\nabla f|^2).$$

Therefore, with $\sigma = 2\tau$, the Riemannian $\mathcal{W}$-functional on $(X^n, g_{i\bar{j}})$ is given by

$$\mathcal{W} = \frac{1}{(2\pi\sigma)^n} \int_X [\sigma(R + |\nabla f|^2) + f - 2n]e^{-f} dV, \qquad (5.76)$$

or, with $u = e^{-f/2}$, by

$$\mathcal{W}(g_{i\bar{j}}, u, \sigma) = \frac{1}{(2\pi\sigma)^n} \int_X [\sigma(Ru^2 + 4|\nabla u|^2) - u^2 \log u^2 - 2nu^2]dV \qquad (5.77)$$

with respect to the Kähler metric $g_{i\bar{j}}$.

The μ-entropy is then given by

$$\mu = \mu(g_{i\bar{j}}, \sigma) = \inf\left\{\mathcal{W}(g_{i\bar{j}}, u, \sigma): (2\pi\sigma)^{-n} \int_X u^2 dV = 1\right\}.$$

For any solution $\hat{g}_{i\bar{j}}(s)$ to the KRF on the maximal time interval $[0, 1)$, by taking $\sigma = 1 - s$, it follows that $\mu(\hat{g}_{i\bar{j}}(s), 1 - s)$ is monotone increasing in s. By the scaling invariance property of μ in (5.65) and the relation between KRF and NKRF as described in Lemma 5.2.4, we get

$$\mu(\hat{g}_{i\bar{j}}(s), 1 - s) = \mu(g_{i\bar{j}}(t), 1). \qquad (5.78)$$

Thus, by the monotonicity of $\mu(\hat{g}_{i\bar{j}}(s), 1 - s)$ and $ds/dt = e^{-t} > 0$, we have

Lemma 5.4.6. *Let $g_{i\bar{j}}(t)$ be a solution to the NKRF on $X^n \times [0, \infty)$. Then,*

$$\mu(g_{i\bar{j}}(t), 1) = \inf\left\{\frac{1}{(2\pi)^n} \int_X (R + |\nabla f|^2 + f - 2n)\right.$$

$$e^{-f} dV: \frac{1}{(2\pi)^n} \int_X e^{-f} dV = 1\right\}$$

$$= \inf\left\{\frac{1}{(2\pi)^n}\right.$$

$$\int_X (Ru^2 + 4|\nabla u|^2 - u^2 \log u^2 - 2nu^2): \frac{1}{(2\pi)^n} \int_X u^2 = 1\right\}$$

is monotone increasing in t.

Finally, we have the corresponding strong no local collapsing theorem for the NKRF:

Theorem 5.4.7 (Strong no local collapsing theorem for NKRF). *Let X^n be a Fano manifold, and let $g_{i\bar{j}}(t)$ be a solution to the NKRF (5.33) on $X^n \times [0, \infty)$. Then there exists a positive constant $\kappa > 0$, depending only the initial metric g_0, such that $g_{i\bar{j}}(t)$ is strongly κ-noncollapsed at very point $(x_0, t_0) \in M \times [0, \infty)$ on all scales less than $e^{t_0/2}$ in the following sense: for any $(x_0, t_0) \in X \times [0, \infty)$ and $0 < r \le e^{t_0/2}$ we have*

$$V_{t_0}(x_0, r) \ge \kappa r^{2n}, \tag{5.79}$$

whenever

$$R(\cdot, t_0) \le r^{-2} \quad on \; B_{t_0}(x_0, r). \tag{5.80}$$

Proof. This is an immediate consequence of Theorem 5.4.5 applied to the KRF on $X^n \times [0, 1)$, and the relation between the KRF and the NKRF as described by Lemma 5.2.4.   □

5.5 Uniform Curvature and Diameter Estimates for NKRF with Nonnegative Bisectional Curvature

Our goal in this section is to prove the uniform diameter and (scalar) curvature estimates by B.L Chen, X.-P. Zhu and the author [CCZ03] for the NKRF with nonnegative holomorphic bisectional curvature. The main ingredients of the proof are the Harnack estimate in Theorem 5.3.6 and the strong non-collapsing estimate in Theorem 5.4.7 for the NKRF.

Theorem 5.5.1. *Let $(X^n, \tilde{g}_{i\bar{j}})$ be a compact Kähler manifold with nonnegative bisectional curvature and let $g_{i\bar{j}}(t)$ be the solution to the NKRF with $g_{i\bar{j}}(0) = \tilde{g}_{i\bar{j}}$. Then, there exist positive constants $C_1 > 0$ and $C_2 > 0$ such that*

(i) $|\mathrm{Rm}|(x, t) \le C_1$ *for all* $(x, t) \in X \times [0, \infty)$;
(ii) $diam(X^n, g_{i\bar{j}}(t)) \le C_2$ *for all* $t \ge 0$.

Proof. By Theorem 5.2.10, we know that $g_{i\bar{j}}(t)$ has nonnegative bisectional curvature for all $t \ge 0$. Thus, it suffices to show the uniform upper bound for the scalar curvature

$$R(x, t) \le C_1$$

on $X \times [0, \infty)$. We divide the proof into several steps:

Step 1: A local uniform bound on R

First of all, we know that the volume $V_t(X^n) = \mathrm{Vol}(X, g_{i\bar{j}}(t))$ and the total scalar curvature $\int_{X^n} R(x, t)dV_t$ are constant along the NKRF. Hence the average scalar curvature is also constant. In fact,

$$\frac{1}{V_t(X^n)} \int_{X^n} R(x, t)dV_t = n, \quad \text{for all } t \geq 0.$$

Now, $\forall\, t > 1$, set $t_1 = t$, $t_2 = t + 1$ and pick a point $y_t \in X$ such that

$$R(y_t, t + 1) = n.$$

Then, $\forall\, x \in X$, by the Harnack inequality in Theorem 5.3.6, and noting that $\forall t \geq 1$,

$$\frac{e^{t+1} - 1}{e^t - 1} \leq e + 1,$$

we have

$$R(x, t) \leq n(e + 1) \exp\left(\frac{e}{4}d_t^2(x, y_t)\right). \tag{5.81}$$

In particular, when $d_t(y_t, x) < 1$, we obtain a uniform upper bound

$$R(\cdot, t) \leq n(e + 1) \exp(e^2/4) \tag{5.82}$$

on the unit geodesic ball $B_t(y_t, 1)$ at time t, for all $t \geq 1$.

Step 2: The uniform diameter bound

Now we have the uniform upper bound (5.82) for the scalar curvature on $B_t(y_t, 1)$. By applying the strong no local collapsing Theorem 5.4.7, there exists a positive constant $\kappa > 0$, depending only on the initial metric g_0, such that we have the following uniform lower bound

$$V_t(y_t, 1) \geq \kappa > 0$$

for the volume of the unit geodesic ball $B_t(y_t, 1)$ for all $t \geq 1$.

Suppose $\mathrm{diam}(X, g_{i\bar{j}}(t))$ is not uniformly bounded from above in t. Then, there exist a sequence of positive numbers $\{D_k\} \to \infty$ and a time sequence $\{t_k\} \to \infty$ such that

$$\mathrm{diam}\,(X, g_{i\bar{j}}(t_k)) > D_k.$$

However, since $g_{i\bar{j}}(t_k)$ has nonnegative Ricci curvature, it follows from an argument of Yau (cf. p. 24 in [ScYau94]) that there exists a universal constant $C = C(n) > 0$ such that

$$V_{t_k}(y_{t_k}, D_k) \geq CV_{t_k}(y_{t_k}, 1)D_k \geq \kappa CD_k \to \infty.$$

But this contradicts the fact that

$$V_{t_k}(y_{t_k}, D_k) \leq V_{t_k}(X^n) = V_0, \qquad k = 1, 2, \cdots.$$

Thus, we have proved the uniform diameter bound: there exists a positive constant $D > 0$ such that for all $t > 0$,

$$\operatorname{diam}(X, g_{i\bar{j}}(t)) \leq D. \tag{5.83}$$

Step 3: The global uniform bound on R

Once we have the uniform diameter upper bound (5.83), the Harnack inequality (5.81) immediately implies the uniform scalar curvature upper bound,

$$R(x, t) \leq n(e + 1)e^{eD^2/4},$$

on $X^n \times [0, \infty)$. $\square$

Remark 5.5.1. As mentioned in the introduction, assuming in addition the existence of K–E metrics, Chen and Tian studied the NKRF with nonnegative bisectional curvature on Del Pezzo surfaces [CheT02] and Fano manifolds in higher dimensions [CheT06].

5.6 Perelman's Uniform Estimates

In the previous section, we saw that when a solution $g_{i\bar{j}}(t)$ to the NKRF has nonnegative bisectional curvature, then the uniform diameter and curvature bounds follow from a nice interplay between the Harnack inequality for the scalar curvature R and the strong no local collapsing theorem. In this section, we shall see Perelman's amazing uniform estimates on the diameter and the scalar curvature for the NKRF on general Fano manifolds (Theorem 5.6.1). In absence of the Harnack inequality, Perelman's proof is much more subtle, yet the monotonicity of the μ-entropy and the ideas used in the proof of the strong non-collapsing estimate played a crucial role.

The material presented in this section follows closely what Perelman gave in a private lecture[3] at MIT in April, 2003. As such, it naturally overlaps considerably with the earlier notes [SeT08] on Perelman's work. The author also presented Perelman's uniform estimates at the Geometry and Analysis seminar at Columbia University in fall 2005.

---

[3]Perelman's private lecture was attended by a very small audience, including this author and the authors of [SeT08].

Theorem 5.6.1. *Let X^n be a Fano manifold and $g_{i\bar{j}}(t)$, $0 \leq t < \infty$, be the solution to the NKRF*

$$\frac{\partial}{\partial t} g_{i\bar{j}} = -R_{i\bar{j}} + g_{i\bar{j}}, \quad g(0) = \tilde{g} \tag{5.84}$$

with the initial metric $g_0 = \tilde{g}$ satisfying $[\omega_0] = \pi c_1(X)$. Let $f = f(t)$ be the Ricci potential of $g_{i\bar{j}}(t)$ satisfying

$$- R_{i\bar{j}}(t) + g_{i\bar{j}}(t) = \partial_i \partial_{\bar{j}} f \tag{5.85}$$

and the normalization

$$\int_{X^n} e^{-f} dV = (2\pi)^n. \tag{5.86}$$

Then there exists a constant $C > 0$ such that

(i) $|R| \leq C$ on $X^n \times [0, \infty)$;
(ii) $\mathrm{diam}(X^n, g_{i\bar{j}}(t)) \leq C$;
(iii) $\|f\|_{C^1} \leq C$ on $X^n \times [0, \infty)$.

The proof will occupy the whole section. First of all, by Lemma 5.2.8, we know that under (5.84) the scalar curvature R evolves according to the equation

$$\frac{\partial}{\partial t} R = \Delta R + |\mathrm{Ric}|^2 - R.$$

Lemma 5.6.2. *There exists a constant $C_1 > 0$ such that the scalar curvature R of the NKRF (5.84) satisfies the estimate*

$$R(x, t) \geq -C_1.$$

for all $t \geq 0$ and all $x \in X^n$.

Proof. Let $R_{\min}(0)$ be the minimum of $R(x, 0)$ on X^n. If $R_{\min}(0) \geq 0$, then by the maximum principle, we have $R(x, t) \geq 0$ for all $t > 0$ and all $x \in X^n$.

Now suppose $R_{\min}(0) < 0$. Set $F(x, t) = R(x, t) - R_{\min}(0)$. Then, $F(x, 0) \geq 0$ and F satisfies

$$\frac{\partial}{\partial t} F = \Delta F + |\mathrm{Ric}|^2 - F - R_{\min}(0) > \Delta F + |\mathrm{Ric}|^2 - F.$$

Hence it follows again from the maximum principle that $F \geq 0$ on $X^n \times [0, \infty)$, i.e.,

$$R(x, t) \geq R_{\min}(0)$$

for all $t > 0$ and all $x \in X^n$.   □

Next, we consider the Ricci potential f satisfying (5.85) and the normalization (5.86). Note that it follows from (5.85) that

$$n - R = \Delta f. \tag{5.87}$$

Also, let $\varphi = \varphi(t)$ be the Kähler potential,

$$g_{i\bar{j}}(t) = \tilde{g}_{i\bar{j}} + \partial_i \partial_{\bar{j}} \varphi,$$

so that φ is a solution to the parabolic scalar equation

$$\varphi_t = \log \frac{\det(\tilde{g}_{i\bar{j}} + \partial_i \partial_{\bar{j}} \varphi)}{\det(\tilde{g}_{i\bar{j}})} + \tilde{f} + \varphi + b(t),$$

where $b(t)$ is a function of t only.

Since $\partial_i \partial_{\bar{j}} \varphi_t = -R_{i\bar{j}} + g_{i\bar{j}}$, by adding a function of t only to φ if necessary, we can assume

$$f = \varphi_t. \tag{5.88}$$

Thus, f satisfies the parabolic equation

$$f_t = \Delta f + f - a(t) \tag{5.89}$$

for some function a(t) of t only.

By differentiating the constraint (5.86), we get

$$\int_{X^n} e^{-f}(-f_t + n - R)dV = 0.$$

Hence, by combining with (5.87) and (5.89), it follows that

$$a(t) = (2\pi)^{-n} \int_{X^n} fe^{-f}dV. \tag{5.90}$$

Lemma 5.6.3. *There exists a constant $C_2 > 0$ such that, for all $t \geq 0$,*

$$-C_2 \leq \int_{X^n} fe^{-f}dV \leq C_2.$$

Proof. The second inequality is easy to see. Now we prove the first inequality. By Lemma 5.4.6 and (5.87), we have

$$A =: \mu(g_{i\bar{j}}(0), 1) \leq \mu(g_{i\bar{j}}(t), 1)$$

$$\leq (2\pi)^{-n} \int_X (R + |\nabla f|^2 + f - 2n)e^{-f} dV$$

$$= (2\pi)^{-n} \int_X (-\Delta f + |\nabla f|^2 + f - n)e^{-f} dV$$

$$= (2\pi)^{-n} \int_X (f - n)e^{-f} dV.$$

Therefore,

$$(2\pi)^{-n} \int_{X^n} fe^{-f} dV \geq A + n. \qquad \qquad \square$$

Lemma 5.6.4. *There exists a constant $C_3 > 0$ such that*

$$f \geq -C_3$$

for all $t \geq 0$ and all $x \in X^n$.

Proof. We argue by contradiction. Suppose the Ricci potential f is very negative at some time $t_0 > 0$ and some point $x_0 \in X^n$ so that

$$f(x_0, t_0) \ll -1.$$

Then, there exists some open neighborhood $U \subset X^n$ of x_0 such that

$$f(x, t_0) \ll -1, \qquad \forall x \in U. \qquad (5.91)$$

On the other hand, by (5.87), (5.89), Lemma 5.6.2, (5.90), and Lemma 5.6.3, we have

$$f_t = n - R + f - a(t) \leq f + C \qquad (5.92)$$

for some uniform constant $C > 0$.

Let us assume $f(\cdot, t)$ and $\varphi(\cdot, t)$ achieve their maximum at x_t and x_t^* respectively. From the constraint (5.86), it is clear that for each $t > 0$, we have a uniform lower estimate

$$f(x_t, t) = \max_X f(\cdot, t) \geq -C$$

for some $C > 0$ independent of t. Moreover, it follows form (5.88) and (5.92) that

$$(f - \varphi)_t \leq C,$$

so

$$f(\cdot,t) - \varphi(\cdot,t) \le \max_X (f - \varphi)(\cdot,t_0) + Ct.$$

Therefore,

$$\varphi(x_t^*,t) \ge \varphi(x_t,t) \ge f(x_t,t) - \max_X (f - \varphi)(\cdot,t_0) - Ct \ge -Ct, \quad \forall t \gg t_0. \tag{5.93}$$

On the other, by (5.92), we have

$$f(x,t) \le e^{t-t_0}(C + f(x,t_0)) \tag{5.94}$$

for $t \ge t_0$ and $x \in X^n$. In particular, by (5.91), we have

$$f(x,t) \le -Ce^{-t_0}e^t, \qquad \forall t > t_0, \forall x \in U. \tag{5.95}$$

Then (5.88) and (5.95) together imply that

$$\varphi(x,t) \le \varphi(x,t_0) - Ce^{-t_0}e^t + C \le -C'e^t, \qquad \forall t \gg t_0, \forall x \in U. \tag{5.96}$$

Next, we claim (5.96) implies

$$\varphi(x_t^*,t) \le -Ce^t + C' \tag{5.97}$$

for some $C' > 0$ independent of $t \gg t_0$. To see this, note that, with respect to the initial metric g_0, we have

$$\varphi(x_t^*,t) = \frac{1}{V_0(X^n)} \int_X \varphi(\cdot,t)dV_0 - \frac{1}{V_0(X^n)} \int_X \Delta_0\varphi(\cdot,t)G_0(x_t^*,\cdot)dV_0, \tag{5.98}$$

where $V_0(X^n) = \mathrm{Vol}(X^n, g_0)$ and $G_0(x_t^*,\cdot)$ denotes a positive Green's function with pole at x_t^*.

Since $n + \Delta_0\varphi = \tilde{g}^{i\bar{j}}g_{i\bar{j}}(t) > 0$, the second term on the RHS of (5.98) can be estimated by

$$-\frac{1}{V_0(X^n)} \int_X \Delta_0\varphi(\cdot,t)G_0(x_t,\cdot)dV_0 \le \frac{n}{V_0(X^n)} \int_X G_0(x_t,\cdot)dV_0 =: C''. \tag{5.99}$$

On the other hand, by using (5.95), it follows that

$$\frac{1}{V_0(X^n)} \int_X \varphi(\cdot,t)dV_0 \le \frac{V_0(X \setminus U)}{V_0(X)}\varphi(x_t^*,t) - \frac{V_0(U)}{V_0(X)}Ce^t. \tag{5.100}$$

Therefore, by (5.98)–(5.100), we have

$$\alpha\varphi(x_t^*, t) \leq C'' - \alpha C e^t$$

for $\alpha = V_0(U)/V_0(X) > 0$. This proves (5.97), a contradiction to (5.93). □

Lemma 5.6.5. *There exists constant $C_4 > 0$ such that, for all $t \geq 0$,*

(a) $|\nabla f|^2 \leq C_4(f + 2C_3)$;
(b) $R \leq C_4(f + 2C_3)$.

Proof. This is essentially a parabolic version of Yau's gradient estimate in [Yau75] (see also [ScYau94]).

First of all, from $|\nabla f|^2 = g^{i\bar{j}} \partial_i f \partial_{\bar{j}} f$, the NKRF, and (5.89), we obtain

$$\frac{\partial}{\partial t}|\nabla f|^2 = (R_{i\bar{j}} - g_{i\bar{j}})\partial_i f \partial_{\bar{j}} f + g^{i\bar{j}}(\partial_i f_t \partial_{\bar{j}} f + \partial_i f \partial_{\bar{j}} f_t)$$

$$= g^{i\bar{j}}[\partial_i(\Delta f)\partial_{\bar{j}} f + \partial_i f \partial_{\bar{j}}(\Delta f)] + \mathrm{Ric}(\nabla f, \nabla f) + |\nabla f|^2.$$

On the other hand, the Bochner formula gives us

$$\Delta|\nabla f|^2 = |\nabla\bar{\nabla} f|^2 + |\nabla\nabla f|^2 + g^{i\bar{j}}[\partial_i(\Delta f)\partial_{\bar{j}} f + \partial_i f \partial_{\bar{j}}(\Delta f)] + \mathrm{Ric}(\nabla f, \nabla f).$$

Hence, we have

$$\frac{\partial}{\partial t}|\nabla f|^2 = \Delta|\nabla f|^2 - |\nabla\bar{\nabla} f|^2 - |\nabla\nabla f|^2 + |\nabla f|^2. \tag{5.101}$$

Also, by (5.85), we have

$$|\mathrm{Ric}|^2 + n - 2R = |\nabla\bar{\nabla} f|^2. \tag{5.102}$$

Thus, from the evolution equation on R, we have

$$\frac{\partial}{\partial t}R \leq \Delta R + |\nabla\bar{\nabla} f|^2 + R$$

Therefore, for any $\alpha \geq 0$, we obtain

$$\frac{\partial}{\partial t}(|\nabla f|^2 + \alpha R) \leq \Delta(|\nabla f|^2 + \alpha R) - (1-\alpha)(|\nabla\bar{\nabla} f|^2 + |\nabla\nabla f|^2) + (|\nabla f|^2 + \alpha R). \tag{5.103}$$

Next, take $B = 2C_3$ so we have $f + B > 1$, and set

$$u = \frac{|\nabla f|^2 + \alpha R}{f + B}. \tag{5.104}$$

Then, we have

$$u_t = \frac{(|\nabla f|^2 + \alpha R)_t}{f + B} - \frac{u}{(f + B)} f_t$$

and

$$\nabla u = \frac{1}{f + B} \nabla(|\nabla f|^2 + \alpha R) - \frac{|\nabla f|^2 + \alpha R}{(f + B)^2} \nabla f. \qquad (5.105)$$

On the other hand, since $|\nabla f|^2 + \alpha R = u(f + B)$, we have

$$\Delta(|\nabla f|^2 + \alpha R) = (f + B)\Delta u + u\Delta f + \nabla u \cdot .\bar{\nabla} f + \bar{\nabla} u \cdot .\nabla f$$

or

$$\Delta u = \frac{\Delta(|\nabla f|^2 + \alpha R)}{f + B} - \frac{u\Delta f}{f + B} - \frac{\nabla u \cdot \bar{\nabla} f + \bar{\nabla} u \cdot \nabla f}{f + B}.$$

Therefore,

$$u_t \le \Delta u - (1 - \alpha)\frac{(|\nabla\bar{\nabla} f|^2 + |\nabla\nabla f|^2)}{f + B} + \frac{\nabla u \cdot \bar{\nabla} f + \bar{\nabla} u \cdot \nabla f}{f + B} + \frac{B + a(t)}{f + B} u. \qquad (5.106)$$

Notice, by (5.105), we have

$$\nabla u \cdot \bar{\nabla} f = \frac{1}{f + B} \nabla(|\nabla f|^2 + \alpha R) \cdot \bar{\nabla} f - \frac{(|\nabla f|^2 + \alpha R)|\nabla f|^2}{(f + B)^2}. \qquad (5.107)$$

Now the trick (see, e.g., p. 19 in [ScYau94]) is to use (5.107) and express

$$\frac{\nabla u \cdot \bar{\nabla} f}{f + B} = (1 - 2\epsilon)\frac{\nabla u \cdot \bar{\nabla} f}{f + B} + \frac{2\epsilon}{f + B}$$

$$\left(\frac{\nabla(|\nabla f|^2 + \alpha R) \cdot \bar{\nabla} f}{f + B} - \frac{|\nabla f|^2(|\nabla f|^2 + \alpha R)}{(f + B)^2} \right). \qquad (5.108)$$

We are ready to conclude the proof of Lemma 5.6.5.
Part (a) Take $\alpha = 0$ so that $u = |\nabla f|^2/(f + B)$. By plugging (5.108) into (5.106), we get

$$u_t \le \Delta u - (1 - 4\epsilon)\frac{|\nabla\bar\nabla f|^2 + |\nabla\nabla f|^2}{f + B} + (1 - 2\epsilon)\frac{\nabla u \cdot \bar\nabla f + \bar\nabla u \cdot \nabla f}{f + B}$$

$$- \frac{\epsilon}{f + B}\left(|2\nabla\bar\nabla f - \frac{\nabla f\bar\nabla f}{f + B}|^2 + |2\nabla\nabla f - \frac{\nabla f\nabla f}{f + B}|^2\right)$$

$$+ \frac{1}{(f + B)}\left(-2\epsilon u^2 + (B + a)u\right).$$

For any $T > 0$, suppose u attains its maximum at (x_0, t_0) on $X^n \times [0, T]$, then

$$u_t(x_0, t_0) \ge 0, \quad \nabla u(x_0, t_0) = 0, \quad \text{and} \quad \Delta u(x_0, t_0) \le 0. \tag{5.109}$$

Thus, by choosing $\epsilon = 1/8$, we arrive at

$$u(x_0, t_0) \le 4(B + a).$$

Therefore, since $T > 0$ is arbitrary, we have shown that

$$\frac{|\nabla f|^2}{f + B} \le 8C_3 + 4C_2 \tag{5.110}$$

on $X^n \times [0, \infty)$.

Part (b) Choose $\alpha = 1/2$ so that

$$u = \frac{|\nabla f|^2 + R/2}{f + B}.$$

Then, from (5.106) and (5.102), we obtain

$$u_t \le \Delta u - \frac{1}{2}\frac{|\mathrm{Ric}|^2 - 2R}{f + B} + \frac{\nabla u \cdot \bar\nabla f + \bar\nabla u \cdot \nabla f}{f + B} + \frac{B + a}{f + B}u.$$

Again, for any $T > 0$, suppose u attains its maximum at (x_0, t_0) on $X^n \times [0, T]$. Then (5.109) holds, and hence

$$0 \le -\frac{1}{2n}\left(\frac{R}{f + B}\right)^2(x_0, t_0) + \frac{R}{f + B}(x_0, t_0)$$

$$\left(1 + \frac{B + a}{2(f + B)}\right) + (8C_3 + 4C_2)(B + a).$$

Here we have used the fact that $|\text{Ric}|^2 \geq R^2/n$, $2f + B \geq 0$, $f + B \geq 1$, and (5.110). It then follows easily that $\frac{R}{f+B}(x_0, t_0)$ is bounded from above uniformly. Therefore, by Part (a), $\frac{R}{f+B}(x, t)$ is bounded uniformly on $X^n \times [0, T]$ for arbitrary $T > 0$. □

Clearly, Lemma 5.6.5 (a) implies that $\sqrt{f + 2C_3}$ is Lipschitz. From now on we assume the Ricci potential $f(\cdot, t)$ attains its minimum at a point $\hat{x} \in X^n$, i.e., $f(\hat{x}, t) = \min_X f(\cdot, t)$. Then, by (5.86), we know

$$f(\hat{x}, t) \leq C$$

for some $C > 0$ independent of t.

Corollary 5.6.6. *There exists a constant $C > 0$ such that $\forall t > 0$ and $\forall x \in X$,*

(i) $f(x, t) \leq C[1 + d_t^2(\hat{x}, x)]$;
(ii) $|\nabla f|^2(x, t) \leq C[1 + d_t^2(\hat{x}, x)]$;
(iii) $R(x, t) \leq C[1 + d_t^2(\hat{x}, x)]$.

Proof. Set $h = f + 2C_2 > 0$. Then, from Lemma 5.6.5 (i), we see that $\sqrt{h}$ is a Lipschitz function satisfying

$$|\nabla \sqrt{h}|^2 \leq C_4.$$

Hence, $\forall x \in X^n$,

$$|\sqrt{h}(x, t) - \sqrt{h}(\hat{x}, t)| \leq Cd_t(\hat{x}, x),$$

or

$$\sqrt{h}(x, t) \leq \sqrt{h}(\hat{x}, t) + Cd_t(\hat{x}, x).$$

Thus, we obtain a uniform upper bound

$$f(x, t) \leq h(x, t) \leq C(d_t^2(\hat{x}, y) + 1)$$

for some constant $C > 0$ independent of t. Now (ii) and (iii) follow immediately from (i) and Lemma 5.6.5. □

By Lemma 5.6.2 and Corollary 5.6.6, it remains to prove the following uniform diameter bound.

Lemma 5.6.7. *There exists a constant $C_5 > 0$ such that*

$$\text{diam}_t(X) =: \text{diam}(X^n, g_{i\bar{j}}(t)) \leq C_5$$

for all $t \geq 0$.

Proof. For each $t > 0$, denote by $A_t(k_1, k_2)$ the annulus region defined by

$$A_t(k_1, k_2) = \{z \in X : 2^{k_1} \le d_t(x, \hat{x}) \le 2^{k_2}\}, \tag{5.111}$$

and by

$$V_t(k_1, k_2) = \text{Vol}(A_t(k_1, k_2)) \tag{5.112}$$

with respect to $g_{i\bar{j}}(t)$.

Note that each annulus $A_t(k, k + 1)$ contains at least 2^{2k} balls B_r of radius $r = 2^{-k}$. Also, for each point $x \in A_t(k, k + 1)$, Corollary 5.6.6 (iii) implies that the scalar curvature is bounded above by $R \le C 2^{2k}$ on $B_t(x, r)$ for some uniform constant $C > 0$. Thus each of these balls B_r has $\text{Vol}(B_r) \ge \kappa(2^{-k})^{2n}$ by Theorem 5.4.7, so we have

$$V_t(k, k + 1) \ge \kappa 2^{2k-1} 2^{-kn}. \tag{5.113}$$

Claim 6.1. For each small $\epsilon > 0$, there exists a large constant $D = D(\epsilon) > 0$ such that if $\text{diam}_t(X) > D$, then one can find large positive constants $k_2 > k_1 > 0$ with the following properties:

$$V_t(k_1, k_2) \le \epsilon \tag{5.114}$$

and

$$V_t(k_1, k_2) \le 2^{10n} V_t(k_1 + 2, k_2 - 2). \tag{5.115}$$

Proof. (a) follows from the fact that $V_t(X^n) = V_0(X^n)$ and the assumption $\text{diam}_t(X) \gg 1$.

Now suppose (a) holds but not (b), i.e.,

$$V_t(k_1, k_2) > 2^{10n} V_t(k_1 + 2, k_2 - 2).$$

Then we consider whether or not

$$V_t(k_1 + 2, k_2 - 2) \le 2^{10n} V_t(k_1 + 4, k_2 - 4).$$

If yes, then we are done. Otherwise we repeat the process.

After j steps, we either have

$$V_t(k_1 + 2(j - 1), k_2 - 2(j - 1)) \le 2^{10nj} V_t(k_1 + 2j, k_2 - 2j), \tag{5.116}$$

or

$$V_t(k_1, k_2) > 2^{10nj} V_t(k_1 + 2j, k_2 - 2j). \tag{5.117}$$

Without loss of generality, we may assume $k_1 + 2j \approx k_2 - 2j$ by choosing a large number $K > 0$ and pick $k_1 \approx K/2, k_2 \approx 3K/2$. Then, when $j \approx K/4$ and using (5.113), this implies that

$$\epsilon \geq V_t(k_1, k_2) \geq 2^{10nK/4} V_t(K, K+1) \geq \kappa 2^{2K(n/4-1)}.$$

So after some finitely many steps $j \approx K(\epsilon)/4$, (5.116) must hold. Therefore, we have found k_1 and $k_2 \approx 3k_1$ satisfying both (5.114) and (5.115).   □

Claim 6.2. There exist constants $r_1 > 0$ and $r_2 > 0$, with $r_1 \in [2^{k_1}, 2^{k_1+1}]$ and $r_2 \in [2^{k_2}, 2^{k_2+1}]$, such that

$$\int_{A_t(r_1, r_2)} R dV_t \leq C V_t(k_1, k_2). \tag{5.118}$$

Proof. First of all, since

$$\frac{d}{dr} \text{Vol}(B(r)) = \text{Vol}(S(r)),$$

we have

$$V(k_1, k_1 + 1) = \int_{2^{k_1}}^{2^{k_1+1}} \text{Vol}(S(r)) dr.$$

Here S_r denotes the geodesic sphere of radius r centered at $\hat{x}$ with respect to $g_{ij}(t)$. Hence, we can choose $r_1 \in [2^{k_1}, 2^{k_1+1}]$ such that

$$\text{Vol}(S_{r_1}) \leq \frac{V_t(k_1, k_2)}{2^{k_1}},$$

for otherwise

$$V(k_1, k_1 + 1) > \frac{V_t(k_1, k_2)}{2^{k_1}} 2^{k_1} = V_t(k_1, k_2),$$

a contradiction because $k_2 > k_1 + 1$. Similarly, there exists $r_2 \in [2^{k_2-1}, 2^{k_2}]$ such that

$$\text{Vol}(S_{r_2}) \leq \frac{V_t(k_1, k_2)}{2^{k_2}}.$$

Next, by integration by parts and Corollary 5.6.6(ii),

$$\left| \int_{A_t(r_1,r_2)} \Delta f \right| \leq \int_{S_{r_1}} |\nabla f| + \int_{S_{r_2}} |\nabla f|$$

$$\leq \frac{V_t(k_1,k_2)}{2^{k_1}} C 2^{k_1+1} + \frac{V_t(k_1,k_2)}{2^{k_2}} C 2^{k_2+1}$$

$$\leq C V_t(k_1,k_2).$$

Therefore, since $R + \Delta f = n$, it follows that

$$\int_{A_t(r_1,r_2)} R dV_t \leq C V_t(k_1,k_2),$$

proving Claim 6.2. □

Now we argue by contradiction to finish the proof: suppose $\text{diam}_t(X^n)$ is unbounded for $0 \leq t < \infty$. Then, for any sequence $\epsilon_i \to 0$, there exists a time sequence $\{t_i\} \to \infty$ and $k_2^{(i)} > k_1^{(i)} > 0$ for which Claim 6.1 holds. Pick smooth cut-off functions $0 \leq \zeta_i(s) \leq 1$ defined on $\mathbb{R}$ such that

$$\zeta_i(s) = \begin{cases} 1, & 2^{k_1^{(i)}+2} \leq s \leq 2^{k_2^{(i)}-2}, \\ 0, & \text{outside } [r_1^{(i)}, r_2^{(i)}], \end{cases}$$

and $|\zeta'| \leq 1$ everywhere. Here $r_1^{(i)} \in [2^{k_1^{(i)}}, 2^{k_1^{(i)}+1}]$ and $r_2^{(i)} \in [2^{k_2^{(i)}-1}, 2^{k_2^{(i)}}]$ are chosen as in Claim 6.2. Define

$$u_i = e^{L_i} \zeta_i(d_{t_i}(x, \hat{x}_i)),$$

where $f(\hat{x}_i, t_i) = \min_X f(\cdot, t_i)$ and the constant L_i is chosen so that

$$(2\pi)^n = \int_X u_i^2 dV_{t_i} = e^{2L_i} \int_{A(r_1^{(i)}, r_2^{(i)})} \zeta_i^2 dV_{t_i}. \tag{5.119}$$

Note that by Claim 6.1, $V_{t_i}(k_1^{(i)}, k_2^{(i)}) \leq \epsilon_i \to 0$. Hence (5.119) implies $L_i \to \infty$.
Now, by Lemma 5.4.6 and similar to the proof of Theorem 5.4.5, we have

$$\mu(g(0), 1) \leq \mu(g(t_i), 1)$$

$$\leq (2\pi)^{-n} \int_X (R u_i^2 + 4|\nabla u_i|^2 - u_i^2 \log u_i^2 - 2n u_i^2) dV_{t_i}$$

$$= (2\pi)^{-n} e^{2L_i} \int_{A_{t_i}(r_1^{(i)}, r_2^{(i)})}$$

$$(R \zeta_i^2 + 4|\zeta_i'|^2 - \zeta_i^2 \log \zeta_i^2 - 2L_i \zeta_i^2 - 2n \zeta_i^2) dV_{t_i}$$

$$= -2(L_i + n) + (2\pi)^{-n} e^{2L_i} \int_{A_{t_i}(r_1^{(i)}, r_2^{(i)})}$$

$$(R\zeta_i^2 + 4|\zeta_i'|^2 - \zeta_i^2 \log \zeta_i^2) dV_{t_i}.$$

Now, by Claim 6.2 and Claim 6.1, we have

$$e^{2L_i} \int_{A_{t_i}(r_1^{(i)}, r_2^{(i)})} R\zeta_i^2 dV_{t_i} \le Ce^{2L_i} V_{t_i}(k_1^{(i)}, k_2^{(i)})$$

$$\le Ce^{2L_i} 2^{10n} V_{t_i}(k_1^{(i)} + 2, k_2^{(i)} - 2)$$

$$\le C2^{10n} \int_{A_{t_i}(r_1^{(i)}, r_2^{(i)})} u_i^2 dV_{t_i} \le C2^{10n}(2\pi)^n.$$

On the other hand, using $|\zeta_i'| \le 1$ and $-s \log s \le e^{-1}$ for $0 \le s \le 1$, we also have

$$e^{2L_i} \int_{A_{t_i}(r_1^{(i)}, r_2^{(i)})} (4|\zeta_i'|^2 - 2\zeta_i^2 \log \zeta_i) dV_{t_i} \le Ce^{2L_i} V(k_1^{(i)}, k_2^{(i)})$$

$$\le C2^{10n}(2\pi)^n.$$

Therefore,

$$\mu(g(0), 1) \le -2(L_i + n) + C$$

for some uniform constant $C > 0$. But this is a contradiction to $\{L_i\} \to \infty$. □

5.7 Remarks on the Formation of Singularities in KRF

Consider a solution $g_{ij}(t)$ to the Ricci flow

$$\frac{\partial g_{ij}(t)}{\partial t} = -2R_{ij}(t)$$

on $M \times [0, T)$, $T \le +\infty$, where either M is compact or at each time t the metric $g_{ij}(t)$ is complete and has bounded curvature. We say that $g_{ij}(t)$ is a *maximal solution* of the Ricci flow if either $T = +\infty$ or $T < +\infty$ and the norm of its curvature tensor $|\text{Rm}|$ is unbounded as $t \to T$. In the latter case, we say $g_{ij}(t)$ is a *singular* solution to the Ricci flow with singular time T. We emphasize that by singular solution $g_{ij}(t)$ we mean the curvature of $g_{ij}(t)$ is not uniformly bounded on $M^n \times [0, T)$, while M^n is a smooth manifold and $g_{ij}(t)$ is a smooth complete metric for each $t < T$.

As in the minimal surface theory and harmonic map theory, one usually tries to understand the structure of a singularity by rescaling the solution (or blow up) to obtain a sequence of solutions and study its limit. For the Ricci flow, the theory was first developed by Hamilton in [Ham95a] and further improved by Perelman [Per02, Per03q].

Now we apply Hamilton's theory to investigate singularity formations of KRF (5.28) on compact Fano manifolds. Consider a (maximal) solution $\hat{g}_{i\bar{j}}(s)$ to KRF (5.28) on $X^n \times [0, 1)$ and the corresponding solution $g_{i\bar{j}}(t)$ to NKRF (5.33) on $X^n \times [0, \infty)$, and let us denote by

$$\hat{K}_{\max}(s) = \max_X |\widehat{\mathrm{Rm}}(\cdot, s)|_{\hat{g}(s)} \quad \text{and} \quad K_{\max}(t) = \max_X |\mathrm{Rm}(\cdot, t)|_{g(t)}.$$

According to Hamilton [Ham95a], one can classify maximal solutions to KRF (5.28) on any compact Fano manifold X^n into Type I and Type II:

Type I: $\limsup_{s \to 1}(1 - s)\hat{K}_{\max}(s) < +\infty$;
Type II: $\limsup_{s \to 1}(1 - s)\hat{K}_{\max}(s) = +\infty$.

On the other hand, by Corollary 5.2.5, $\hat{K}_{\max}(s)$ and $K_{\max}(t)$ are related by

$$(1 - s)\hat{K}_{\max}(s) = K_{\max}(t(s)).$$

Thus, we immediately get

Lemma 5.7.1. *Let $\hat{g}_{i\bar{j}}(s)$ be a solution to KRF (5.28) on $X^n \times [0, 1)$ and $g_{i\bar{j}}(t)$ be the corresponding solution to NKRF (5.33) on $X^n \times [0, \infty)$. Then,*

(a) $\hat{g}_{i\bar{j}}(s)$ is a Type I solution if and only if $g_{i\bar{j}}(t)$ is a nonsingular solution, i.e., $K_{\max}(t) \leq C$ for some constant $C > 0$ for all $t \in [0, \infty)$;
(b) $\hat{g}_{i\bar{j}}(s)$ is a Type II solution if and only if $g_{i\bar{j}}(t)$ is a singular solution.

For each type of (maximal) solutions $\hat{g}_{i\bar{j}}(s)$ to KRF (5.28) or the corresponding solutions $g_{i\bar{j}}(t)$ for NKRF (5.33), following Hamilton [Ham95a] (see also Chap. 4 of [CZ06]) we define a corresponding type of limiting singularity models.

Definition 5.7.2. A solution $g_{i\bar{j}}^\infty(t)$ to KRF on a complex manifold X_∞^n with complex structure J_∞, where either X_∞^n is compact or at each time t the Kähler metric $g_{i\bar{j}}^\infty(t)$ is complete and has bounded curvature, is called a Type I or Type II **singularity model** if it is not flat and of one of the following two types:

Type I: $g_{i\bar{j}}^\infty(t)$ exists for $t \in (-\infty, \Omega)$ for some Ω with $0 < \Omega < +\infty$ and

$$|\mathrm{Rm}^\infty|(x, t) \leq \Omega/(\Omega - t)$$

everywhere on $X_\infty^n \times (-\infty, \Omega)$ with equality somewhere at $t = 0$;
Type II: $g_{i\bar{j}}^\infty(t)$ exists for $t \in (-\infty, +\infty)$ and

$$|\mathrm{Rm}^\infty|(x, t) \leq 1$$

everywhere on $X_\infty^n \times (-\infty, \Omega)$ with equality somewhere at $t = 0$.

With the help of the strong κ-noncollapsing theorem, we can apply Hamilton's Type I and Type II blow up arguments to get the following result, a Kähler analog of Theorem 16.2 in [Ham95a]:

Theorem 5.7.3. *For any (maximal) solution $\hat{g}_{i\bar{j}}(s)$, $0 \leq s < 1$, to KRF (5.28) on compact Fano manifold X^n (or the corresponding solution $g_{i\bar{j}}(t)$ to NKRF (5.33) on $X^n \times [0, \infty)$), which is of either Type I or Type II, there exists a sequence of dilations of the solution which converges in C^∞_{loc} topology to a singularity model $(X^n_\infty, J_\infty, g^\infty(t))$ of the corresponding Type. Moreover, the Type I singularity model $(X^n_\infty, J_\infty, g^\infty(t))$ is compact with $X^n_\infty = X^n$ as a smooth manifold, while the Type II singularity model $(X^n_\infty, J_\infty, g^\infty(t))$ is complete noncompact.*

Proof. **Type I case:** Let

$$\Omega =: \limsup_{t \to 1}(1 - s)\hat{K}_{\max}(s) < +\infty.$$

First we note that $\Omega > 0$. Indeed by the evolution equation of curvature,

$$\frac{d}{ds}\hat{K}_{\max}(s) \leq \text{Const} \cdot \hat{K}^2_{\max}(s).$$

This implies that

$$\hat{K}_{\max}(s) \cdot (1 - s) \geq \text{Const} > 0,$$

because

$$\limsup_{t \to 1} \hat{K}_{\max}(s) = +\infty.$$

Thus Ω must be positive.

Next we choose a sequence of points x_k and times s_k such that $s_k \to 1$ and

$$\lim_{k \to \infty}(1 - s_k)|\widehat{\text{Rm}}|(x_k, s_k) = \Omega.$$

Denote by

$$Q_k = |\widehat{\text{Rm}}|(x_k, s_k).$$

Now translate the time so that s_k becomes 0 in the new time, and dilate in space-time by the factor Q_k (time like distance squared) to get the rescaled solution

$$\hat{g}^{(k)}_{i\bar{j}}(\hat{t}) = Q_k \hat{g}_{i\bar{j}}(s_k + Q_k^{-1}\hat{t})$$

to the KRF

$$\frac{\partial}{\partial \hat{t}} \hat{g}_{ij}^k = -2 \hat{R}_{i\bar{j}}^{(k)},$$

where $\hat{R}_{i\bar{j}}^{(k)}$ is the Ricci tensor of $\hat{g}_{i\bar{j}}^{(k)}$, on the time interval $[-Q_k s_k, Q_k(1-s_k))$, with

$$Q_k s_k = s_k |\widehat{\mathrm{Rm}}|(x_k, s_k) \to \infty \quad \text{and} \quad Q_k(1-s_k) = (1-s_k)|\widehat{\mathrm{Rm}}|(x_k, s_k) \to \Omega.$$

For any $\epsilon > 0$ we can find a time $\tau < 1$ such that for $s \in [\tau, 1)$,

$$|\widehat{\mathrm{Rm}}| \leq (\Omega + \epsilon)/(1-s)$$

by the assumption. Then for $\hat{t} \in [Q_k(\tau - s_k), Q_k(1-s_k))$, the curvature of $\hat{g}_{i\bar{j}}^{(k)}(\hat{t})$ is bounded by

$$|\widehat{\mathrm{Rm}^{(k)}}| = Q_k^{-1}|\widehat{\mathrm{Rm}(\hat{g})}|$$

$$\leq \frac{\Omega + \epsilon}{Q_k(1-s)} = \frac{\Omega + \epsilon}{Q_k(1-s_k) + Q_k(s_k - s)}$$

$$\to (\Omega + \epsilon)/(\Omega - \hat{t}), \qquad \text{as } k \to +\infty.$$

With the above curvature bound and the injectivity radius estimates coming from κ-noncollapsing, one can apply Hamilton's compactness theorem (cf. [Ham95a] or Theorem 4.1.5 in [CZ06]) to get a subsequence of $\hat{g}_{i\bar{j}}^{(k)}(\hat{t})$ which converges in the C_{loc}^∞ topology to a limit metric $g_{i\bar{j}}^{(\infty)}(t)$ in the Cheeger sense on (X^n, J_∞) for some complex structure J_∞ such that $g_{i\bar{j}}^{(\infty)}(t)$ is a solution to the KRF with $t \in (-\infty, \Omega)$ and its curvature satisfies the bound

$$|\mathrm{Rm}^{(\infty)}| \leq \Omega/(\Omega - t)$$

everywhere on $X_\infty^n \times (-\infty, \Omega)$ with the equality somewhere at $t = 0$.

   Type II: Take a sequence $S_k \to 1$ and pick space-time points (x_k, s_k) such that, as $k \to +\infty$,

$$Q_k(S_k - s_k) = \max_{x \in X, s \leq S_k} (S_k - s)|\widehat{\mathrm{Rm}}|(x, s) \to +\infty,$$

where again we denote by $Q_k = |\widehat{\mathrm{Rm}}|(x_k, s_k)$. Now translate the time and dilate the solution as before to get

$$\hat{g}_{i\bar{j}}^{(k)}(\hat{t}) = Q_k \hat{g}_{i\bar{j}}(s_k + Q_k^{-1}\hat{t}),$$

which is a solution to the KRF and satisfies the curvature bound

$$|\widehat{Rm}^{(k)}| = Q_k^{-1}|\widehat{Rm}(\hat{g})| \leq \frac{(S_k - s_k)}{(S_k - s)}$$

$$= \frac{Q_k(S_k - s_k)}{Q_k(S_k - s_k) - \hat{t}} \quad \text{for } \hat{t} \in [-Q_k s_k, Q_k(S_k - s_k)).$$

Then as before, by applying Hamilton's compactness theorem, there exists a subsequence of $\hat{g}_{i\bar{j}}^{(k)}(\hat{t})$ which converges in the C_{loc}^∞ topology to a limit metric $g_{i\bar{j}}^{(\infty)}(t)$ in the Cheeger sense on a limiting complex manifold (X_∞^n, J_∞) such that $g_{i\bar{j}}^{(\infty)}(t)$ is a complete solution to the KRF with $t \in (-\infty, +\infty)$, and its curvature satisfies

$$|Rm^{(\infty)}| \leq 1$$

everywhere on $X_\infty^n \times (-\infty, +\infty)$ and the equality holds somewhere at $t = 0$. □

Remark 5.7.1. The injectivity radius bound needed in Hamilton's compactness theorem is satisfied due to the "Little Loop Lemma" (cf. Theorem 4.2.4 in [CZ06]), which is a consequence of Perelman's κ-noncollapsing theorem.

Thanks to Perelman's monotonicity of μ-entropy and the uniform scalar curvature bound in Theorem 5.6.1, we can say more about the singularity models in Theorem 5.7.3.

First of all, the following result on Type I singularity models of KRF (5.28) is well-known.

Theorem 5.7.4. *Let $\tilde{g}_{i\bar{j}}(s)$ be a Type I solution to KRF (5.28) on $X^n \times [0, 1)$ and $g_{i\bar{j}}(t)$ be the corresponding nonsingular solution to NKRF (5.33) on $X^n \times [0, \infty)$. Then there exists a sequence $\{t_k\} \to \infty$ such that $g_{i\bar{j}}^{(k)}(t) =: g_{i\bar{j}}(t + t_k)$ converges in the Cheeger sense to a gradient shrinking Kähler–Ricci soliton $g^\infty(t)$ on (X^n, J_∞), where J_∞ is a certain complex structure on X^n, possibly different from J.*

Proof. This is a consequence of Theorem 5.7.3, and the fact that every compact Type I singularity model is necessarily a shrinking gradient Ricci soliton (see [Se05, SeT08] or p. 662 of [PSSW08b]; also Corollary 1.2 in [CCZ03]). □

Next, for Type II solutions to the KRF we have the following two results, which were known to Hamilton and the author (Cao and Hamilton, 2004, unpublished work on the formation of singularities in KRF) back in 2004,[4] and were also observed independently by Ruan–Zhang–Zhang [RZZ09].

---

[4] Theorems 5.7.5 and 5.7.6 were observed by Hamilton and the author during the IPAM conference "Workshop on Geometric Flows: Theory and Computation" in February, 2004.

Theorem 5.7.5. *Let $g_{i\bar{j}}(t)$ be a singular solution to NKRF (5.33) on $X^n \times [0, \infty)$. Then there exists a sequence $\{t_k\} \to \infty$ and rescaled solution metrics $g^{(k)}(t)$ to KRF such that $(X^n, J, g^{(k)}(t))$ converges in the Cheeger sense to some noncompact limit $(X^n_\infty, J_\infty, g_\infty(t))$, $-\infty < t < \infty$, with the following properties:*

(i) $g_\infty(t)$ is Calabi–Yau (i.e, Ricci flat Kähler);

(ii) $|Rm|_{g_\infty(t)}(x, t) \leq 1$ everywhere and with equality somewhere at $t = 0$;

(iii) $(X^n_\infty, g_\infty(t))$ has maximal volume growth: for any $x_0 \in X^n_\infty$ there exists a positive constant $c > 0$ such that

$$\text{Vol}(B(x_0, r)) \geq cr^{2n}, \qquad \text{for all } r > 0.$$

Proof. This is an immediate consequence of Theorems 5.7.3 and 5.6.1 (i). Indeed, Theorem 5.7.3 implies the existence of a noncompact Type II singularity model $(X^n_\infty, J_\infty, g_\infty(t))$ satisfying property (ii). Property (iii) follows from the fact that the κ-noncollapsing property for KRF or NKRF in Theorem 5.4.7 is dilation invariant, hence (5.79) and (5.80) holds for each rescaled solution on larger and larger scales for the same $\kappa > 0$, hence the maximal volume growth in the limit of dilations. Finally, for property (i), note that the scalar curvature R of $g_{i\bar{j}}(t)$ is uniformly bounded on $X \times [0, \infty)$ by Theorem 5.6.1 and the rescaling factors go to infinite, so we have $R^\infty = 0$ everywhere in the limit of dilations. On the other hand, since $g^\infty_{i\bar{j}}(t)$ is a solution to KRF, R^∞ satisfies the evolution equation

$$\frac{\partial}{\partial t} R^\infty = \Delta R^\infty + |\text{Ric}^\infty|^2.$$

Thus, we have $|\text{Ric}^\infty|^2 = 0$ everywhere hence g_∞ is Ricci-flat. $\quad\square$

Theorem 5.7.6. *Let X^2 be a Del Pezzo surface (i.e., a Fano surface) and let $g_{i\bar{j}}(t)$ be a singular solution to NKRF (5.33) on $X^2 \times [0, \infty)$. Then the Type II limit space $(X^2_\infty, J_\infty, g_\infty)$ in Theorem 5.7.5 is a non-compact Calabi–Yau space satisfies the following properties:*

(a) $|Rm|_{g_\infty} \leq 1$ everywhere on X^2_∞ and with equality somewhere;

(b) (X^2_∞, g_∞) has maximal volume growth: for any $x_0 \in X^2_\infty$ there exists a positive constant $c > 0$ such that

$$\text{Vol}(B(x_0, r)) \geq cr^4, \qquad \text{for all } r > 0;$$

(c) $\int_{X^2_\infty} |Rm(g_\infty)|^2 dV_\infty < \infty$.

Proof. Clearly, we only need to verify property (c). But this follows from the facts the integral

$$\int_{X^2} |Rm|^2(x, t) dV_t$$

is dilation invariant in complex dimension $n = 2$ (real dimension 4); that it differs from $\int_X R^2 dV_t$ up to a constant depending only on the Kähler class of $g(0)$ and the Chern classes $c_1(X)$ and $c_2(X)$ (cf. Proposition 1.1 in [Cal82]); and that, before the dilations, $\int_X R^2 dV_t$ is uniformly bounded for all $t \in [0, \infty)$ by Theorem 5.6.1 (i).

$\square$

Remark 5.7.2. The work of Bando–Kasue–Nakajima [BKN89] implies that Calabi–Yau surfaces satisfying conditions (b) and (c) are asymptotically locally Euclidean (ALE) of order 4.

Remark 5.7.3. Kronheimer [Kron89] has classified ALE Hyper–Kähler surfaces (i.e., simply connected ALE Calabi–Yau surfaces).

Acknowledgements This article is based on a mini-course on KRF delivered at University of Toulouse III in February 2010, a talk on Perelman's uniform estimates for NKRF at Columbia University's Geometry and Analysis Seminar in Fall 2005, and several conference talks, including "Einstein Manifolds and Beyond" at CIRM (Marseille—Luminy, fall 2007), "Program on Extremal Kähler Metrics and Kähler–Ricci Flow" at the De Giorgi Center (Pisa, spring 2008), and "Analytic Aspects of Algebraic and Complex Geometry" at CIRM (Marseille—Luminy, spring 2011). This article also served as the lecture notes by the author for a graduate course at Lehigh University in spring 2012, as well as a short course at the Mathematical Sciences Center of Tsinghua University in May, 2012. I would like to thank Philippe Eyssidieux, Vincent Guedj, and Ahmed Zeriahi for inviting me to give the mini-course in Toulouse, and especially Vincent Guedj for inviting me to write up the notes for a special volume. I also wish to thank the participants in my courses, especially Qiang Chen, Xin Cui, Chenxu He, Xiaofeng Sun, Yingying Zhang and Meng Zhu, for their helpful suggestions. Finally, I would like to take this opportunity to express my deep gratitude to Professors E. Calabi, R. Hamilton, and S.-T. Yau for teaching me the Kähler geometry, the Ricci flow, and geometric analysis over the years. Partially supported by NSF grant DMS-0909581.

References

[Aub78] T. Aubin, Equation de type Monge-Ampère sur les variétés kählériennes compactes. Bull. Sci. Math. **102**, 63–95 (1978)

[Baker10] C. Baker, The mean curvature flow of submanifolds of high codimension. Ph.D. thesis, Australian National University, 2010 [arXiv:1104.4409v1]

[Bando84] S. Bando, On the classification of three-dimensional compact Kaehler manifolds of nonnegative bisectional curvature. J. Differ. Geom. **19**(2), 283–297 (1984)

[BKN89] S. Bando, A. Kasue, H. Nakajima, On a construction of coordinates at infinity on manifolds with fast curvature decay and maximal volume growth. Invent. Math. **97**, 313–349 (1989)

[BM87] S. Bando, T. Mabuchi, Uniqueness of Einstein Kähler metrics modulo connected group actions, in *Algebraic Geometry* (Sendai, 1985), ed. by T. Oda. Advanced Studies in Pure Mathematics, vol. 10 (Kinokuniya, 1987), pp. 11–40 (North-Holland, Amsterdam, 1987)

[BS09] S. Brendle, R. Schoen, Manifolds with 1/4-pinched curvature are space forms. J. Am. Math. Soc. **22**(1), 287–307 (2009)

[Cal82] E. Calabi, Extremal Kähler metrics, in *Seminar on Differential Geometry*. Annals of Mathematics Studies, vol. 102 (Princeton University Press, Princeton, 1982), pp. 259–290

[Cao85] H.D. Cao, Deformation of Kähler metrics to Kähler-Einstein metrics on compact Kähler manifolds. Invent. Math. **81**(2), 359–372 (1985)

[Cao92] H.D. Cao, On Harnack's inequalities for the Kähler–Ricci flow. Invent. Math. **109**(2), 247–263 (1992)

[Cao94] H.-D. Cao, Existence of gradient Kähler–Ricci solitons, in *Elliptic and Parabolic Mathords in Geometry* (Minneapolis, MN, 1994) (A.K. Peters, Wellesley, 1996), pp. 1–16

[Cao97] H.-D. Cao, Limits of Solutions to the Kähler–Ricci flow. J. Differ. Geom. **45**, 257–272 (1997)

[Cao10] H.-D. Cao, Recent progress on Ricci solitons, in *Recent Advances in Geometric Analysis*. Advanced Lectures in Mathematics (ALM), vol. 11 (International Press, Somerville, 2010), pp. 1–38

[CCZ03] H.-D. Cao, B.-L. Chen, X.-P. Zhu, Ricci flow on compact Kähler manifolds of positive bisectional curvature. C. R. Math. Acad. Sci. Paris **337**(12), 781–784 (2003)

[CaoChow86] H.-D. Cao, B. Chow, Compact Kähler manifolds with nonnegative curvature operator. Invent. Math. **83**(3), 553–556 (1986)

[CZ09] H.-D. Cao, M. Zhu, A note on compact Kähler–Ricci flow with positive bisectional curvature. Math. Res. Lett. **16**(6), 935–939 (2009)

[CZ06] H.-D. Cao, X.-P. Zhu, A complete proof of the Poincaré and geometrization conjectures- application of the Hamilton-Perelman theory of the Ricci flow. Asian J. Math. **10**(2), 165–492 (2006)

[ChauT06] A. Chau, L.-F. Tam, On the complex structure of Kähler manifolds with nonnegative curvature. J. Differ. Geom. **73**(3), 491–530 (2006)

[ChauT08] A. Chau, L.-F. Tam, A survey on the Kähler–Ricci flow and Yau's uniformization conjecture, in *Surveys in Differential Geometry*, vol. XII. Geometric Flows. [Surv. Differ. Geom. vol. 12] (International Press, Somerville, 2008), pp. 21–46

[CTZ04] B.-L. Chen, S.-H. Tang, X.-P. Zhu, A uniformization theorem for non complete non-compact Kähler surfaces with positive bisectional curvature. J. Differ. Geom. **67**(3), 519–570 (2004)

[CheT02] X.X. Chen, G. Tian, Ricci flow on Kähler-Einstein surfaces. Invent. Math. **147**(3), 487–544 (2002)

[CheT06] X.X. Chen, G. Tian, Ricci flow on Kähler-Einstein manifolds. Duke Math. J. **131**(1), 17–73 (2006)

[Chow91] B. Chow, The Ricci flow on the 2-sphere. J. Differ. Geom. **33**(2), 325–334 (1991)

[DW11] A.S. Dancer, M.Y. Wang, On Ricci solitons of cohomogeneity one. Ann. Glob. Anal. Geom. **39**(3), 259–292 (2011)

[Det83] D.M. DeTurck, Deforming metrics in the direction of their Ricci tensors. J. Differ. Geom. **18**(1), 157–162 (1983)

[DT92] W.-Y. Ding, G. Tian, Kähler-Einstein metrics and the generalized Futaki invariant. Invent. Math. **110**(2), 315–335 (1992)

[Don02] S.K. Donaldson, Scalar curvature and stability of toric varieties. J. Differ. Geom. **62**, 289–349 (2002)

[FIK03] M. Feldman, T. Ilmanen, D. Knopf, Rotationally symmetric shrinking and expanding gradient Kähler–Ricci solitons. J. Differ. Geom. **65**(2), 169–209 (2003)

[Fut83] A. Futaki, An obstruction to the existence of Einstein Kähler metrics. Invent. Math. **73**, 437–443 (1983)

[FutW11] A. Futaki, M.-T. Wang, Constructing Kähler–Ricci solitons from Sasaki-Einstein manifolds. Asian J. Math. **15**(1), 33–52 (2011)

[Gu09] H.-L. Gu, A new proof of Mok's generalized Frankel conjecture theorem. Proc. Am. Math. Soc. **137**(3), 1063–1068 (2009)

[Ham82] R.S. Hamilton, Three-manifolds with positive Ricci curvature. J. Differ. Geom. **17**(2), 255–306 (1982)

[Ham86] R.S. Hamilton, Four-manifolds with positive curvature operator. J. Differ. Geom. **24**(2), 153–179 (1986)

[Ham88] R. Hamilton, The Ricci flow on surfaces, in *Mathematics and General Relativity (Santa Cruz, CA, 1986)*. Contemporary Mathematics, vol. 71 (American Mathematical Society, Providence, 1988), pp. 237–262

[Ham93a] R. Hamilton, The Harnack estimate for the Ricci flow. J. Differ. Geom. **37**, 225–243 (1993)

[Ham93b] R. Hamilton, Eternal solutions to the Ricci flow. J. Differ. Geom. **38**(1), 1–11 (1993)

[Ham95a] R.S. Hamilton, The formation of singularities in the Ricci flow, in *Surveys in Differential Geometry*, vol. II (Cambridge, MA, 1993) (International Press, Cambridge, 1995), pp. 7–136

[Ham95b] R. Hamilton, A compactness property for solution of the Ricci flow. Am. J. Math. **117**, 545–572 (1995)

[Ham97] R.S. Hamilton, Four-manifolds with positive isotropic curvature. Comm. Anal. Geom. **5**(1), 1–92 (1997)

[Ham99] R.S. Hamilton, Non-singular solutions to the Ricci flow on three manifolds. Comm. Anal. Geom. **7**, 695–729 (1999)

[Koiso90] N. Koiso, On rotationally symmetric Hamilton's equation for Kähler-Einstein metrics, in *Advanced Studies in Pure Mathematics*, vol. 18-I. Recent Topics in Differential and Analytic Geometry, pp. 327–337 (Academic, Boston, 1990)

[Kron89] B.P. Kronheimer, The construction of ALE spaces as hyper-Kähler quotients. J. Differ. Geom. **29**, 665–683 (1989)

[LiYau86] P. Li, S.-T. Yau, On the parabolic kernel of the Schrödinger operator. Acta Math. **156**(3–4), 153–201 (1986)

[Mab86] T. Mabuchi, K-energy maps integrating Futaki invariants. Tohoku Math. J. (2) **38**(4), 575–593 (1986)

[Mok88] N. Mok, The uniformization theorem for compact Kähler manifolds of nonnegative holomorphic bisectional curvature. J. Differ. Geom. **27**(2), 179–214 (1988)

[Mori79] S. Mori, Projective manifolds with ample tangent bundles. Ann. Math. **100**, 593–606 (1979)

[MSz09] O. Munteanu, G. Székelyhidi, On convergence of the Kähler–Ricci flow. Commun. Anal. Geom. **19**(5), 887–903 (2011)

[Ng07] H. Nguyen, Invariant curvature cones and the Ricci flow. Ph.D. thesis, Australian National University, 2007

[Ni05] L. Ni, Ancient solutions to Kähler–Ricci flow. Math. Res. Lett. **12**(5–6), 633–653 (2005)

[Pal08] N. Pali, Characterization of Einstein-Fano manifolds via the Kähler–Ricci flow. Indiana Univ. Math. J. **57**(7), 3241–3274 (2008)

[Per02] G. Perelman, The entropy formula for the Ricci flow and its geometric applications (2002). Preprint [arXiv: math.DG/0211159]

[Per03q] G. Perelman, Ricci flow with surgery on three-manifolds (2003). Preprint [arXiv:math.DG/0303109]

[Per03b] G. Perelman, Finite extinction time for the solutions to the Ricci flow on certain three-manifolds (2003). Preprint [arXiv:math.DG/0307245]

[PSSW08b] D.H. Phong, J. Song, J. Sturm, B. Weinkove, The Kähler–Ricci flow with positive bisectional curvature. Invent. Math. **173**(3), 651–665 (2008)

[PSSW09] D.H. Phong, J. Song, J. Sturm, B. Weinkove, The Kähler–Ricci flow and the $\bar{\partial}$ operator on vector fields. J. Differ. Geom. **81**(3), 631–647 (2009)

[PSSW11] D.H. Phong, J. Song, J. Sturm, B. Weinkove, On the convergence of the modified Kähler–Ricci flow and solitons. Comment. Math. Helv. **86**(1), 91–112 (2011)

[PS06] D.H. Phong, J. Sturm, On stability and the convergence of the Kähler–Ricci flow. J. Differ. Geom. **72**(1), 149–168 (2006)

[Rot81] O.S. Rothaus, Logarithmic Sobolev inequalities and the spectrum of Schrödinger opreators. J. Funct. Anal. **42**(1), 110–120 (1981)

[RZZ09] W.-D. Ruan, Y. Zhang, Z. Zhang, Bounding sectional curvature along the Kähler–Ricci flow. Comm. Contemp. Math. **11**(6), 1067–1077 (2009)

[ScYau94] R. Schoen, S.T. Yau, Lectures on differential geometry, in *Conference Proceedings and Lecture Notes in Geometry and Topology*, vol. 1 (International Press Publications, Somerville, 1994)

[Se05] N. Šešum, Curvature tensor under the Ricci flow. Am. J. Math. **127**(6), 1315–1324 (2005)

[SeT08] N. Sesum, G. Tian, Bounding scalar curvature and diameter along the Kähler–Ricci flow (after Perelman). J. Inst. Math. Jussieu **7**(3), 575–587 (2008)

[Shi90] W.X. Shi, Complete noncompact Kähler manifolds with positive holomorphic bisectional curvature. Bull. Am. Math. Soc. **23**, 437–440 (1990)

[Shi97] W.X. Shi, Ricci flow and the uniformization on complete noncompact Kähler manifolds. J. Differ. Geom. **45**(1), 94–220 (1997)

[SW] J. Song, B. Weinkove, Lecture Notes on the Kähler–Ricci Flow, in *An Introduction to the Kähler–Ricci Flow*, ed. by S. Boucksom, P. Eyssidieux, V. Guedj. Lecture Notes in Mathematics (Springer, Heidelberg, 2013)

[Sz10] G. Székelyhidi, The Kähler–Ricci flow and K-stability. Am. J. Math. **132**(4), 1077–1090 (2010)

[Tian97] G. Tian, Kähler–Einstein metrics with positive scalar curvature. Invent. Math. **130**, 239–265 (1997)

[Tzha06] G. Tian, Z. Zhang, On the Kähler–Ricci flow on projective manifolds of general type. Chin. Ann. Math. Ser. B **27**(2), 179–192 (2006)

[TZ00] G. Tian, X. Zhu, Uniqueness of Kähler–Ricci solitons. Acta Math. **184**(2), 271–305 (2000)

[TZ02] G. Tian, X. Zhu, A new holomorphic invariant and uniqueness of Kähler–Ricci solitons. Comment. Math. Helv. **77**(2), 297–325 (2002)

[Tos10a] V. Tosatti, Kähler–Ricci flow on stable Fano manifolds. J. Reine Angew. Math. **640**, 67–84 (2010)

[Tsu88] H. Tsuji, Existence and degeneration of Kähler-Einstein metrics on minimal algebraic varieties of general type. Math. Ann. **281**(1), 123–133 (1988)

[WZ04] X.J. Wang, X. Zhu, Kähler–Ricci solitons on toric manifolds with positive first Chern class. Adv. Math. **188**(1), 87–103 (2004)

[Wil10] B. Wilking, A Lie algebraic approach to Ricci flow invariant curvature conditions and Harnack inequalities (2010). Preprint [arXiv:1011.3561v2]

[Yau75] S.-T. Yau, Harmonic functions on complete Riemannian manifolds. Comm. Pure Appl. Math. **28**, 201–228 (1975)

[Yau78] S.T. Yau, On the Ricci curvature of a compact Kähler manifold and the complex Monge-Ampère equation, I. Comm. Pure Appl. Math. **31**(3), 339–411 (1978)

[Yau93] S.-T. Yau, Open problems in geometry. Proc. Symp. Pure Math. **54**, 1–28 (1993)

[Zha11] Z.L. Zhang, Kähler Ricci flow on Fano manifolds with vanished Futaki invariants. Math. Res. Lett. **18**, 969–982 (2011)

Chapter 6
Convergence of the Kähler–Ricci Flow on a Kähler–Einstein Fano Manifold

Vincent Guedj

Abstract The goal of these notes is to sketch the proof of the following result, due to Perelman and Tian–Zhu: on a Kähler–Einstein Fano manifold with discrete automorphism group, the normalized Kähler–Ricci flow converges smoothly to the unique Kähler–Einstein metric. We also explain an alternative approach due to Berman–Boucksom–Eyssidieux–Guedj–Zeriahi, which only yields weak convergence but also applies to Fano varieties with log terminal singularities.

Introduction

Let X be a Fano manifold, i.e. a compact (connected) complex projective algebraic manifold whose first Chern class $c_1(X)$ is positive, i.e. can be represented by a Kähler form. It has been an open question for decades to understand when such a manifold admits a Kähler–Einstein metric, i.e. if we can find a Kähler form $\omega_{KE} \in c_1(X)$ such that

$$\mathrm{Ric}(\omega_{KE}) = \omega_{KE}.$$

By comparison with the cases when $c_1(X) < 0$ (or $c_1(X) = 0$) treated in Chap. 3, there is neither existence nor uniqueness in general of Kähler–Einstein metrics in the Fano case.

After the spectacular progress in Ricci flow techniques, it has become a natural question to wonder whether the Ricci flow could help in understanding this problem. The goal of this series of lectures is to sketch the proof of an important result in this direction, which is due to Perelman:

V. Guedj (✉)
Institut de Mathématiques de Toulouse and Institut Universitaire de France, Université
Paul Sabatier, 118 route de Narbonne, 31062, Toulouse Cedex 9, France
e-mail: vincent.guedj@math.univ-toulouse.fr

S. Boucksom et al. (eds.), *An Introduction to the Kähler–Ricci Flow*,
Lecture Notes in Mathematics 2086, DOI 10.1007/978-3-319-00819-6_6,
© Springer International Publishing Switzerland 2013

Theorem 6.0.7 (Perelman, seminar talk at MIT, 2003). *Let X be a Fano manifold which admits a unique Kähler–Einstein metric ω_{KE}. Fix $\omega_0 \in c_1(X)$ an arbitrary Kähler form. Then the normalized Kähler–Ricci flow*

$$\frac{\partial}{\partial t}\omega_t = -\mathrm{Ric}(\omega_t) + \omega_t$$

converges, as $t \to +\infty$, in the C^∞-sense to ω_{KE}.

In other words, the normalized Kähler–Ricci flow detects the (unique) Kähler–Einstein metric if it exists.

This result has been generalized by Tian and Zhu [TZ07] to the case of Kähler–Ricci soliton. Other generalizations by Phong and his collaborators can be found in [PS10]. We follow here a slightly different path, using pluripotential techniques to establish a uniform C^0-a priori estimate along the flow.

All proofs rely on deep estimates due to Perelman. These are explained in Chap. 5, to which we refer the reader.

Nota Bene. These notes are written after the lectures the author delivered at the third ANR-MACK meeting (24–27 October 2011, Marrakech, Morocco). There is no claim of originality. As the audience consisted of non specialists, we have tried to make these lecture notes accessible with only few prerequisites.

6.1 Background

6.1.1 The Kähler–Einstein Equation on Fano Manifolds

Let X be an n-dimensional Fano manifold and fix $\omega \in c_1(X)$ an arbitrary Kähler form. If we write locally

$$\omega = \sum \omega_{\alpha\beta}\frac{i}{\pi}dz_\alpha \wedge d\bar{z}_\beta,$$

then the Ricci form of ω is

$$\mathrm{Ric}(\omega) := -\sum \frac{\partial^2 \log\left(\det \omega_{pq}\right)}{\partial z_\alpha \partial \bar{z}_\beta}\frac{i}{\pi}dz_\alpha \wedge d\bar{z}_\beta.$$

Observe that $\mathrm{Ric}(\omega)$ is a closed $(1,1)$-form on X such that for any other Kähler form ω' on X, the following holds globally:

$$\mathrm{Ric}(\omega') = \mathrm{Ric}(\omega) - dd^c\left[\log \omega'^n/\omega^n\right].$$

Here $d = \partial + \bar{\partial}$ and $d^c = (\partial - \bar{\partial})/2i\pi$ are both real operators.

In particular $\mathrm{Ric}(\omega')$ and $\mathrm{Ric}(\omega)$ represents the same cohomology class, which turns out to be $c_1(X)$.

The Associated Complex Monge–Ampère Equation

Since we have picked $\omega \in c_1(X)$, it follows from the dd^c-lemma that

$$\mathrm{Ric}(\omega) = \omega - dd^c h$$

for some smooth function $h \in C^\infty(X, \mathbb{R})$ which is uniquely determined, up to an additive constant. We normalize h by asking for

$$\int_X e^{-h} \omega^n = V := \int_X \omega^n = c_1(X)^n.$$

We look for $\omega_{\mathrm{KE}} = \omega + dd^c \varphi_{\mathrm{KE}}$ a Kähler form such that $\mathrm{Ric}(\omega_{\mathrm{KE}}) = \omega_{\mathrm{KE}}$. Since $\mathrm{Ric}(\omega_{\mathrm{KE}}) = \mathrm{Ric}(\omega) - dd^c \log(\omega_{\mathrm{KE}}^n / \omega^n)$, an easy computation shows that

$$dd^c \{ \log(\omega_{\mathrm{KE}}^n / \omega^n) + \varphi_{\mathrm{KE}} + h \} = 0.$$

Since pluriharmonic functions are constant on X (by the maximum principle), we infer

$$(\omega + dd^c \varphi_{\mathrm{KE}})^n = e^{-\varphi_{\mathrm{KE}}} e^{-h+C} \omega^n \tag{MA}$$

for some normalizing constant $C \in \mathbb{R}$. Solving $\mathrm{Ric}(\omega_{\mathrm{KE}}) = \omega_{\mathrm{KE}}$ is thus equivalent to solving the above complex Monge–Ampère equation (MA).

Known Results

When $n = 1$, X is the Riemann sphere $\mathbb{CP}^1$ and (a suitable multiple of) the Fubini–Study Kähler form is a Kähler–Einstein metric.

When $n = 2$ it is not always possible to solve (MA). In this case X is a *DelPezzo surface*, biholomorphic either to $\mathbb{CP}^1 \times \mathbb{CP}^1$ or $\mathbb{CP}^2$ which both admit the (product) Fubini–Study metric as a Kähler–Einstein metric, or else to X_r, the blow up of $\mathbb{CP}^2$ at r points in general position, $1 \leq r \leq 8$. Various authors (notably Yau, Siu, Tian, Nadel) have studied the Kähler–Einstein problem on DelPezzo surfaces in the eighties. The final and difficult step was done by Tian who proved the following:

Theorem 6.1.1 ([Tian90]). *The DelPezzo surface X_r admits a Kähler–Einstein metric if and only if $r \neq 1, 2$.*

The interested reader will find an up-to-date proof of this result in [Tos12].

The situation becomes much more difficult and largely open in higher dimension. There is a finite but long list (105 families) of deformation classes of Fano threefolds. It is unknown, for most of them, whether they admit or not a Kähler–Einstein metric. Among them, the Mukai–Umemura manifold is particularly interesting: this manifold admits a Kähler–Einstein metric as was shown by Donaldson [Don08], and there are arbitrary small deformations of it which do (resp. do not) admit a Kähler–Einstein metric as shown by Donaldson (resp. Tian).

There are even more families in dimension $n \geq 4$. Those which are *toric* admit a Kähler–Einstein metric if and only if the Futaki invariant vanishes (see [WZ04]), the non-toric case is essentially open and has motivated an important conjecture of Yau–Tian–Donaldson (see [PS10]).

Uniqueness Issue

Bando and Mabuchi have shown in [BM87] that any two Kähler–Einstein metrics on a Fano manifold can be connected by the holomorphic flow of a holomorphic vector field. This result has been generalized recently by Berndtsson [Bern11]. We shall make in the sequel the simplifying assumption that X does not admit non-zero holomorphic vector field, so that it admits a unique Kähler–Einstein metric, if any.

6.1.2 The Analytic Criterion of Tian

Given $\varphi : X \to \mathbb{R} \cup \{-\infty\}$ an upper semi-continuous function, we say that φ is ω-plurisubharmonic (ω-psh for short) and write $\varphi \in \mathrm{PSH}(X, \omega)$ if φ is locally given as the sum of a smooth and a plurisubharmonic function, and $\omega + dd^c \varphi \geq 0$ in the weak sense of currents. Set

$$E(\varphi) := \frac{1}{n+1} \sum_{j=0}^{n} V^{-1} \int_X \varphi (\omega + dd^c \varphi)^j \wedge \omega^{n-j}.$$

We let the reader check, by using Stokes formula, that

$$\frac{d}{dt} E(\varphi + tv)_{|t=0} = \int_X v \, \mathrm{MA}(\varphi), \quad \text{where } \mathrm{MA}(\varphi) := V^{-1}(\omega + dd^c \varphi)^n.$$

The functional E is thus a primitive of the complex Monge–Ampère operator, in particular $\varphi \mapsto E(\varphi)$ is non-decreasing since $E' = \mathrm{MA} \geq 0$.

Definition 6.1.2. The *Ding functional*[1] is defined as

$$\mathrm{Ding}(\varphi) := -E(\varphi) - \log \left[\int_X e^{-\varphi - h} \omega^n \right].$$

The reader will check that φ is a critical point of the Ding functional if and only if

$$\mathrm{MA}(\varphi) = \frac{e^{-\varphi - h} \omega^n}{\int_X e^{-\varphi - h} \omega^n}$$

---

[1]This functional seems to have been first explicitly considered by W.Y. Ding in [Ding88, p. 465], hence the chosen terminology.

so that $\omega + dd^c\varphi$ is Kähler–Einstein. Observe that $\mathrm{Ding}(\varphi + c) = \mathrm{Ding}(\varphi)$, for all $c \in \mathbb{R}$, thus Ding is actually a functional acting on the metrics $\omega_\varphi := \omega + dd^c\varphi$. It is natural to try and extremize the Ding functional. This motivates the following:

Definition 6.1.3. We say that the Ding functional is *proper* if $\mathrm{Ding}(\varphi_j) \to +\infty$ whenever $\varphi_j \in \mathrm{PSH}(X, \omega) \cap C^\infty(X)$ is such that $E(\varphi_j) \to -\infty$ and $\int_X \varphi_j \omega^n = 0$.

The importance of this notion was made clear in a series of works by Ding and Tian in the 1990s, culminating with the following deep result of [Tian97]:

Theorem 6.1.4 ([Tian97]). *Let X be a Fano manifold with no holomorphic vector field. There exists a Kähler–Einstein metric if and only if the Ding functional is proper.*

6.1.3 The Kähler–Ricci Flow Approach

The Ricci flow is the parabolic evolution equation

$$\frac{\partial}{\partial t}\omega_t = -\mathrm{Ric}(\omega_t) \quad \text{with initial data } \omega_0. \tag{KRF}$$

When ω_0 is a Kähler form, so is ω_t, $t > 0$ hence it is called the Kähler–Ricci flow.

Long Time Existence

The short time existence is guaranteed by standard parabolic theory (see Chap. 2 of the present volume). In the Kähler context, this translates into a parabolic scalar equation as we explain below.

It is more convenient to analyze the long time existence by considering the normalized Kähler–Ricci flow, namely

$$\frac{\partial}{\partial t}\omega_t = -\mathrm{Ric}(\omega_t) + \omega_t. \tag{NKRF}$$

One passes from (KRF) to (NKRF) by changing $\omega(t)$ in $e^t\omega(1 - e^{-t})$. At the level of cohomology classes,

$$\frac{d\{\omega_t\}}{dt} = -c_1(X) + \{\omega_t\} \in H^{1,1}(X, \mathbb{R})$$

therefore $\{\omega_t\} \equiv c_1(X)$ is constant if we start from $\omega_0 \in c_1(X)$. This justifies the name (normalized KRF) since in this case

$$\mathrm{vol}_{\omega_t}(X) = \mathrm{vol}_{\omega_0}(X) = c_1(X)^n$$

is constant. Note that the volume blows up exponentially fast if $\{\omega_0\} > c_1(X)$.

Theorem 6.1.5 ([Cao85]). *Let X be a Fano manifold and pick a Kähler form $\omega_0 \in$ $c_1(X)$. Then the normalized Kähler–Ricci flow exists for all times $t > 0$.*

We will outline a proof of this result, although it is already essentially contained in Chap. 3.

The main issue is then whether (ω_t) converges as $t \to +\infty$. Hopefully, we should have $\frac{\partial}{\partial t}\omega_t \to 0$ and $\omega_t \to \omega_{KE}$ such that $\text{Ric}(\omega_{KE}) = \omega_{KE}$. We can now formulate Perelman's result as follows:

Theorem 6.1.6 (Perelman 03). *Let X be a Fano manifold and pick an arbitrary Kähler form $\omega_0 \in c_1(X)$. If the Ding functional is proper, then the normalized Kähler–Ricci flow (ω_t) converges, as $t \to +\infty$, towards the unique Kähler–Einstein metric ω_{KE}.*

Remark 6.1.7. It turns out that the properness assumption ensures that there can be no holomorphic vector field, hence the Kähler–Einstein metric (which exists by Tian's result) is unique (by Bando–Mabuchi's result).

The situation is much more delicate in the presence of holomorphic vector fields. For $n = 1$, the problem is already non-trivial and was settled by Hamilton in [Ham88] and Chow in [Chow91]. For $n \geq 2$ we refer the reader to [CSz12] for up-to-date references.

Reduction to a Scalar Parabolic Equation

Let $\omega = \omega_0 \in c_1(X)$ denote the initial data. Since ω_t is cohomologous to ω, we can find $\varphi_t \in \text{PSH}(X, \omega)$ a smooth function such that $\omega_t = \omega + dd^c\varphi_t$. The function φ_t is defined up to a time dependent additive constant. Then

$$\frac{d\{\omega_t\}}{dt} = dd^c\dot{\varphi}_t = -\text{Ric}(\omega_t) + \omega + dd^c\varphi_t,$$

where $\dot{\varphi}_t := \frac{\partial}{\partial t}\varphi_t$. Let $h \in C^\infty(X, \mathbb{R})$ be the unique function such that

$$\text{Ric}(\omega) = \omega - dd^c h, \text{ normalized so that } \int_X e^{-h}\omega^n = V.$$

We also consider $h_t \in C^\infty(X, \mathbb{R})$ the unique function such that

$$\text{Ric}(\omega_t) = \omega_t - dd^c h_t, \text{ normalized so that } \int_X e^{-h_t}\omega_t^n = V.$$

It follows that $\text{Ric}(\omega_t) = \omega - dd^c h - dd^c \log(\omega_t^n/\omega^n)$, hence

$$dd^c\left[\log\left(\frac{\omega_t^n}{\omega^n}\right) + h + \varphi_t - \dot{\varphi}_t\right] = 0,$$

therefore

$$(\omega + dd^c \varphi_t)^n = e^{\dot{\varphi}_t - \varphi_t - h + \beta(t)} \omega^n,$$

for some normalizing constant $\beta(t)$.

Observe also that $dd^c \dot{\varphi}_t = -\text{Ric}(\omega_t) + \omega_t = dd^c h_t$ hence $\dot{\varphi}_t(x) = h_t(x) + \alpha(t)$ for some time dependent constant $\alpha(t)$. Our plan is to show the convergence of the metrics $\omega_t = \omega + dd^c \varphi_t$ by studying the properties of the potentials φ_t, so we should be very careful in the way we normalize the latter.

6.1.4 Plan of the Proof

Step 1: Choice of Normalization

We will first explain two possible choices of normalizing constants. Chen and Tian have proposed in [CheT02] a normalization which has been most commonly used up to now. We will emphasize an alternative normalization, which is most likely the one used by Perelman.[2]

Step 2: Uniform C^0-Estimate

Once φ_t has been suitably normalized, we will use the properness assumption to show that there exists $C_0 > 0$ such that

$$|\varphi_t(x)| \leq C_0, \quad \text{for all } (x,t) \in X \times \mathbb{R}^+.$$

This C^0-uniform estimate along the flow is the one that fails when there is no Kähler–Einstein metric. It is considered by experts as the core of the proof. We will indicate an alternative argument using pluripotential techniques to deduce it from Perelman's estimate.

Step 3: Uniform Estimate for $\dot{\varphi}_t$

We will explain how to bound $|\dot{\varphi}_t|$ uniformly in finite time, i.e. on $X \times [0, T]$. To get a uniform bound for $|\dot{\varphi}_t|$ on $X \times \mathbb{R}^+$, one needs to invoke Perelman's deep estimates: the latter will not be explained here, but are sketched in Chap. 5.

---

[2]In his seminar talk, Perelman apparently focused on his key estimates and did not say much about the remaining details.

Step 4: Uniform C^2-Estimate

We will then show that $|\Delta_\omega \varphi_t| \leq C_2$ independent of $(x,t) \in X \times \mathbb{R}^+$, by a clever use of the maximum principle for the Heat operator $\frac{\partial}{\partial t} - \Delta_{\omega_t}$. This is a parabolic analogue of Yau's celebrated Laplacian estimate. The constant C_2 depends on uniform bounds for φ_t and $\dot\varphi_t$, hence on Steps 2, 3.

Step 5: Higher Order Estimate

At this stage one can either establish a parabolic analogue of Calabi's C^3-estimates (global reasoning, see [PSS07]), or a complex version of the parabolic Evans–Krylov theory (local arguments) to show that there exists $\alpha > 0$ and $C_{2,\alpha} > 0$ such that

$$\|\varphi_t\|_{C^{2,\alpha}(X\times\mathbb{R}+)} \leq C_{2,\alpha},$$

where the Sobolev norm has to be taken with respect to the parabolic distance

$$d\left((x,y),(t,s)\right) := \max\{D(x,y), \sqrt{|t-s|}\}.$$

We won't say a word about these estimates in these notes. The reader will find a neat treatment of the C^3-estimates in Chap. 3, and an idea of the Evans–Krylov approach in the real setting in Chap. 2 (see [Gill11, ShW11] for the complex case).

With these estimates in hands, one can try and estimate the derivatives of the curvature as in Chap. 3, or simply invoke the parabolic Schauder theory to conclude (using a bootstrapping argument) that there exists $C_k > 0$ such that

$$\|\varphi_t\|_{C^k(X\times\mathbb{R}+)} \leq C_k.$$

Step 6: Convergence of the Flow

At this point, we know that (φ_t) is relatively compact in C^∞ and it remains to show that it converges.

For the first normalization, an easy argument shows that $\dot\varphi_t \to 0$. A differential Harnack inequality (à la Li–Yau) allows then to show that the flow converges exponentially fast towards a Kähler–Einstein potential, which is thus the unique cluster point by Bando–Mabuchi's result.

For Perelman's normalization, one can conclude by using the variational characterization of the Kähler–Einstein metric: it is the unique minimizer of the Ding functional.

6.2 Normalization of Potentials

Recall that ω_t is a solution of the normalized Kähler–Ricci flow (NKRF),

$$\frac{\partial}{\partial t}\omega_t = -\mathrm{Ric}(\omega_t) + \omega_t \qquad\text{(NKRF)}$$

with initial data $\omega = \omega_0 \in c_1(X)$. We let $\varphi_t \in \mathrm{PSH}(X,\omega) \cap C^\infty(X)$ denote a potential for ω_t, $\omega_t = \omega + dd^c\varphi_t$ which is uniquely determined up to a time dependent additive constant. It satisfies the complex parabolic Monge–Ampère flow

$$\dot{\varphi}_t := \frac{\partial}{\partial t}\varphi_t = \log\left(\frac{\omega_t^n}{\omega^n}\right) + \varphi_t + h - \beta(t)$$

for some normalizing constant $\beta(t) \in \mathbb{R}$.

6.2.1 First Normalization

Observe that $dd^c\varphi_0 = \omega_0 - \omega = 0$, hence $\varphi_0(x) \equiv c_0$ is a constant. The choice of c_0 will turn out to be crucial.

It is somehow natural to adjust the normalization of φ_t so that $\beta(t) \equiv 0$. This amounts to replace φ_t by $\varphi_t + B(t)$, where B solves the ODE $B' - B = -\beta$. Now

$$\dot{\varphi}_t := \frac{\partial}{\partial t}\varphi_t = \log\left(\frac{\omega_t^n}{\omega^n}\right) + \varphi_t + h$$

with $\varphi_0(x) \equiv c_0' = c_0 + B(0)$. Since we can choose $B(0)$ arbitrarily without affecting this complex Monge–Ampère flow (in other words the transformation $\varphi_t \mapsto \varphi_t + B(0)e^t$ leaves the flow invariant), we can still choose the value of $c_0' \in \mathbb{R}$. This choice is now clearly crucial, since two different choices lead to a difference in potentials which blows up exponentially in time.

The Mabuchi Functional

Recall that the scalar curvature of a Kähler form ω is the trace of the Ricci curvature,

$$\mathrm{Scal}(\omega) := n\frac{\mathrm{Ric}(\omega) \wedge \omega^{n-1}}{\omega^n}.$$

Its mean value is denoted by

$$\overline{\mathrm{Scal}(\omega)} := V^{-1} \int_X \mathrm{Scal}(\omega)\omega^n = n\frac{c_1(X) \cdot \{\omega\}^{n-1}}{\{\omega\}^n}.$$

The *Mabuchi energy* is defined by its derivative: if $\omega_t = \omega + dd^c\psi_t$ is any path of Kähler forms within the cohomology class $\{\omega\}$, then

$$\frac{d}{dt}\mathrm{Mab}(\psi_t) := V^{-1} \int_X \dot{\psi}_t \left[\overline{\mathrm{Scal}(\omega_t)} - \mathrm{Scal}(\omega_t)\right]\omega_t^n.$$

As we work here with $\omega \in c_1(X)$, we obtain $\overline{\mathrm{Scal}(\omega_t)} = n$. Since

$$\mathrm{Ric}(\omega_t) = \omega_t - dd^c h_t,$$

we observe that

$$\overline{\mathrm{Scal}(\omega_t)} - \mathrm{Scal}(\omega_t) = \Delta_{\omega_t} h_t := n\frac{dd^c h_t \wedge \omega_t^{n-1}}{\omega_t^n}.$$

Recall now that $dd^c\dot{\varphi}_t = dd^c h_t$. Therefore along the normalized Kähler–Ricci flow,

$$\frac{d}{dt}\mathrm{Mab}(\varphi_t) = V^{-1} \int_X \dot{\varphi}_t \Delta_{\omega_t}(\dot{\varphi}_t)\omega_t^n = -nV^{-1} \int_X d\dot{\varphi}_t \wedge d^c\dot{\varphi}_t \wedge \omega_t^{n-1} \le 0.$$

We have thus proved the following important property:

Lemma 6.2.1. *The Mabuchi energy is non-increasing along the normalized Kähler–Ricci flow. More precisely,*

$$\frac{d}{dt}\mathrm{Mab}(\varphi_t) = -nV^{-1} \int_X d\dot{\varphi}_t \wedge d^c\dot{\varphi}_t \wedge \omega_t^{n-1} \le 0.$$

We explain hereafter (see Proposition 6.2.5) that the Mabuchi functional is bounded below if and only if the Ding functional introduced above is so. The previous computation therefore yields

$$\int_0^{+\infty} \|\nabla_t\dot{\varphi}_t\|_{L^2(X)}^2 dt < +\infty.$$

One chooses c_0 so as to guarantee that

$$a(t) := V^{-1} \int_X \dot{\varphi}_t\omega_t^n \overset{t\to+\infty}{\longrightarrow} 0.$$

This convergence will be necessary to show the convergence of the flow (see the discussion before Lemma 1 in [PSS07]).

Lemma 6.2.2. *The function $a(t)$ converges to zero as $t \to +\infty$ iff we choose*

$$\varphi_0(x) \equiv c_0 := \int_0^{+\infty} \|\nabla_t \dot{\varphi}_t\|^2_{L^2(X)} e^{-t} dt - V^{-1} \int_X h_0 \, \omega^n.$$

Proof. Observe that

$$a'(t) = V^{-1} \int_X \ddot{\varphi}_t \omega_t^n + nV^{-1} \int_X \dot{\varphi}_t dd^c \dot{\varphi}_t \wedge \omega_t^{n-1} = a(t) + \frac{d}{dt} \mathrm{Mab}(\varphi_t).$$

Indeed

$$\ddot{\varphi}_t = \frac{d}{dt} \left[\log \left(\frac{\omega_t^n}{\omega^n} \right) + \varphi_t + h_0 \right] = \dot{\varphi}_t + \Delta_{\omega_t} \dot{\varphi}_t$$

hence $\int_X \ddot{\varphi}_t \omega_t^n = \int_X \dot{\varphi}_t \omega_t^n$. We can integrate this ODE and obtain

$$a(t) = \left[a_0 + \int_0^t k'(s) e^{-s} ds \right] e^t,$$

where $k(s) := \mathrm{Mab}(\varphi_s)$. Since k is non-increasing and bounded below, the function $k'(s) e^{-s}$ is integrable on $\mathbb{R}^+$ and $a(t) \to 0$ as $t \to +\infty$ if and only if

$$a(0) = -\int_0^{+\infty} k'(s) e^{-s} ds.$$

Now $a(0) = V^{-1} \int_X \dot{\varphi}_0 \omega^n = c_0 + V^{-1} \int_X h_0 \omega^n$. The result follows. $\qquad\square$

Conclusion

The first normalization amounts to considering the parabolic flow of potentials

$$\dot{\varphi}_t := \frac{\partial}{\partial t} \varphi_t = \log \left(\frac{\omega_t^n}{\omega^n} \right) + \varphi_t + h_0$$

with constant initial potential

$$\varphi_0(x) \equiv c_0 := \int_0^{+\infty} \|\nabla_t \dot{\varphi}_t\|^2_{L^2(X)} e^{-t} dt - V^{-1} \int_X h_0 \, \omega^n.$$

This choice of initial potential being possible only when the Mabuchi functional is bounded below, which is the case under our assumptions.

6.2.2 Perelman's Normalization?

There is another choice of normalization which is perhaps more natural from a variational point of view. Namely we choose

$$\beta(t) = \log\left[V^{-1} \int_X e^{-\varphi_t - h_0} \omega^n \right]$$

so that

$$\dot{\varphi}_t = \log\left[\frac{MA(\varphi_t)}{\mu_t} \right],$$

where $MA(\varphi_t) = (\omega + dd^c\varphi_t)^n / V$ and

$$\mu_t := \frac{e^{-\varphi_t - h_0}\omega^n}{\int_X e^{-\varphi_t - h_0}\omega^n}$$

are both probability measures. This is the normalization used in [BBEGZ11].

Observe that changing further $\varphi_t(x)$ in $\varphi_t(x) + B(t)$ leaves both $MA(\varphi_t)$ and μ_t unchanged, but modifies $\dot{\varphi}_t(x)$ into $\dot{\varphi}_t(x) + B'(t)$. Thus we can only afford replacing φ_t by $\varphi_t - c_0$ so that $\varphi_0 = 0$.

The Ricci Deviation

Recall that we have set $Ric(\omega_t) = \omega_t - dd^c h_t$, with

$$V^{-1} \int_X e^{-h_t} \omega_t^n = 1.$$

We have observed that $\dot{\varphi}_t(x)$ and $h_t(x)$ only differ by a constant (in space). Now

$$V^{-1} \int_X e^{-\dot{\varphi}_t} \omega_t^n = \int_X e^{-\dot{\varphi}_t} MA(\varphi_t) = \mu_t(X) = 1,$$

so that $\dot{\varphi}_t \equiv h_t$ with this choice of normalization. As we recall below, Perelman has succeeded in getting uniform estimates on the Ricci deviations h_t, these estimates therefore apply immediately to the function $\dot{\varphi}_t$ with our present choice of normalization.

Monotonicity of the Functionals Along the Flow

We have observed previously that the Mabuchi functional is non-decreasing along the normalized Kähler–Ricci flow. Since this functional acts on metrics (rather than on potentials), this property is independent of the chosen normalization. The same holds true for the Ding functional:

Lemma 6.2.3. *The Ding functional is non-increasing along the normalized Kähler–Ricci flow. More precisely,*

$$-\frac{d}{dt}\mathrm{Ding}(\varphi_t) = H_{\mathrm{MA}(\varphi_t)}(\mu_t) + H_{\mu_t}(\mathrm{MA}(\varphi_t)) \geq 0.$$

Here $H_\mu(\nu)$ denotes the relative entropy of the probability measure ν with respect to the probability measure μ. It is defined by

$$H_\mu(\nu) = \int_X \log\left(\frac{\nu}{\mu}\right) d\nu \in [0, +\infty]$$

if ν is absolutely continuous with respect to μ, and $H_\mu(\nu) = +\infty$ otherwise. It follows from the concavity of the logarithm that

$$H_\mu(\nu) = -\int_X \log\left(\frac{\mu}{\nu}\right) d\nu \geq -\log(\mu(X)) = 0,$$

with strict inequality unless $\nu = \mu$.

Proof. Recall that $\mathrm{Ding}(\varphi) = -E(\varphi) - \log\left[\int_X e^{-\varphi - h_0} \omega^n\right]$, where E is a primitive of the complex Monge–Ampère operator. We thus obtain along (NKRF)

$$\frac{d}{dt}E(\varphi_t) = \int_X \dot{\varphi}_t \mathrm{MA}(\varphi_t) = \int_X \log\left(\frac{\mathrm{MA}(\varphi_t)}{\mu_t}\right) \mathrm{MA}(\varphi_t) = H_{\mu_t}(\mathrm{MA}(\varphi_t)),$$

while

$$\frac{d}{dt}\log\left[\int_X e^{-\varphi_t - h_0} \omega^n\right] = -\int_X \dot{\varphi}_t d\mu_t = H_{\mathrm{MA}(\varphi_t)}(\mu_t).$$

This proves the lemma. □

Recall that in the first normalization, the initial constant c_0 has been chosen so that

$$a(t) := V^{-1}\int_X \dot{\varphi}_t \omega_t^n = \int_X \dot{\varphi}_t \mathrm{MA}(\varphi_t)$$

converges to zero as $t \to +\infty$. We relate this quantity to the above functionals:

Lemma 6.2.4. *Along the normalized Kähler–Ricci flow, one has*

$$\mathrm{Ding}(\varphi_t) + V^{-1}\int_X \dot{\varphi}_t \omega_t^n = \mathrm{Mab}(\varphi_t) + V^{-1}\int_X h_0 \omega^n.$$

Observe that the right hand side only depends on ω_t, while the left hand side depends on the choice of normalization for φ_t. It is understood here that this identity holds under the Perelman normalization.

Proof. Recall that

$$\dot{\varphi}_t = \log(\omega_t^n/\omega^n) + \varphi_t + h_0 + \beta(t), \text{ with } \beta(t) = \log\left[V^{-1}\int_X e^{-\varphi_t - h_0}\omega^n\right].$$

We let $a(t) = \int_X \dot{\varphi}_t \mathrm{MA}(\varphi_t)$ denote the left hand side and get as before

$$a'(t) = a(t) + \beta'(t) + \frac{d}{dt}\mathrm{Mab}(\varphi_t) = \frac{d}{dt}[-\mathrm{Ding}(\varphi_t) + \mathrm{Mab}(\varphi_t)],$$

noting that $a(t) = \frac{d}{dt}E(\varphi_t)$.

The conclusion follows since $a(0) = V^{-1}\int_X h_0\,\omega^n$ while $\mathrm{Ding}(\varphi_0) = \mathrm{Mab}(\varphi_0) = 0$.   □

Mabuchi vs Ding

We now show that the Mabuchi and the Ding functionals are bounded below simultaneously. This seems to have been noticed only recently (see [Li08, CLW09]).

Proposition 6.2.5. *Let X be a Fano manifold. The Mabuchi functional is bounded below along (NKRF) if and only if the Ding functional is so. If such is the case, then*

$$\inf_{t>0} \mathrm{Ding}(\varphi_t) = \inf_{t>0} \mathrm{Mab} + V^{-1}\int_X h_0\,\omega^n.$$

Proof. We have noticed in previous lemma, using Perelman' normalization, that

$$\mathrm{Ding}(\varphi_t) + V^{-1}\int_X \dot{\varphi}_t\omega_t^n = \mathrm{Mab}(\varphi_t) + V^{-1}\int_X h_0\,\omega^n.$$

It follows from Perelman's estimates that $\dot{\varphi}_t$ is uniformly bounded along the flow. Thus $\mathrm{Mab}(\varphi_t)$ is bounded if and only if $\mathrm{Ding}(\varphi_t)$ is so. We assume such is the case.

The error term $a(t) = V^{-1}\int_X \dot{\varphi}_t\omega_t^n$ is non-negative, with

$$0 \le a(t) = \frac{d}{dt}E(\varphi_t).$$

Since $\mathrm{Ding}(\varphi_t) = -E(\varphi_t) - \beta(t)$ is bounded from above and $t \mapsto \beta(t)$ is increasing, $E(\varphi_t)$ is bounded above as well. Thus $\int^{+\infty} a(t)dt < +\infty$, hence there exists $t_j \to +\infty$ such that $a(t_j) \to 0$. We infer

$$\inf_{t>0} \mathrm{Ding}(\varphi_t) = \inf_{t>0} \mathrm{Mab}(\varphi_t) + V^{-1} \int_X h_0 \, \omega^n. \qquad \square$$

Conclusion

The Perelman normalization amounts to consider the parabolic flow of potentials

$$\dot{\varphi}_t := \log\left(\frac{\omega_t^n}{\omega^n}\right) + \varphi_t + h_0 + \log\left[V^{-1} \int_X e^{-\varphi_t - h_0} \omega^n\right],$$

with initial potential $\varphi_0 \equiv 0$. Our plan is to show that if the Ding functional is proper and $H^0(X, TX) = 0$, then $\tilde{\varphi}_t := \varphi_t - V^{-1} \int_X \varphi_t \omega^n$ converges, in the C^∞-sense, towards the unique function φ_{KE} such that

$$\mathrm{MA}(\varphi_{\mathrm{KE}}) = \frac{e^{-\varphi_{\mathrm{KE}} - h} \omega^n}{\int_X e^{-\varphi_{\mathrm{KE}} - h} \omega^n}$$

and $\int_X \varphi_{\mathrm{KE}} \omega^n = 0$. This will imply that ω_t smoothly converges towards the unique Kähler–Einstein metric $\omega_{\mathrm{KE}} = \omega + dd^c \varphi_{\mathrm{KE}}$.

6.2.3 Perelman's Estimates

We first explain how a uniform control on $|\varphi_t(x)|$ in finite time easily yields a uniform control in finite time on $|\dot{\varphi}_t(x)|$:

Proposition 6.2.6. *Assume $\varphi_t \in \mathrm{PSH}(X, \omega) \cap C^\infty(X)$ satisfies*

$$\dot{\varphi}_t = \log\left[\frac{(\omega + dd^c \varphi_t)^n}{\omega^n}\right] + \varphi_t + h_0 + \beta(t),$$

with $\varphi_0 = 0$, $\beta(t) = \log\left[V^{-1} \int_X e^{-\varphi_t - h_0} \omega^n\right]$. Then $\forall (x, t) \in X \times [0, T]$,

$$e^{2T} \inf_X h_0 \le \dot{\varphi}_t(x) \le \mathrm{osc}_X(\varphi_t) + (n+1)T + \sup_X h_0.$$

Proof. Consider

$$H(x, t) := \dot{\varphi}_t(x) - \varphi_t(x) - (n+1)t - \beta(t),$$

and let $(x_0, t_0) \in X \times [0, T]$ be a point at which H realizes its maximum.

Set $\Delta_t := \Delta_{\omega_t}$. Observe that $\ddot{\varphi}_t = \dot{\varphi}_t + \Delta_t \dot{\varphi}_t + \beta'(t)$ and estimate

$$\left(\frac{\partial}{\partial t} - \Delta_t\right) H = \Delta_t \varphi_t - (n+1) \le -1,$$

where the latter inequality comes from the identity

$$\Delta_t \varphi_t = n - n\frac{\omega \wedge \omega_t^{n-1}}{\omega_t^n} \le n.$$

We infer that $t_0 = 0$, hence for all $(x, t) \in X \times [0, T]$,

$$H(x, t) \le H(x_0, 0) = h_0(x_0) \le \sup_X h_0,$$

thus

$$\dot{\varphi}_t(x) \le [\sup_X \varphi_t + \beta(t)] + (n+1)T + \sup_X h_0.$$

The desired upper-bound follows by observing that $\beta(t) \le -\inf_X \varphi_t$.

We use a similar reasoning to obtain the lower-bound, using the minimum principle for the Heat operator $\frac{\partial}{\partial t} - \Delta_t$, instead of the maximum principle. Indeed observe that

$$\left(\frac{\partial}{\partial t} - \Delta_t\right)(\dot{\varphi}_t) = \dot{\varphi}_t + \beta'(t) \ge \dot{\varphi}_t,$$

hence

$$\left(\frac{\partial}{\partial t} - \Delta_t\right)(e^{-2t}\dot{\varphi}_t) \ge -e^{-2t}\dot{\varphi}_t.$$

Let $(x, 0, t_0) \in X \times [0, T]$ be a point where $e^{-2t}\dot{\varphi}_t(x)$ realizes its minimum. If $t_0 > 0$, then

$$0 \ge \left(\frac{\partial}{\partial t} - \Delta_t\right)(e^{-2t}\dot{\varphi}_t)_{|(x_0, t_0)} \ge -e^{-2t_0}\dot{\varphi}_{t_0}(x_0)$$

hence $\dot{\varphi}_t(x) \ge 0$ for all (x, t). If $t_0 = 0$, then

$$e^{-2t}\dot{\varphi}_t(x) \ge \dot{\varphi}_0(x_0) = \inf_X h_0 + \beta(0) = \inf_X h_0.$$

The desired lower-bound follows, as $\inf_X h_0 \le 0$ since $\int_X e^{-h_0}\omega^n = V$. $\square$

We let the reader check that similar bounds can be obtained for the first normalization. These bounds are sufficient to prove Cao's result [Cao85] (the normalized Kähler–Ricci flow exists in infinite time), however they blow up as $t \to +\infty$ hence are too weak to study the convergence of the NKRF.

By using the monotonicity of his $\mathcal{W}$-functional, together with a non-collapsing argument, Perelman was able to prove the following deep estimate:

Theorem 6.2.7. *There exists $C_1 > 0$ such that for all $(x, t) \in X \times \mathbb{R}^+$,*

$$|\dot{\varphi}_t(x)| \leq C_1.$$

We refer the reader to [SeT08] for a detailed proof. An outline is also provided in the appendix of [TZ07], and more information can be found in Chap. 5.

6.3 C^0-Estimate

The main purpose of this section is to explain how to derive a uniform estimate on $|\varphi_t(x)|$. We first show that this is an elementary task in finite time, and then use the properness assumption and pluripotential tools to derive a uniform estimate on $X \times \mathbb{R}^+$. The latter estimate can not hold on Fano manifolds which do not admit a Kähler–Einstein metric.

6.3.1 Control in Finite Time

Proposition 6.3.1. *Assume $\varphi_t \in \mathrm{PSH}(X, \omega) \cap C^\infty(X)$ satisfies*

$$\dot{\varphi}_t = \log\left(\frac{(\omega + dd^c \varphi_t)^n}{\omega^n}\right) + \varphi_t + h_0 + \beta(t),$$

with $\varphi_0 = 0$, $\beta(t) = \log\left[V^{-1} \int_X e^{-\varphi_t - h_0} \omega^n\right]$. Then $\forall (x, t) \in X \times [0, T]$,

$$e^{2T} \inf_X h_0 \leq \varphi_t(x) \leq e^{4T} \operatorname{osc}_X(h_0).$$

Proof. Let $(x_0, t_0) \in X \times [0, T]$ be a point at which the function $(x, t) \mapsto F(x, t) = e^{-2t} \varphi_t(x)$ realizes its maximum. If $t_0 = 0$, we obtain

$$e^{-2t} \varphi_t(x) \leq \varphi_0(x_0) = 0, \text{ hence } \varphi_t(x) \leq 0.$$

If $t_0 > 0$, then at (x_0, t_0) we have $dd^c F = e^{-2t_0} dd^c \varphi_{t_0}(x_0) \leq 0$ hence

$$\dot{\varphi}_{t_0}(x_0) \leq \varphi_{t_0}(x_0) + \sup_X h_0 + \beta(t_0),$$

while

$$0 \leq \frac{\partial}{\partial t} F = e^{-2t_0} \left[\dot{\varphi}_{t_0}(x_0) - 2\varphi_{t_0}(x_0) \right] \leq e^{-2t_0} \left[\sup_X h_0 + \beta(t_0) - \varphi_{t_0}(x_0) \right].$$

The upper-bound follows by recalling that β is non-decreasing and

$$\beta(T) \leq -\inf_X \varphi_T \leq e^{2T} (-\inf_X h_0),$$

assuming the lower-bound holds true.

The latter is proved along the same lines: looking at the point where F realizes its minimum, we end up with a lower-bound

$$\varphi_t(x) \geq e^{2T} \inf_X h_0 + \beta(0) = e^{2T} \inf_X h_0,$$

since β vanishes at the origin. $\square$

6.3.2   Uniform Bound in Infinite Time

Theorem 6.3.2. *Let X be a Fano manifold such that the Ding functional is proper. Let $\omega_t := \omega + dd^c \psi_t$ be the solution of the normalized Kähler–Ricci flow with initial data $\omega \in c_1(X)$, where $\psi_t \in \mathrm{PSH}(X, \omega)$ is normalized so that $\int_X \psi_t \omega^n = 0$. There exists $C_0 > 0$ such that*

$$\forall (x, t) \in X \times \mathbb{R}^+, \ |\psi_t(x)| \leq C_0.$$

Proof. Observe that $\psi_t = \varphi_t - \int_X \varphi_t \omega^n$, where φ_t satisfies

$$\dot{\varphi}_t = \log \left[\frac{\mathrm{MA}(\varphi_t)}{\mu_t} \right],$$

with

$$\mathrm{MA}(\varphi_t) = V^{-1}(\omega + dd^c \varphi_t)^n \text{ and } \mu_t = \frac{e^{-\varphi_t - h_0} \omega^n}{\int_X e^{-\varphi_t - h_0} \omega^n}.$$

We have observed that the Ding functional is translation invariant and non-increasing along the NKRF. Since it is proper, we infer that $E(\psi_t)$ is uniformly bounded below. But the mean value property shows that $\sup_X \psi_t \leq \int_X \psi_t \omega^n + C_\omega = C_\omega$ (see [GZ05, Proposition 1.7]), and it follows that ψ_t belong to

$$\mathcal{E}_C^1(X, \omega) := \{ u \in \mathrm{PSH}(X, \omega) \mid u \leq C \text{ and } E(u) \geq -C \},$$

for some fixed $C > 0$. This is a compact set (for the L^1-topology) of functions which have zero Lelong numbers at all points (see below). It follows therefore from Skoda's uniform integrability theorem [Zer01] that there exists $A > 0$ such that

$$\sup_{t \geq 0} \int_X e^{-2\psi_t - 2h_0} \omega^n \leq A.$$

Note that $\int_X e^{-\psi_t - h_0} \omega^n \geq V e^{-\sup_X \psi_t} \geq \delta_0 > 0$ and recall that $\dot{\varphi}_t(x) \leq C_1$ by Perelman's fundamental estimate to conclude that

$$\mathrm{MA}(\psi_t) = e^{\dot{\varphi}_t} \frac{e^{-\psi_t - h_0} \omega^n}{\int_X e^{-\psi_t - h_0} \omega^n} = f_t \omega^n,$$

where the densities $0 \leq f_t$ are uniformly in $L^2(X)$, $\|f_t\|_{L^2(\omega^n)} \leq A'$. It follows therefore from Theorem 6.3.8 that ψ_t is uniformly bounded.   □

Remark 6.3.3. The reader will find a rather different approach in [TZ07, PSS07, PS10], using the first normalization, Moser iterative process and a uniform Sobolev inequality along the flow. It takes some efforts to check that the two normalizations are uniformly comparable along the flow, give it a try!

6.3.3   Pluripotential Tools

We explain here some of the pluripotential tools that have been used in the above proof.

6.3.3.1   Finite Energy Classes

Recall that X is an n-dimensional Fano manifold, ω is a fixed Kähler form in $c_1(X)$, and $V = c_1(X)^n = \int_X \omega^n$. The energy $E(\psi)$ of a smooth ω-plurisubharmonic function,

$$E(\psi) := \frac{1}{n+1} \sum_{j=0}^{n} V^{-1} \int_X \psi(\omega + dd^c \psi)^j \wedge \omega^{n-j},$$

is non-decreasing in ψ. It can thus be extended to any $\varphi \in \mathrm{PSH}(X, \omega)$ by setting

$$E(\varphi) := \inf_{\psi \geq \varphi} E(\psi),$$

where the infimum runs over all smooth ω-psh functions ψ that lie above φ.

Definition 6.3.4. We set

$$\mathcal{E}^1(X, \omega) := \{\varphi \in \mathrm{PSH}(X, \omega) \mid E(\varphi) > -\infty\}.$$

and

$$\mathcal{E}^1_C(X, \omega) := \{\varphi \in \mathcal{E}^1(X, \omega) \mid E(\varphi) \geq -C \text{ and } \varphi \leq C\}.$$

The following properties are established in [GZ07, BEGZ10]:

- The complex Monge–Ampère operator MA(·) is well defined on the class $\mathcal{E}^1(X, \omega)$, since the Monge–Ampère measure of a function $\varphi \in \mathcal{E}^1(X, \omega)$ is very well approximated (in the Borel sense) by the Monge–Ampère measures $\mathrm{MA}(\varphi_j)$ of its canonical approximants $\varphi_j := \max\{\varphi, -j\}$;
- The maximum and comparison principles hold, namely if $\varphi, \psi \in \mathcal{E}^1(X, \omega)$,

$$1_{\{\varphi > \psi\}} \mathrm{MA}(\max\{\varphi, \psi\}) = 1_{\{\varphi > \psi\}} \mathrm{MA}(\varphi)$$

and

$$\int_{\{\varphi < \psi\}} \mathrm{MA}(\psi) \leq \int_{\{\varphi < \psi\}} \mathrm{MA}(\varphi).$$

- The functions with finite energy have zero Lelong number at all points, as follows by observing that the class $\mathcal{E}^1(X, \omega)$ is stable under the max-operation, while $\chi \log \mathrm{dist}(\cdot, x)$ is ω-plurisubharmonic for each fixed point $x \in X$ and a suitable cut-off function χ, while it does not belong to $\mathcal{E}^1(X, \omega)$;
- The sets $\mathcal{E}^1_C(X, \omega)$ are compact subsets of $L^1(X)$: this easily follows from the upper semi-continuity property of the energy, together with the fact that the set

$$\{\varphi \in \mathrm{PSH}(X, \omega) \mid -C' \leq \sup_X \varphi \leq C\}$$

is compact in $L^1(X)$.

Recall now the following uniform version of Skoda's integrability theorem [Zer01]:

Theorem 6.3.5. *Let $\mathcal{B} \subset \mathrm{PSH}(X, \omega)$ be a compact family of ω-psh functions, set*

$$\nu(\mathcal{B}) := \sup\{\nu(\varphi, x) \mid x \in X \text{ and } \varphi \in \mathcal{B}\}.$$

For every $A < 2/\nu(\mathcal{B})$, there exists $C_A > 0$ such that

$$\forall \varphi \in \mathcal{B}, \quad \int_X e^{-A\varphi} \omega^n \leq C_A.$$

It follows from this result that functions from $\mathcal{E}_C^1(X, \omega)$ satisfy such a uniform integrability property with $A > 0$ as large as we like.

6.3.3.2 Capacities and Volume

For a Borel set $K \subset X$, we consider

$$M_\omega(K) := \sup_X V_{K,\omega}^* \in [0, +\infty], \tag{6.1}$$

where

$$V_{K,\omega} := \sup\{\varphi \in \mathrm{PSH}(X, \omega) \mid \varphi \leq 0 \text{ on } K\}.$$

One checks that $M_\omega(K) = +\infty$ if and only if K is pluripolar. We also set

$$\mathrm{Cap}(K) := \sup\left\{\int_K \mathrm{MA}(u) \mid 0 \leq u \leq 1\right\}.$$

This is the Monge–Ampère capacity. It vanishes on pluripolar sets.

Lemma 6.3.6. *For every non-pluripolar compact subset K of X, we have*

$$1 \leq \mathrm{Cap}(K)^{-1/n} \leq \max\{1, M_\omega(K)\}.$$

Proof. The left-hand inequality is trivial. In order to prove the right-hand inequality we consider two cases. If $M_\omega(K) \leq 1$, then $V_{K,\omega}^*$ is a candidate in the definition of $\mathrm{Cap}(K)$. One checks that $\mathrm{MA}(V_{K,\omega}^*)$ is supported on K, thus

$$\mathrm{Cap}(K) \geq \int_K \mathrm{MA}(V_{K,\omega}^*) = \int_X \mathrm{MA}(V_{K,\omega}^*) = 1$$

and the desired inequality holds in that case.

On the other hand if $M := M_\omega(K) \geq 1$ we have $0 \leq M^{-1} V_{K,\omega}^* \leq 1$ and it follows by definition of the capacity again that

$$\mathrm{Cap}(K) \geq \int_K \mathrm{MA}(M^{-1} V_{K,\omega}^*).$$

Since $\mathrm{MA}(M^{-1} V_{K,\omega}^*) \geq M^{-n} \mathrm{MA}(V_{K,\omega}^*)$ we deduce that

$$\int_K \mathrm{MA}(M^{-1} V_{K,\omega}^*) \geq M^{-n} \int_X \mathrm{MA}(V_{K,\omega}^*) = M^{-n}$$

and the result follows. $\qquad\square$

Proposition 6.3.7. *Let* $\mu = fdV$ *be a positive measure with* L^p *density with respect to Lebesgue measure, with* $p > 1$. *Then there exists* $C > 0$ *such that*

$$\mu(B) \leq C \cdot \mathrm{Cap}(B)^2$$

for all Borelian $B \subset X$, *where* $C := (p-1)^{-2n} A \|f\|_{L^{1+\varepsilon}(dV)}$, *and* $A = A(\omega, dV)$.

Proof. It is enough to consider the case where $B = K$ is compact. We can also assume that K is non-pluripolar since $\mu(K) = 0$ otherwise and the inequality is then trivial. Set

$$\nu(X) := \sup_{T, x} \nu(T, x) \tag{6.2}$$

the supremum ranging over all positive currents $T \in c_1(X)$ and all $x \in X$, and $\nu(T, x)$ denoting the Lelong number of T at x. Since all Lelong numbers of $\nu(X)^{-1} T$ are < 2 for each positive current $T \in c_1(X)$, Skoda's uniform integrability theorem yields $C_\omega > 0$ only depending on dV and ω such that

$$\int_X \exp(-\nu(X)^{-1} \psi) dV \leq C_\omega$$

for all ω-psh functions ψ normalized by $\sup_X \psi = 0$. Applying this to $\psi = V_{K,\omega}^* - M_\omega(K)$ [which has the right normalization by (6.1)] we get

$$\int_X \exp(-\nu(X)^{-1} V_{K,\omega}^*) dV \leq C_\omega \exp(-\nu(X)^{-1} M_\omega(K)).$$

On the other hand $V_{K,\omega}^* \leq 0$ on K a.e. with respect to Lebesgue measure, hence

$$\mathrm{vol}(K) \leq C_\omega \exp(-\nu(X)^{-1} M_\omega(K)). \tag{6.3}$$

Now Hölder's inequality yields

$$\mu(K) \leq \|f\|_{L^p(dV)} \mathrm{vol}(K)^{1/q}, \tag{6.4}$$

where q denotes the conjugate exponent. We may also assume that $M_\omega(K) \geq 1$. Otherwise Lemma 6.3.6 implies $\mathrm{Cap}(K) = 1$, and the result is thus clear in that case. By Lemma 6.3.6, (6.3) and (6.4) together we thus get

$$\mu(K) \leq C_\omega^{1/q} \|f\|_{L^p(dV)} \exp\left(-\frac{1}{q\nu(X)} \mathrm{Cap}(K)^{-1/n}\right)$$

and the result follows since $\exp(-t^{-1/n}) = O(t^2)$ when $t \to 0_+$. $\qquad\square$

6.3.3.3 Kolodziej's Uniform A Priori Estimate

We are now ready to prove the following celebrated result of Kolodziej [Kol98]:

Theorem 6.3.8. *Let* $\mu = \mathrm{MA}(\varphi) = f\,dV$ *be a probability Monge–Ampère measure with density* $f \in L^p$, $p > 1$. *Then*

$$\mathrm{osc}_X \varphi \leq C$$

where C *only depends on* $\omega, dV, \|f\|_{L^p}$.

Proof. We can assume φ is normalized so that $\sup_X \varphi = 0$. Consider

$$g(t) := (\mathrm{Cap}\{\varphi < -t\})^{1/n}.$$

Our goal is to show that $g(M) = 0$ for some M under control. Indeed we will then have $\varphi \geq -M$ on $X \setminus P$ for some Borel subset P such that $\mathrm{Cap}(P) = 0$. It then follows from Proposition 6.3.7 (applied to the Lebesgue measure itself) that P has Lebesgue measure zero hence $\varphi \geq -M$ will hold everywhere.

Since $\mathrm{MA}(\varphi) = \mu$ it follows from Proposition 6.3.7 and Lemma 6.3.9 that

$$g(t + \delta) \leq \frac{C^{1/n}}{\delta} g(t)^2 \quad \text{for all } t > 0 \text{ and } 0 < \delta < 1.$$

We can thus apply Lemma 6.3.10 below which yields $g(M) = 0$ for $M := t_0 + 5C^{1/n}$. Here $t_0 > 0$ has to be chosen so that

$$g(t_0) < \frac{1}{5C^{1/n}}.$$

Now Lemma 6.3.9 (with $\delta = 1$) implies that

$$g(t)^n \leq \mu\{\varphi < -t + 1\} \leq \frac{1}{t-1} \int_X |\varphi| f\,dV \leq \frac{1}{t-1} \|f\|_{L^p(dV)} \|\varphi\|_{L^q(dV)}$$

by Hölder's inequality. Since φ belongs to the compact set of ω-psh functions normalized by $\sup_X \varphi = 0$, its $L^q(dV)$-norm is bounded by a constant C_2 only depending on ω, dV and p. It is thus enough to take

$$t_0 > 1 + 5^{n-1} C_2 C \|f\|_{L^p(dV)}. \qquad \square$$

Lemma 6.3.9. *Fix* $\varphi \in \mathcal{E}^1(X, \omega)$. *Then for all* $t > 0$ *and* $0 < \delta < 1$ *we have*

$$\mathrm{Cap}\{\varphi < -t - \delta\} \leq \delta^{-n} \int_{\{\varphi < -t\}} \mathrm{MA}(\varphi).$$

Proof. Let ψ be a ω-psh function such that $0 \le \psi \le 1$. We then have

$$\{\varphi < -t - \delta\} \subset \{\varphi < \delta\psi - t - \delta\} \subset \{\varphi < -t\}.$$

Since $\delta^n MA(\psi) \le MA(\delta\psi)$ and $\varphi \in \mathcal{E}^1(X, \omega)$ it follows from the comparison principle that

$$\delta^n \int_{\{\varphi < -t-\delta\}} MA(\psi) \le \int_{\{\varphi < \delta\psi - t-\delta\}} MA(\delta\psi)$$

$$\le \int_{\{\varphi < \delta\psi - t-\delta\}} MA(\varphi) \le \int_{\{\varphi < -t\}} MA(\varphi)$$

and the proof is complete. $\square$

Lemma 6.3.10. *Let $g : \mathbb{R}^+ \to [0, 1]$ be a decreasing function such that $g(+\infty) = 0$ and*

$$g(t + \delta) \le \frac{C^{1/n}}{\delta} g(t)^2 \quad \text{for all } t > 0 \text{ and } 0 < \delta < 1.$$

Then $g(t) = 0$ for all $t \ge t_0 + 5C^{1/n}$, where

$$t_0 = \inf\{s > 0 \mid g(s) \le e^{-1}C^{-1/n}\}.$$

Proof. Set $f(t) = -\log g(t)$ so that $f : \mathbb{R}^+ \to \mathbb{R}^+$ is increasing with

$$f(t + \delta) \ge 2f(t) - \log\left(\delta/C^{1/n}\right).$$

By induction we define an increasing sequence t_j such that

$$t_{j+1} = t_j + \delta_j, \quad \text{with } \delta_j = eC^{1/n}\exp(-f(t_j)) = eC^{1/n}g(t_j).$$

Observe that $0 < \delta_0$ is smaller than 1 if we choose t_0 as indicated. Since (t_j) is increasing and g is decreasing, this ensures that δ_j is smaller than 1 for all $j \in \mathbb{N}$. We can thus use the growth estimate and obtain

$$f(t_{j+1}) = f(t_j + \delta_j) \ge f(t_j) + 1.$$

Since $f \ge 0$, we infer $f(t_j) \ge j$ for all j. Now

$$t_\infty := t_0 + \sum_{j \ge 0}(t_{j+1} - t_j) \le t_0 + C^{1/n}e\sum_{j \ge 0}\exp(-j) \le t_0 + 5C^{1/n}.$$

The proof is thus complete since $f(t) \ge f(t_\infty) = +\infty$ for all $t \ge t_\infty$. $\square$

6.4 Higher Order Estimates

6.4.1 Preliminaries

We shall need two auxiliary results.

Lemma 6.4.1. *Let* α, β *be positive* $(1, 1)$-*forms. Then*

$$n \left(\frac{\alpha^n}{\beta^n} \right)^{\frac{1}{n}} \leq \operatorname{tr}_\beta(\alpha) \leq n \left(\frac{\alpha^n}{\beta^n} \right) \cdot (\operatorname{tr}_\alpha(\beta))^{n-1}.$$

The proof is elementary (see Lemma 4.1.1) Applying these inequalities to $\alpha = \omega_t := \omega + dd^c \varphi_t$ and $\beta = \omega$, we obtain:

Corollary 6.4.2. *There exists* $C > 0$ *which only depends on* $\|\dot\varphi_t\|_{L^\infty}$ *such that*

$$\frac{1}{C} \leq \operatorname{tr}_\omega(\omega_t) \leq C \left[\operatorname{tr}_{\omega_t}(\omega) \right]^{n-1}.$$

The second result we need is the following estimate which goes back to the work of Aubin [Aub78] and Yau [Yau78]; in this form it is due to Siu [Siu87].

Lemma 6.4.3. *Let* ω, ω' *be arbitrary Kähler forms. Let* $-B \in \mathbb{R}$ *be a lower bound on the holomorphic bisectional curvature of* (X, ω). *Then*

$$\Delta_{\omega'} \log \operatorname{tr}_\omega(\omega') \geq -\frac{\operatorname{tr}_\omega(\operatorname{Ric}(\omega'))}{\operatorname{tr}_\omega(\omega')} - B \operatorname{tr}_{\omega'}(\omega).$$

We refer the reader to Proposition 4.1.2 for a proof.

6.4.2 C^2-Estimate

Theorem 6.4.4. *Let* X *be a Fano manifold such that* $\mathcal{F}$ *is proper. Let* ω_t *be the solution of the normalized Kähler–Ricci flow with initial data* $\omega \in c_1(X)$. *There exists* $C_2 > 0$ *such that for all* $(x, t) \in X \times \mathbb{R}^+$,

$$0 \leq \operatorname{tr}_\omega(\omega_t) \leq C_2.$$

Proof. Set $\alpha(x, t) := \log \operatorname{tr}_\omega(\omega_t) - (B + 1)\varphi_t$, were $-B$ denotes a lower bound on the holomorphic bisectional curvature of (X, ω) (as in Lemma 6.4.3). Fix $T > 0$ and let $(x_0, t_0) \in X \times [0, T]$ be a point at which α realizes its maximum.

Either $t_0 = 0$, in which case $\alpha(x,t) \le \alpha(x_0,0) = \log n$ yields

$$\mathrm{tr}_{\omega}(\omega_t)(x) \le n \exp([B+1]\varphi_t(x)) \le C_2' = n \exp([B+1]C_0),$$

since φ_t is uniformly bounded from above.

Or $t_0 > 0$. In this case it follows from Lemma 6.4.5 that at point (x_0, t_0),

$$0 \le \left(\frac{\partial}{\partial t} - \Delta_t\right)\alpha \le -\mathrm{tr}_{\omega_{t_0}}(\omega)(x_0) + \kappa$$

so that

$$\mathrm{tr}_{\omega}(\omega_t)(x) \le C_2'' = \kappa \exp(2[B+1]C_0).$$

The conclusion follows since both C_2' and C_2'' are independent of T. $\square$

Lemma 6.4.5. *Set* $\alpha(x,t) := \log \mathrm{tr}_{\omega}(\omega_t) - (B+1)\varphi_t$. *There exists* $\kappa > 0$ *such that*

$$\forall (x,t) \in X \times \mathbb{R}^+, \quad \left(\frac{\partial}{\partial t} - \Delta_t\right)\alpha \le -\mathrm{tr}_{\omega_t}(\omega) + \kappa.$$

Here $-B$ denotes a lower bound on the holomorphic bisectional curvature of (X, ω) (as in Lemma 6.4.3).

Proof. It follows from Perelman's estimate that

$$\frac{\partial}{\partial t}\alpha = \frac{\Delta_{\omega}\dot{\varphi}_t}{\mathrm{tr}_{\omega}(\omega_t)} - (B+1)\dot{\varphi}_t \le \frac{\Delta_{\omega}\dot{\varphi}_t}{\mathrm{tr}_{\omega}(\omega_t)} + C.$$

Now $\dot{\varphi}_t = \log(\omega_t^n/\omega^n) + \varphi_t + h_0 + \beta(t)$ thus

$$\Delta_{\omega}\dot{\varphi}_t = \Delta_{\omega}\log\left(\frac{\omega_t^n}{\omega^n}\right) + \mathrm{tr}_{\omega}(\omega_t) - n + \Delta_{\omega}h_0$$

$$\le \Delta_{\omega}\log\left(\frac{\omega_t^n}{\omega^n}\right) + \mathrm{tr}_{\omega}(\omega_t) + C'.$$

Since $dd^c \log\left(\frac{\omega_t^n}{\omega^n}\right) = \mathrm{Ric}(\omega) - \mathrm{Ric}(\omega_t)$, we infer

$$\Delta_{\omega}\dot{\varphi}_t \le -\mathrm{tr}_{\omega}(\mathrm{Ric}(\omega_t)) + \mathrm{tr}_{\omega}(\omega_t) + C'',$$

hence

$$\frac{\partial}{\partial t}\alpha \le -\frac{\mathrm{tr}_{\omega}(\mathrm{Ric}(\omega_t))}{\mathrm{tr}_{\omega}(\omega_t)} + \frac{C''}{\mathrm{tr}_{\omega}(\omega_t)} + C + 1.$$

We now estimate $\Delta_{\omega_t}\alpha = \Delta_t\alpha$ from below. It follows from Lemma 6.4.3 that

$$\Delta_t\alpha = \Delta_t \log \mathrm{tr}_\omega(\omega_t) - (B+1)[n - \mathrm{tr}_{\omega_t}(\omega)]$$
$$\geq -\frac{\mathrm{tr}_\omega(\mathrm{Ric}(\omega_t))}{\mathrm{tr}_\omega(\omega_t)} + \mathrm{tr}_{\omega_t}(\omega) - n(B+1).$$

Therefore

$$\left(\frac{\partial}{\partial t} - \Delta_t\right)(\alpha) \leq -\mathrm{tr}_{\omega_t}(\omega) + \frac{C''}{\mathrm{tr}_\omega(\omega_t)} + C'''.$$

The conclusion follows since $\mathrm{tr}_\omega(\omega_t)$ is uniformly bounded from below away from zero, as we have observed in the preliminaries.   □

Remark 6.4.6. The reader can go through the above proof and realize that one can obtain similarly a uniform upper bound for $\mathrm{tr}_\omega(\omega_t)$ on any finite interval of time, without assuming the properness of the functional $\mathcal{F}$.

6.4.3 Complex Parabolic Evans–Krylov Theory and Schauder Estimates

At this stage, it follows from local arguments that one can obtain higher order uniform a priori estimates. We won't dwell on these techniques here and rather refer the reader to Chap. 2 for the real theory. The latter can not be directly applied in the complex setting, but the technique can be adapted as was done for instance in [Gill11].

6.5 Convergence of the Flow

6.5.1 Asymptotic of the Time-Derivatives

Proposition 6.5.1. *The time-derivatives $\dot\psi_t$ converge to zero in $C^\infty(X)$.*

Proof. Note that $\int_X \psi_t \omega^n = 0$ hence $\int_X \dot\psi_t \omega^n = 0$ in the Perelman normalization, while for the first normalization, φ_t has been so normalized that

$$\int_X \dot\varphi_t \omega_t^n \overset{t\to+\infty}{\longrightarrow} 0.$$

It therefore suffices to check that

$$\int_X d\dot{\varphi}_t \wedge d^c\dot{\varphi}_t \wedge \omega_t^{n-1} \longrightarrow 0,$$

since ω_t and ω are uniformly equivalent, by Theorem 6.4.4.

To check the latter convergence, we follow some arguments by Phong and Sturm [PS06]. Set

$$Y(t) := \int_X |\nabla_t \dot{\varphi}_t|^2 \omega_t^n = n \int_X d\dot{\varphi}_t \wedge d^c\dot{\varphi}_t \wedge \omega_t^{n-1}.$$

Recall that the Mabuchi functional is bounded below and non-increasing along the flow, with

$$-\frac{d}{dt}\text{Mab}(\varphi_t) = Y(t) \geq 0, \quad \text{thus} \quad \int_0^{+\infty} Y(t)dt < +\infty.$$

We cannot of course immediately deduce that $Y(t) \to 0$ as $t \to +\infty$, however Phong–Sturm succeed, by using a Bochner–Kodaira type formula and a uniform control of the curvatures along the flow, in showing that $Y' \leq CY$ for some uniform positive constant $C > 0$.

The reader will easily check that this further estimate allows to conclude. We refer to Chap. 3 for the controls on the curvatures along the flow, and to [PS06] for the remaining details. We propose in Lemma 6.5.2 a slightly weaker, but economical control that is also sufficient, as the reader will check.   □

Lemma 6.5.2. *Set*

$$Z(t) := n \int_X d\dot{\varphi}_t \wedge d^c\dot{\varphi}_t \wedge \omega^{n-1}.$$

Then $Z'(t) \leq 2Z(t) + C$ for some uniform constant $C > 0$.

Proof. Observe that

$$Z'(t) = -2n \int_X \ddot{\varphi}_t dd^c\dot{\varphi}_t \wedge \omega^{n-1} \quad \text{with} \quad \dot{\varphi}_t = \log\left(\frac{\omega_t^n}{\omega^n}\right) + \varphi_t + h_0.$$

We use here the first normalization, this clearly does not affect the value of $Z(t)$. Since $\ddot{\varphi}_t = \Delta_t \dot{\varphi}_t + \dot{\varphi}_t$, we infer

$$Z'(t) = 2Z(t) - 2\int_X \Delta_t \dot{\varphi}_t \Delta_\omega \dot{\varphi}_t \omega^n \leq 2Z(t) + C,$$

since the latter quantities are uniformly bounded along the flow.   □

6.5.2 Conclusion

We are now in position to conclude.

First Normalization

It follows from previous sections that the family (φ_t) is relatively compact in $C^\infty(X \times [0, +\infty])$. Let $\varphi_\infty = \lim_{j \to +\infty} \varphi_{t_j}$ be a cluster point of $(\varphi_t)_{t>0}$. It follows from Proposition 6.5.1 that $\dot{\varphi}_{t_j} \to 0$ hence

$$(\omega + dd^c \varphi_\infty)^n = e^{-\varphi_\infty} e^{-h_0} \omega^n, \qquad (\dagger)$$

hence $\omega + dd^c \varphi_\infty$ is a Kähler–Einstein metric. Since we have assumed that X has no holomorphic vector field, it follows from Bando–Mabuchi's uniqueness result [BM87] that φ_∞ coincides with the Kähler–Einstein potential φ_{KE}, which is the unique solution of $(\dagger)$. There is thus a unique cluster point for (φ_t) as $t \to +\infty$, hence the whole family converges in the C^∞-sense towards φ_{KE}.

It turns out that the above convergence holds at an exponential speed. We refer the interested reader to [PSSW08a, PS10] for a proof of this fact.

Perelman Normalization

A similar argument could be used for the potentials $\psi_t = \varphi_t - V^{-1} \int_X \varphi_t \omega^n$ if we could show the convergence of $\int_X \varphi_t \omega^n$ as $t \to +\infty$. To get around this difficulty, we can proceed as follows: let $\mathcal{K}$ denote the set of cluster values of $(\omega_t)_{t>0}$. Observe that $\mathcal{K}$ is invariant under the normalized Kähler–Ricci flow and the Ding functional is constant on $\mathcal{K}$.

It follows now from Lemma 6.2.3 that the Ding functional is strictly increasing along the NKRF, unless we start from a fixed point ω_0. Thus $\mathcal{K}$ consists in fixed points for the NKRF. There is only one such fixed point, the unique Kähler–Einstein metric. Therefore ω_t converges to ω_{KE} and ψ_t converges to the unique Kähler–Einstein potential ψ_{KE} such that $\omega_{\mathrm{KE}} = \omega + dd^c \psi_{\mathrm{KE}}$ and $\int_X \psi_{\mathrm{KE}} \omega^n = 0$.

6.6 An Alternative Approach

We finally briefly mention an alternative approach to the weak convergence of the normalized Kähler–Ricci flow, as recently proposed in [BBEGZ11].

The convergence of ω_t towards ω_{KE} is only proved in the weak sense of (positive) currents, but without using Perelman's deep estimates: this allows us in [BBEGZ11] to extend Perelman's convergence result to singular settings (weak Fano varieties and pairs), where these estimates are not available.

6.6.1 The Variational Characterization of K–E Currents

The alternative approach we propose in [BBEGZ11] relies on the variational characterization of Kähler–Einstein currents established in [BBGZ13].

Let X be a Fano manifold and fix $\omega \in c_1(X)$ a Kähler form. A positive current $T = \omega + dd^c \psi \in c_1(X)$ is said to have finite energy if $E(\psi) > -\infty$. We then introduce the J-functional

$$J(T) := V^{-1} \int_X \psi \omega^n - E(\psi).$$

We let $\mathcal{E}^1(c_1(X))$ denote the set of currents with finite energy in $c_1(X)$ and

$$\mathcal{E}_C^1(c_1(X)) := \{ T \in \mathcal{E}^1(c_1(X)) \mid J(T) \le C \}$$

the compact convex set of those positive closed currents in $c_1(X)$ whose energy is uniformly bounded from below by C.

A combination of [BM87, Tian97] and [BBGZ13, Theorems D,E] yields the following criterion:

Theorem 6.6.1. *Let X be a Fano manifold with $H^0(X, TX) = 0$. Let $T \in c_1(X)$ be a closed positive current of finite energy. The following are equivalent:*

1. *T minimizes the Ding functional.*
2. *T is a Kähler–Einstein current;*
3. *T is the unique Kähler–Einstein metric;*

We say here that a current $T = \omega + dd^c \varphi \in \mathcal{E}^1(c_1(X))$ is Kähler–Einstein if it satisfies $T^n = e^{-\varphi - h_0} \omega^n$, where as previously $\mathrm{Ric}(\omega) = \omega - dd^c h_0$.

It was realized by Ding–Tian [DT92] that the Kähler–Einstein metric is the unique Kähler metric maximizing Ding. This result being extended to the class of finite energy currents allows to use the soft compacity criteria available in these Sobolev-like spaces:

Corollary 6.6.2. *Let X be a Fano manifold whose Ding functional is proper. If $\omega_t \in c_1(X)$ is a family of Kähler forms such that*

$$\mathrm{Ding}(\omega_t) \to \inf_{\mathcal{E}^1(X, \omega)} \mathrm{Ding},$$

then

$$\omega_t \longrightarrow \omega_{\mathrm{KE}}$$

in the weak sense of (positive) currents.

6.6.2 Maximizing Subsequences

We let the potential $\varphi_t \in \mathrm{PSH}(X, \omega) \cap C^\infty(X)$ evolve according to the complex Monge–Ampère flow,

$$\dot{\varphi}_t = \log \left[\frac{\mathrm{MA}(\varphi_t)}{\mu_t} \right] = \log \left(\frac{\omega_t^n}{\omega^n} \right) + \varphi_t + h_0 + \beta(t),$$

where

$$\beta(t) = \log \left[\int_X e^{-\varphi_t - h_0} \omega^n \right],$$

with initial condition $\varphi_0 = 0$. We set $\psi_t := \varphi_t - \int_X \varphi_t \, \omega^n$.

Recall that the Ding functional is non-increasing along this flow. It follows more precisely from Lemma 6.2.3 and Pinsker's inequality (see [Villani, Remark 22.12]) that for all $0 < s < t$,

$$\mathrm{Ding}(\varphi_t) - \mathrm{Ding}(\varphi_s) \leq - \int_s^t \| \mathrm{MA}(\varphi_r) - \mu_r \|^2 \, dr, \tag{P}$$

where $\| \nu - \mu \|$ denotes the total variation of the signed measure $\nu - \mu$.

Since the Ding functional is assumed to be proper, it follows from the monotonicity property that the ψ_t's have uniformly bounded energy, hence form a relatively compact family. Let ψ_∞ be any cluster point. If we could show that

$$\mathrm{Ding}(\varphi_t) \searrow \inf_{\mathcal{E}^1(X,\omega)} \mathrm{Ding},$$

it would follow from the lower semicontinuity of Ding that $\mathrm{Ding}(\psi_\infty) = \inf \mathrm{Ding}$, hence ψ_∞ is the only minimizer of Ding, the Kähler–Einstein potential normalized by $\int_X \psi_\infty \, \omega^n = 0$. Thus the whole family $(\psi_t)_{t>0}$ actually converges towards ψ_∞ (see Corollary 6.6.2). Note that this convergence is easy when (ψ_t) is known to be relatively compact in C^∞. The delicate point here is that we only have weak compactness.

It thus remains to check that $\mathrm{Ding}(\varphi_t) \searrow \inf_{\mathcal{E}^1(X,\omega)} \mathrm{Ding}$. By (P), we can find $r_j \to +\infty$ such that

$$\mathrm{MA}(\varphi_{r_j}) - \mu_{r_j} \longrightarrow 0,$$

since Ding is bounded from above. By compactness we can further assume that $\psi_{r_j} \to \psi_\infty$ in $L^1(X)$, almost everywhere, and *in energy* (see below), so that

$$\mathrm{MA}(\psi_\infty) = \mu(\psi_\infty).$$

Thus $\omega + dd^c \psi_\infty$ is a Kähler–Einstein current. It follows again from the variational characterization that it minimizes Ding, hence the infimum along the flow coincides with the absolute infimum, and we are done.

6.6.3 Convergence in Energy

As explained above, the last step to be justified is that $\mathrm{Ding}(\varphi_t)$ decreases towards the absolute minimum of Ding when ω_t evolves along the normalized Kähler–Ricci flow, without assuming high order a priori estimates.

We already know that the normalized potentials $\omega + dd^c \psi_t = \omega_t$, $\int_X \psi_t \omega^n = 0$, have uniformly bounded energies hence form a relatively compact family. Using (P) we have selected a special subsequence $\psi_{t_j} \to \psi_\infty$ (convergence in L^1 and almost everywhere) such that

$$\mathrm{MA}(\psi_{t_j}) \longrightarrow \mu(\psi_\infty) = \frac{e^{-\psi_\infty}\mu}{\int_X e^{-\psi_\infty} d\mu}, \quad \text{where } \mu = e^{-h_0}\omega^n / V$$

We would be done if we could justify that $\mathrm{MA}(\psi_{t_j}) \to \mathrm{MA}(\psi_\infty)$.

The delicate problem is that the complex Monge–Ampère operator is not continuous for the L^1-topology. A slightly stronger notion of convergence (convergence in energy) is necessary. We refer the reader to [BBEGZ11] for its precise definition; suffice it to say here that it is equivalent to checking that

$$\int_X |\psi_{t_j} - \psi_\infty| \mathrm{MA}(\psi_{t_j}) \longrightarrow 0.$$

Set

$$f_t := e^{\dot{\psi}_t} \frac{e^{-\varphi_t}}{\int_X e^{-\varphi_t} d\mu} \quad \text{so that } \mathrm{MA}(\psi_t) = f_t \mu.$$

If the densities f_t were uniformly in L^p for some $p > 1$, we could conclude by using Hölder inequality, since

$$\int_X |\psi_{t_j} - \psi_\infty| \mathrm{MA}(\psi_{t_j}) \le \|f_{t_j}\|_{L^p(\mu)} \cdot \|\psi_{t_j} - \psi_\infty\|_{L^q(\mu)}.$$

We cannot prove such a strong uniform bound in general, however our next lemma provides us with a weaker bound that turns out to be sufficient:

Lemma 6.6.3. *Set* $\mu := e^{-h_0}\omega^n / V$. *Then*

$$\mathrm{Mab}(\psi_t) = H_\mu(\mathrm{MA}(\psi_t)) + \int_X \psi_t \mathrm{MA}(\psi_t) - E(\psi_t) - V^{-1} \int_X h_0 \omega^n.$$

Therefore there exists $C > 0$ such that for all $t > 0$,

$$0 \leq \int_X f_t \log f_t \, d\mu \leq C.$$

Proof. Recall that $\psi_t = \varphi_t - \int_X \varphi_t \omega^n / V$. It follows from Lemma 6.2.4 that

$$\mathrm{Mab}(\varphi_t) = \mathrm{Ding}(\varphi_t) + \int_X \dot{\varphi}_t \mathrm{MA}(\varphi_t) - \int_X h_0 \omega^n / V$$

$$= -E(\varphi_t) - \beta(t) + \int_X \dot{\varphi}_t \mathrm{MA}(\varphi_t) - \int_X h_0 \omega^n / V,$$

where $\beta(t) = \log\left[\int_X e^{-\varphi_t} d\mu\right]$, while

$$H_\mu(\mathrm{MA}(\psi_t)) = \int_X \log\left(\frac{\mathrm{MA}(\varphi_t)}{\mu}\right) \mathrm{MA}(\varphi_t)$$

$$= \int_X \dot{\varphi}_t \mathrm{MA}(\varphi_t) - \int_X \varphi_t \mathrm{MA}(\varphi_t) - \beta(t).$$

The equality follows.

Recall now that the Mabuchi functional Mab is bounded along the flow, as well as the energy $E(\psi_t)$. Since the latter are uniformly comparable to $\int_X \psi_t \mathrm{MA}(\psi_t)$, we infer that the entropies $H_\mu(\mathrm{MA}(\psi_t))$ are uniformly bounded, i.e.

$$0 \leq H_\mu(\mathrm{MA}(\psi_t)) = \int_X f_t \log f_t \, d\mu \leq C. \qquad \square$$

We can thus use the Hölder–Young inequality to deduce that

$$\int_X |\psi_{t_j} - \psi_\infty| \, f_{t_j} \, d\mu \leq C' \|\psi_{t_j} - \psi_\infty\|_{L^\chi(\mu)},$$

where $\chi : t \in \mathbb{R}^+ \mapsto e^t - t - 1 \in \mathbb{R}^+$ denotes the convex weight conjugate to the weight $t \in \mathbb{R}^+ \mapsto (t+1)\log(t+1) - t \in \mathbb{R}^+$ naturally associated to the entropy, and $\|\cdot\|_{L^\chi(\mu)}$ denotes the Luxembourg norm on $L^\chi(\mu)$,

$$\|g\|_{L^\chi(\mu)} := \inf\left\{\alpha > 0 \mid \int_X \chi\left(\alpha^{-1}|g|\right) d\mu \leq 1\right\}.$$

It remains to check that $\|\psi_{t_j} - \psi_\infty\|_{L^\chi(\mu)} \to 0$. By definition, this amounts to verifying that for all $\alpha > 0$,

$$\int_X \chi \left(\alpha^{-1} |\psi_{t_j} - \psi_\infty| \right) d\mu \longrightarrow 0.$$

Since $\chi(t) \leq te^t$ and the functions (ψ_{t_j}) have uniformly bounded energies, the latter convergence follows from Hölder's inequality and Skoda's uniform integrability theorem.

Acknowledgements It is a pleasure to thank D.H.Phong for patiently explaining several aspects of the proof of this result.

References

[Aub78] T. Aubin, Equation de type Monge-Ampère sur les variétés kählériennes compactes. Bull. Sci. Math. **102**, 63–95 (1978)

[BM87] S. Bando, T. Mabuchi, Uniqueness of Einstein Kähler metrics modulo connected group actions, in *Algebraic Geometry* (Sendai, 1985), ed. by T. Oda. Advanced Studies in Pure Mathematics, vol. 10 (Kinokuniya, 1987), pp. 11–40 (North-Holland, Amsterdam, 1987)

[BBGZ13] R. Berman, S. Boucksom, V. Guedj, A. Zeriahi, A variational approach to complex Monge-Ampère equations. Publ. Math. I.H.E.S. **117**, 179–245 (2013)

[BBEGZ11] R. Berman, S. Boucksom, P.Eyssidieux, V. Guedj, A. Zeriahi, Kähler–Ricci flow and Ricci iteration on log-Fano varieties (2011). Preprint [arXiv]

[Bern11] B. Berndtsson, A Brunn-Minkowski type inequality for Fano manifolds and the Bando-Mabuchi uniqueness theorem (2011). Preprint [arXiv:1103.0923]

[BEGZ10] S. Boucksom, P. Eyssidieux, V. Guedj, A. Zeriahi, Monge-Ampère equations in big cohomology classes. Acta Math. **205**, 199–262 (2010)

[Cao85] H.D. Cao, Deformation of Kähler metrics to Kähler-Einstein metrics on compact Kähler manifolds. Invent. Math. **81**(2), 359–372 (1985)

[CLW09] X.X. Chen, H. Li, B. Wang, Kähler–Ricci flow with small initial energy. Geom. Funct. Anal. **18** (5), 1525–1563 (2009)

[CheT02] X.X. Chen, G. Tian, Ricci flow on Kähler-Einstein surfaces. Invent. Math. **147**(3), 487–544 (2002)

[Chow91] B. Chow, The Ricci flow on the 2-sphere. J. Differ. Geom. **33**(2), 325–334 (1991)

[CSz12] T. Collins, G. Székelyhidi, The twisted Kähler–Ricci flow (2012). Preprint [arXiv:1207.5441]

[Ding88] W.-Y. Ding, Remarks on the existence problem of positive Kähler-Einstein metrics. Math. Ann. **282**, 463–471 (1988)

[DT92] W.-Y. Ding, G. Tian, Kähler-Einstein metrics and the generalized Futaki invariant. Invent. Math. **110**(2), 315–335 (1992)

[Don08] S.K. Donaldson, Kähler geometry on toric manifolds, and some other manifolds with large symmetry, in *Handbook of Geometric Analysis*, No. 1. Advanced Lectures in Mathematics (ALM), vol. 7 (International Press, Somerville, 2008), pp. 29–75

[Gill11] M. Gill, Convergence of the parabolic complex Monge-Ampère equation on compact Hermitian manifolds. Comm. Anal. Geom. **19**(2), 277–303 (2011)

[GZ05] V. Guedj, A. Zeriahi, Intrinsic capacities on compact Kähler manifolds. J. Geom. Anal. **15**(4), 607–639 (2005)

[GZ07] V. Guedj, A. Zeriahi, The weighted Monge-Ampère energy of quasiplurisubharmonic functions. J. Funct. Anal. **250**, 442–482 (2007)

[Ham88] R. Hamilton, The Ricci flow on surfaces, in *Mathematics and General Relativity (Santa Cruz, CA, 1986)*. Contemporary Mathematics, vol. 71 (American Mathematical Society, Providence, 1988), pp. 237–262

[Kol98] S. Kołodziej, The complex Monge-Ampère equation. Acta Math. **180**(1), 69–117 (1998)

[Li08] H. Li, On the lower bound of the K-energy and F-functional. Osaka J. Math. **45**(1), 253–264 (2008)

[PSS07] D.H. Phong, N. Sesum, J. Sturm, Multiplier ideal sheaves and the Kähler–Ricci flow. Comm. Anal. Geom. **15**(3), 613–632 (2007)

[PSSW08a] D.H. Phong, J. Song, J. Sturm, B. Weinkove, The Moser-Trudinger inequality on Kähler-Einstein manifolds. Am. J. Math. **130**(4), 1067–1085 (2008)

[PS06] D.H. Phong, J. Sturm, On stability and the convergence of the Kähler–Ricci flow. J. Differ. Geom. **72**(1), 149–168 (2006)

[PS10] D.H. Phong, J. Sturm, Lectures on stability and constant scalar curvature, in *Handbook of Geometric Analysis*, No. 3. Advanced Lectures in Mathematics (ALM), vol. 14 (International Press, Somerville, 2010), pp. 357–436

[SeT08] N. Sesum, G. Tian, Bounding scalar curvature and diameter along the Kähler–Ricci flow (after Perelman). J. Inst. Math. Jussieu **7**(3), 575–587 (2008)

[ShW11] M. Sherman, B. Weinkove, Interior derivative estimates for the Kähler–Ricci flow (2011). Preprint [arXiv]

[Siu87] Y.T. Siu, in *Lectures on Hermitian-Einstein Metrics for Stable Bundles and Kähler-Einstein Metrics*. DMV Seminar, vol. 8 (Birkhäuser, Basel, 1987)

[Tian90] G. Tian, On Calabi's conjecture for complex surfaces with positive first Chern class. Invent. Math. **101**(1), 101–172 (1990)

[Tian97] G. Tian, Kähler-Einstein metrics with positive scalar curvature. Invent. Math. **130**, 239–265 (1997)

[TZ07] G. Tian, X. Zhu, Convergence of Kähler–Ricci flow. J. Am. Math. Soc. **20**(3), 675–699 (2007)

[Tos12] V. Tosatti, Kähler-Einstein metrics on Fano surfaces Expo. Math. **30**(1), 11–31 (2012)

[Villani] C. Villani, Optimal transport. Old and new. Grundlehren der Mathematischen Wissenschaften, vol. 338 (Springer-Verlag, Berlin, 2009), xxii+973 pp.

[WZ04] X.J. Wang, X. Zhu, Kähler–Ricci solitons on toric manifolds with positive first Chern class. Adv. Math. **188**(1), 87–103 (2004)

[Yau78] S.T. Yau, On the Ricci curvature of a compact Kähler manifold and the complex Monge-Ampère equation, I. Comm. Pure Appl. Math. **31**(3), 339–411 (1978)

[Zer01] A. Zeriahi, Volume and capacity of sublevel sets of a Lelong class of psh functions. Indiana Univ. Math. J. **50**(1), 671–703 (2001)

LECTURE NOTES IN MATHEMATICS

Edited by J.-M. Morel, B. Teissier; P.K. Maini

Editorial Policy (for Multi-Author Publications: Summer Schools / Intensive Courses)

1. Lecture Notes aim to report new developments in all areas of mathematics and their applications - quickly, informally and at a high level. Mathematical texts analysing new developments in modelling and numerical simulation are welcome. Manuscripts should be reasonably selfcontained and rounded off. Thus they may, and often will, present not only results of the author but also related work by other people. They should provide sufficient motivation, examples and applications. There should also be an introduction making the text comprehensible to a wider audience. This clearly distinguishes Lecture Notes from journal articles or technical reports which normally are very concise. Articles intended for a journal but too long to be accepted by most journals, usually do not have this "lecture notes" character.

2. In general SUMMER SCHOOLS and other similar INTENSIVE COURSES are held to present mathematical topics that are close to the frontiers of recent research to an audience at the beginning or intermediate graduate level, who may want to continue with this area of work, for a thesis or later. This makes demands on the didactic aspects of the presentation. Because the subjects of such schools are advanced, there often exists no textbook, and so ideally, the publication resulting from such a school could be a first approximation to such a textbook. Usually several authors are involved in the writing, so it is not always simple to obtain a unified approach to the presentation.

   For prospective publication in LNM, the resulting manuscript should not be just a collection of course notes, each of which has been developed by an individual author with little or no coordination with the others, and with little or no common concept. The subject matter should dictate the structure of the book, and the authorship of each part or chapter should take secondary importance. Of course the choice of authors is crucial to the quality of the material at the school and in the book, and the intention here is not to belittle their impact, but simply to say that the book should be planned to be written by these authors jointly, and not just assembled as a result of what these authors happen to submit.

   This represents considerable preparatory work (as it is imperative to ensure that the authors know these criteria before they invest work on a manuscript), and also considerable editing work afterwards, to get the book into final shape. Still it is the form that holds the most promise of a successful book that will be used by its intended audience, rather than yet another volume of proceedings for the library shelf.

3. Manuscripts should be submitted either online at www.editorialmanager.com/lnm/ to Springer's mathematics editorial, or to one of the series editors. Volume editors are expected to arrange for the refereeing, to the usual scientific standards, of the individual contributions. If the resulting reports can be forwarded to us (series editors or Springer) this is very helpful. If no reports are forwarded or if other questions remain unclear in respect of homogeneity etc, the series editors may wish to consult external referees for an overall evaluation of the volume. A final decision to publish can be made only on the basis of the complete manuscript; however a preliminary decision can be based on a pre-final or incomplete manuscript. The strict minimum amount of material that will be considered should include a detailed outline describing the planned contents of each chapter.

   Volume editors and authors should be aware that incomplete or insufficiently close to final manuscripts almost always result in longer evaluation times. They should also be aware that parallel submission of their manuscript to another publisher while under consideration for LNM will in general lead to immediate rejection.

4. Manuscripts should in general be submitted in English. Final manuscripts should contain at least 100 pages of mathematical text and should always include

   – a general table of contents;
   – an informative introduction, with adequate motivation and perhaps some historical remarks: it should be accessible to a reader not intimately familiar with the topic treated;
   – a global subject index: as a rule this is genuinely helpful for the reader.

   Lecture Notes volumes are, as a rule, printed digitally from the authors' files. We strongly recommend that all contributions in a volume be written in the same LaTeX version, preferably LaTeX2e. To ensure best results, authors are asked to use the LaTeX2e style files available from Springer's web-server at

   ftp://ftp.springer.de/pub/tex/latex/svmonot1/ (for monographs) and
   ftp://ftp.springer.de/pub/tex/latex/svmultt1/ (for summer schools/tutorials).

   Additional technical instructions, if necessary, are available on request from:
   lnm@springer.com.

5. Careful preparation of the manuscripts will help keep production time short besides ensuring satisfactory appearance of the finished book in print and online. After acceptance of the manuscript authors will be asked to prepare the final LaTeX source files and also the corresponding dvi-, pdf- or zipped ps-file. The LaTeX source files are essential for producing the full-text online version of the book. For the existing online volumes of LNM see:

   http://www.springerlink.com/openurl.asp?genre=journal&issn=0075-8434.

   The actual production of a Lecture Notes volume takes approximately 12 weeks.

6. Volume editors receive a total of 50 free copies of their volume to be shared with the authors, but no royalties. They and the authors are entitled to a discount of 33.3 % on the price of Springer books purchased for their personal use, if ordering directly from Springer.

7. Commitment to publish is made by letter of intent rather than by signing a formal contract. Springer-Verlag secures the copyright for each volume. Authors are free to reuse material contained in their LNM volumes in later publications: a brief written (or e-mail) request for formal permission is sufficient.

Addresses:

Professor J.-M. Morel, CMLA,
École Normale Supérieure de Cachan,
61 Avenue du Président Wilson, 94235 Cachan Cedex, France
E-mail: morel@cmla.ens-cachan.fr

Professor B. Teissier, Institut Mathématique de Jussieu,
UMR 7586 du CNRS, Équipe "Géométrie et Dynamique",
175 rue du Chevaleret,
75013 Paris, France
E-mail: teissier@math.jussieu.fr

For the "Mathematical Biosciences Subseries" of LNM:

Professor P. K. Maini, Center for Mathematical Biology,
Mathematical Institute, 24-29 St Giles,
Oxford OX1 3LP, UK
E-mail: maini@maths.ox.ac.uk

Springer, Mathematics Editorial I,
Tiergartenstr. 17,
69121 Heidelberg, Germany,
Tel.: +49 (6221) 4876-8259
Fax: +49 (6221) 4876-8259
E-mail: lnm@springer.com